D1064079

Tropical Agroecosystems

Advances in Agroecology

Series Editor: Clive A. Edwards

Agroecosystem Sustainability: Developing Practical Strategies,
Stephen R. Gliessman
Agroforestry in Sustainable Agricultural Systems,
Louise E. Buck, James P. Lassoie, and Erick C.M. Fernandes
Biodiversity in Agroecosystems,
Wanda Williams Collins and Calvin O. Qualset
Interactions Between Agroecosystems and Rural Communities
Cornelia Flora
Landscape Ecology in Agroecosystems Management
Lech Ryszkowski
Soil Ecology in Sustainable Agricultural Systems,
Lijbert Brussaard and Ronald Ferrera-Cerrato
Soil Tillage in Agroecosystems
Adel El Titi
Structure and Function in Agroecosystem Design and Management
Masae Shiyomi and Hiroshi Koizumi

Tropical Agroecosystems

Edited by

John H. Vandermeer

CRC PRESS

Boca Raton London New York Washington, D.C.

Front and back covers: Rustic coffee production in Chiapas, Mexico. (Photos courtesy of John Vandermeer.)

Library of Congress Cataloging-in-Publication Data

Tropical agroecosystems / edited by John H. Vandermeer.
p. cm. -- (Advances in agroecology)
Includes bibliographical references (p.).
ISBN 0-8493-1581-6 (alk. paper)
1. Agricultural ecology--Tropics. 2. Agricultural systems--Tropics. I. Vandermeer, John H. II. Series.

S589.76.T73 T76 2002
630′.913—dc21 2002031318

Direct all inquiries to CRC Press LLC, 2000 N.W. Corporate Blvd., Boca Raton, Florida 33431.

Visit the CRC Press Web site at www.crcpress.com

International Standard Book Number 0-8493-1581-6
Library of Congress Card Number 2002031318
Printed in the United States of America 1 2 3 4 5 6 7 8 9 0
Printed on acid-free paper

The Editor

John H. Vandermeer Ph.D., is a Margaret Davis Collegiate Professor in the Department of Ecology and Evolutionary Biology at the University of Michigan, Ann Arbor. His work has been in tropical agroecosystem ecology, tropical forest ecology, and theoretical ecology. He is the author of over 150 scientific articles and 9 books.

Professor Vandermeer was born in 1940 in Chicago, Illinois. He received his BS in zoology from the University of Illinois, Champaign/Urbana, and his masters in zoology from the University of Kansas. His doctorate was awarded in biology from the University of Michigan in 1969 based on laboratory and theoretical work with protozoa in a thesis entitled *The Competitive Structure of Communities: A Theoretical and Empirical Approach with Protozoa.* He did postdoctoral work at the University of Chicago with Richard Levins and was assistant professor at State University of New York at Stony Brook before returning to the University of Michigan in 1971 as an assistant professor.

Professor Vandermeer has been a visiting scholar at Princeton University; CATIE, Costa Rica; INPA, Brazil; ECOSUR, Mexico; UCA, UNA, and URACCAN, Nicaragua, and has taught or coordinated courses for OTS. He has received two Fullbright Awards, one to support a year as visiting professor in the National Agricultural University of Nicaragua and one to support a year as a visiting scholar in the Department of Ecological Agriculture at Wageningen University, The Netherlands. He is a faculty associate in the Program in American Culture, Program in Latin American and Caribbean Studies, Program in Science, Technology and Society, and Michigan Consortium for Theoretical Physics. He was an Alfred Thurneau Distinguished Professor from 1994 to 1997 and was awarded the Sokal Prize in 1996. In 2002 he was appointed a Margaret Davis Collegiate Professor.

Contributors

Sonia Altizer
Department of Environmental Studies
Emory University
Atlanta, Georgia
altizer@emory.edu

Carlos M. Araya
Escuela de Ciencias Agrarias
Universidad Nacional
Heredia, Costa Rica
caraya@una.ac.cr

Inge Armbrecht
Department de Biología
Universidad del Valle
Cali, Columbia
inge@biologus.univalle.edu.co

Hendrien Beukema
Department of Plant Biology
Biol Centrum Rijksuniversiteit
Groningen, The Netherlands
H.Beukema@biol.rug.nl

Andrew P. Dobson
Department of Ecology and Evolution
Princeton University
Princeton, New Jersey
adobson@princeton.edu

Johannes Foufopoulos
School of Natural Resources and Environment
University of Michigan
Ann Arbor, Michigan
jfoufop@umich.edu

Luis García-Barrios
Colegio de la Frontera Sur
San Cristobal de las Casas, Mexico
lgarcia@sclc.ecosur.mx

Luko Hilje
Unidad de Fitoprotección
Centro Agronómico Tropical de Investigación y Enseñanza
Turrialba, Costa Rica
lhilje@cati.e.,ac.cr

Laxman Joshi
International Centre for Research in Agroforestry
Jl. CIFOR, Sindang Barang
Bogor, Indonesia
L.Joshi@cgiar.org

Ivette Perfecto
School of Natural Resources and Environment
University of Michigan
Ann Arbor, Michigan
perfecto@umich.edu

Chris Picone
The Land Institute
Salina, Kansas
picone@landinstitute.org

Robert A. Rice
Smithsonian Migratory Bird Center
National Zoological Park
Washington, D.C.
rarice@igc.org

Bernal E. Valverde
Royal Veterinary and Agricultural University
Department of Agricultural Sciences
Frederiksberg, Denmark
bev@kvl.dk

John H. Vandermeer
Department of Ecology and Evolutionary Biology
University of Michigan
Ann Arbor, Michigan
jvander@umich.edu

Meine van Noordwijk
International Centre for Research in Agroforestry
Jl. CIFOR, Sindang Barang
Bogor, Indonesia
M.van-noordwijk@cgiar.org

Gede Wibawa
Indonesian Rubber Research Institute
Palembang, Indonesia
gdwibawa@mdp.co.id

Sandy Williams
Swansea, U.K.
s.e.williams@scriptoria.co.uk

Mark L. Wilson
Departments of Epidemiology and of Ecology and Evolutionary Biology
University of Michigan
Ann Arbor, Michigan
wilsonml@umich.edu

Contents

CHAPTER 1

Introduction

John Vandermeer

CONTENTS

THE SCOPE OF TROPICAL AGROECOSYSTEMS

In the classical word association test of psychologists, what comes to mind when I say tropical agriculture? I tried this on several friends and got answers like extreme rural poverty, deforestation, banana plantations, and the like. The answer mainly depended on the person taking the test. To the conservationist, the answer was deforestation, to the Nicaraguan farmer "my only option" was the response, and poor soils, intransigent pests, and slave labor were others — responses as varied and eclectic as the people giving them. It is, indeed, an eclectic subject. That eclectic nature is both intellectually more satisfying and probably in the end more relevant to the problems of today's tropics than a more traditional approach that might be expected from a collection with the title *Tropical Agroecosystems.*

Consider, for example, the traditional "increasing agricultural production," a sacrosanct goal of almost all classically trained agronomists. Examining this goal

0-8493-1581-6/03/$0.00+$1.50

from a broad perspective, we see articulations with seemingly unrelated issues. Conservationists, for instance, would emphasize that tropical agriculture competes for land with conservation goals (not really an accurate point of view, but a popular one — see, for example, Vandermeer and Perfecto, 1999). So entrenched is this idea that it is sometimes even difficult to bring it up in serious conversation — agriculture bad, natural forest good, case closed, no argument permitted. However, a serious examination of this issue reveals considerable complication, as noted perceptively in the recent volume by Angelsen and Kaimowitz (2001). They note problems with the simple idea that an increase in agricultural productivity (produce per hectare) is positive for conservation goals (with better technology you can produce the same amount on less land, thereby leaving more land for conservation/preservation). Indeed, such a position is at odds with standard economic theory. Standard theory posits that with increased production, more producers are attracted to the productive activity. Thus the unintended consequence of greater productivity could be the transformation of even more natural forest to agriculture (e.g., once pesticides were available to control the boll worm, forests could be cut down to make cotton fields, something that no one would have done before the technology was available). So, does increased agricultural productivity encourage more conservation or does it promote more agricultural transformation? Angelsen and Kaimowitz, after examining several case studies, conclude that one cannot generalize, but the trend is actually in the direction of more land clearing as agricultural efficiency increases, quite opposite to the standard assumption of the conservationists.

It is also somewhat standard, or at least was so in the recent past, to note that the majority of poverty exists in the tropics, and that in order to feed the burgeoning populations in tropical lands, better agricultural technology needs to be developed. Yet world food production has been more than adequate to feed the entire world population ever since the Malthusian position was first articulated in the 19th century. Clearly the problem of hunger is not a problem of agricultural productivity, as has been repeatedly noted in the past (e.g., Lappé et al., 2000).

Many other subjects could be cited in which tropical agriculture is intimately involved, albeit at times quite indirectly. To repeat, it is an eclectic subject. So what ought to be the focus of a volume entitled *Tropical Agroecosystems,* embedded in a series entitled *Advances in Agroecology*?

IN THE SPIRIT OF THE AGROECOLOGICAL APPROACH

In the spirit of the word "agroecology," certain foci should be emphasized and, by implication, others ignored. In the agroecological approach, ecological knowledge takes center stage. Yet ecological knowledge itself is eclectic and voluminous. What methods might be used to categorize and systematize the immense amount of ecological knowledge that exists, from professional ecology journals to oral traditions?

The Nature of Ecological Knowledge

First, ecological processes are general, but their particulars are local. Weeds compete with crops, a general ecological phenomenon. But, for example, in a Nicaraguan backyard garden it is the sedges that are most competitive against maize, whereas the morning glory vines and *Heliconia* plants that may appear as weeds at first glance are really beneficial because they maintain the field free of sedges over the long run (Schraeder, 1999). Predators eat their prey in all ecosystems, but which predator eats which prey is dependent on local conditions. It is itself almost a general rule that local particulars may override general rules in ecology, a fact that continues to frustrate the attempts by ecologists to devise meaningful general rules.

Second, in much the same way that ecological forces are simultaneously general and local, our knowledge about those forces is also both local and general. Intimate experience of local farmers cannot be matched by generalized knowledge of the ecologist, yet sophisticated training of the ecologist cannot be matched by experiential knowledge of the local farmer. Thus, for example, residents in a small valley in Cuba observe that trees grow toward the wind (Levins, pers. comm.). This particular valley is arranged such that the surrounding mountains block out the sun most of the day, except when it appears through the same mountain pass that allows the daily breezes access to the valley. Consequently, the trees that strain for more light, according to ecological principles, in fact do grow toward the wind according to local knowledge. Such stories could be multiplied a thousandfold across the globe. Local residents may have intimate knowledge about the ecological forces that surround them. However, their experience is limited to a relatively small geographical and intellectual setting, preventing them from seeing their knowledge in the larger context that a professional ecologist may automatically assume. On the other hand, the ecologist may find the deviance of local circumstances baffling in relation to that same larger context and be unable to appreciate the rich texture that comes from detailed particular knowledge that the local farmer automatically assumes. If local knowledge is to be part of the process of agricultural transformation, a clear prerequisite to the development of a truly ecological agriculture, the people who own that knowledge must be part of the planning process. This implies a great deal about equality — equality in education, equality in economic and political power.

A New Green Revolution: The Contour Pathway

A new agricultural revolution seems to be brewing. It is the revolutionary change to more ecological forms of agriculture. This change is neither unwelcomed nor unexpected. But change, even revolutionary change, is not new to agriculture, with revolutions being announced at rather regular intervals recently. The forces that actually direct and promote change are not always evident.

The consequence of the original agricultural (Neolithic) revolution was that human cultural evolution invaded the natural world in a major way. While any ecosystem is subject to the changes wrought by genetic change in component species, the agroecosystem is unique among ecosystems in that changes can be extremely rapid because they are driven by cultural rather than organic evolution. We may have genetically modified wild teosinte to produce domesticated maize, but we plant maize not because we have a maize-planting gene inside of us, but because of a series of historical events in our evolving culture. Domestication may have been a genetically based evolutionary change, but agroecosystems evolve through the process of cultural evolution.

Because cultural evolution is the driving force behind change, changes in agroecosystems in the past have not invariably been in the direction of improving the system, but rather have emerged from whatever happened to be the forces shaping cultural evolutionary changes of the times. Thus, for example, the expansion of sugar beet production in Napoleonic France was not for the purpose of producing sugar for the French people. It was done to secure independence from English-controlled cane sugar markets as part of Napoleon's imperial strategy. Many other examples could be cited in which economic, cultural, or political forces, the forces that in fact shape cultural evolution, created the conditions under which agroecosystems underwent dramatic changes.

The contemporary world is no different, except perhaps in that change happens far more quickly than in the past, and its effects are felt worldwide. But agroecosystems remain under pressure to change. That changes in agroecosystem structure and function should be brought under control and directed rationally is part of the philosophy of the agroecological approach. Rather than allowing skewed economic interests to dictate the direction of change, rather than encouraging the externalization of real costs of production, rather than letting a philosophy born of other interests (the unrelenting search for increased profits, for example) dictate the direction of change, there ought to be a concerted effort to design agroecosystems in a rational fashion. It is here that the ecological agriculture movement has been fairly clear in its philosophical distinction from industrial agriculture. A useful metaphor is the hunter in unfamiliar terrain with a topographic map for guidance. To get from point A to point B one might draw a straight line between the two points on the map and proceed to follow that line, climbing hill and valley, perhaps scrambling up cliff faces and rappelling down steep gullies, eventually getting to point B in the "most efficient manner possible." Indeed it would be the most efficient in the sense that the line on the map was the shortest possible. Another way of getting from A to B would be to follow a pathway along the contours provided on the topographic map. While this will be a longer absolute distance, one will likely arrive at point B faster and more efficiently than with the straight-line approach. In the first case we ignore the topographic contours and insist that our peculiar notion of efficiency dictates that a straight line is the shortest distance between two points, that we can ignore the contours. In the second case we view the contours as our signposts to guide us to the goal. The philosophy of industrial agriculture has been akin to drawing a straight line on a topographic map. The metaphorical contours are the myriad ecological interactions that inevitably exist in an agroecosystem, the interactions that industrial agriculture has sought to ignore. Ecological agriculture philosophy acknowledges the contours not only as extant barriers to the straight line approach, but as useful signposts as to where to construct the contour pathway that will most efficiently get us to point B (Swift et al., 1996).

Ecological Knowledge and a Gentle, Thought-Intensive Technology

In response to the crisis in modern industrial agriculture, conventional agricultural researchers clamor for more funds to do more of the same, seeking a new technological fix to each problem that arises. Even more troubling, at the other end of the political spectrum a surprisingly reactionary force seems to have emerged. If the modern industrial system is a consequence of hard-nosed scientific research applied to the problems of ever higher production at whatever environmental cost, they argue, it is this hard-nosed scientific approach that needs to be changed. This position frequently takes on a romantic air, and the techniques of our forebearers become models to be emulated, even venerated, regardless of their potentially negative environmental effects. I have always been in opposition to those who locate the problem of industrial agriculture in the application of Western science. Indeed, my position is not that we need less science and more old-fashioned tradition, but rather that we need more and better science.

Given that the industrial model still dominates, we might think of agriculture, as so many in the developed world yet think of it, as simply another form of industrial production, with inputs to and outputs from the factory, with expenditures, gross revenues, and profits, and so forth. In this model, if agrochemicals are inputs to the factory and wheat or barley are the outputs from the factory, where is the factory? Is it the farm? But where are the machines in the factory? Are they simply the tractors, plows, cultivators, pickup trucks, and the like? Perhaps. But there is surely a more important collection of "machines" on the farm — the soil and its dynamic biological inhabitants, the insect pests and their natural enemies, the sun and shade and weeds and water, and all the other factors that are usually referred to as ecological factors. These are the machines that do the real producing on the farm, and the trucks and tractors only function to bring the inputs to and take the outputs away from the real machines.

But there is a profound difference between these machines and the machines of a regular factory (Carroll, Vandermeer, and Rossett, 1990). The machines in a regular factory were designed and built by men and women, and their exact function and operation are known with great precision — they do pretty much what their designers intended them to do. But no human being designed the machines in the factory that is the farm. They were molded by hundreds of millions of years of evolution, and their exact operation remains enigmatic to anyone who minimally understands our current state of scientific knowledge about how ecosystems work. We are, in effect, making products with machines we understand only superficially. It is as if we send a blind worker into a factory filled with machines, tell her nothing about what the machines are or what they are intended to do, and then tell her to produce something useful.

The industrial model has been able to avoid understanding much about the machines, at least temporarily, by a brute-force approach. It is as if the blind worker began by knocking over all the machines that were not immediately familiar to her upon first touch. It is increasingly obvious to most observers that the science of ecology must play a larger role. The machines in the factory are the ecological forces that dictate what can and cannot be done on the farm, and understanding those machines is the goal of agricultural ecology, whether applied to conventional agriculture or to a more ecological agriculture. The mobilization of the science of ecology in service of understanding agroecosystems is critical to future developments in agriculture if we are to avoid the mistakes of the past.

Yet most applications of ecological principles to agroecosystems remain either hopelessly naive or are even mislabeled. For example, the recent textbook titled *Crop Ecology* (Loomis and Connor, 1992) provides excellent material about the physiology of crop plants, but by most standards of what the science of ecology actually covers, the book probably should not have the word ecology in the title. And the many alternative, organic, holistic, or ecological agriculture texts and guidebooks are frequently guilty of substituting popular visions of ecology for what really ought to be concrete scientific knowledge. It is not that popular visions of ecology are irrelevant, nor that local experiential knowledge about particular ecosystem function is unimportant, but rather that there in fact is a body of scientific knowledge called ecology that should be mobilized in an effort to make a more ecologically sound agriculture. Just as physics represents the scientific foundation of mechanical engineering and chemistry that of chemical engineering, the science of ecology should, and I predict one day will, represent at least one pillar of the scientific foundation of ecological agriculture.

Richard Levins has provided a useful guiding philosophy for the future development of an agriculture based on ecological principles. What we want is not the bull-in-the-china-shop technological developments we have seen in the past in industrial agriculture. Yet we also do not want a return-to-the-good-old-days philosophy that rejects modern scientific approaches. What we want is an approach to technological development that takes a holistic view, incorporates both the knowledge we have of academic ecological science and the local knowledge farmers have of functioning ecology, and seeks to develop an agriculture that produces for the good of people. It is a philosophy that rejects standard economic accounting (assuming that what is profitable is what is good) and seeks to internalize all environmental costs. In short, it is a philosophy that looks to the contours on the map as signposts for development, gently maneuvering through the ecological realities rather than trying to bully them into submission, and it is based on general human knowledge, not restricted knowledge gained from reductionist science. In short, it is gentle, thought-intensive technology.

A FOCUS ON THE TROPICS

The observations in the above paragraphs are general. Their application to tropical situations is special. Tropical areas present ecological, cultural, and political problems that demand separate analysis. Tropical ecology has long been recognized as a distinct science, with its own professional journals and academic positions. The environment of the tropics is thought to be special in many ways, from the lack of a biological down season (winter), to generally poor soil conditions, to extremely high biodiversity. All of these ecological factors are of major import to agricultural development and practice.

Similarly, the cultures of the tropics are themselves complicated and diverse. Still not completely transformed by Western consumer culture, many tropical cultures retain their traditional methods of resource management, including agriculture. It is not only a moral imperative that we tread lightly as we seek transformation of agroecosystems in such areas, here is much of the raw material for future rational development. Consider, for example, the recent work of Morales and Perfecto (2000). In seeking to understand and study traditional methods of pest control among the highland Maya of Guatemala, they began by asking the question, "What are your pest problems?" Surprisingly they found almost unanimity in the attitude of most of the farmers they interviewed — "We have no pest problems." Taken aback, they reformulated their questionnaire and asked, "What kind of insects do you have?", to which they received a large number of answers, including all the main characteristic pests of maize and beans in the region. They then asked why these insects were not pests, and again received all sorts of answers, always in the form of how the agroecosystem was managed. The farmers were certainly aware that these insects could present problems, but they also had ways of managing the agroecosystem such that the insects remained below levels that would be categorized as pests. The initial approach taken by these researchers probably was influenced by their original training in agronomy and classical entomology, but interactions with the Mayan farmers caused them to change. Rather than study how Mayan farmers solve their problems, they focused on why the Mayan farmers do not have problems. The existence of this very special tropical culture drove the research of these two Western scientists.

Political forces are likewise very special in most of the world's tropics. Modern agriculture, like the rest of the modern capitalist economy, is really an international affair. And most important in that international structure is the relationship between the developed and underdeveloped world, the North and the South. The decolonizations of this century have created what is in effect a different form of colonialism. The former colonies are now members of the so-called South (in the recent past called the Third World) and retain important remnants of their colonial structure. As difficult as life may be for some citizens of the developed nations, one can hardly fail to notice a dramatic difference in conditions of life in the South. It is in the South that agricultural production has seemingly not been able to keep up with population growth (although this is only an illusion), where dangerous production processes are located, where raw materials and labor are supplied to certain industries at ridiculously low cost, where air and water pollution run rampant, where people live in desperation, where talk of bettering the state of the environment is frequently met with astonishment — "How can you expect me to worry about tropical deforestation when I must spend all my worry-time on where I will find the next meal for myself and my children?"

HOW THIS BOOK IS ORGANIZED

The chapters in this book are a small subset of chapters that could have been written. They were chosen to represent a broad range of approaches to agroecosystem analysis, focusing on the special problems of the tropics. To speak of blatant omissions first, there are no chapters on tropical soils, although Picone's chapter on mycorrhizae treats the subject indirectly. I had planned on a chapter discussing tropical soils, but all potential authors refused, mainly because the subject is too large for a single chapter. Clearly, a separate volume on tropical soils and agroecology is warranted.

The chapters that do form the body of the book are roughly organized in three main sections. Chapters 2 through 4 treat specific ecological issues associated with production, Chapters 5 and 6 examine two case studies of agricultural transformation and its effect on biodiversity, and Chapters 7 through 9 treat some key landscape issues.

In Chapter 2, García-Barrios presents a classification of tropical agroecosystems in a new and challenging manner. The underlying principles of plant ecology are presented as a framework, and tropical agroecosystem types are organized within the framework. In this way the normally confusing cacophony of classification systems is significantly reduced.

While the chapter by García-Barrios is based at the level of plant-to-plant competition, the incorporation of trophic dynamics is represented in the offering by Hilje and colleagues in Chapter 3. In agroecosystems generally, the ecological force of herbivory is normally cast as the problem of pests and the solution as pest management. Hilje et al. provide an important historical focus to the development of IPM programs in Mesoamerica. They describe an impressive number of IPM programs that are currently underway, but go on to emphasize the barriers to further development. Structural constraints derived largely from the special historical circumstances of the Mesoamerican countries severely constrain the development of IPM programs. They go on to analyze the role of agrichemical industries in the developing paradigms of IPM and discuss the possible costs and benefits of transgenic crops in pest management.

The recent understanding of mycorrhizal biology as described in great detail in Chapter 4 by Picone is in the tradition of detailed ecological knowledge. Picone summarizes recent mycorrhizal research generally and then applies it to the particular situations of tropical environments, finally providing some insight as to the potential for improving on the sustainability of agroecosystems.

The next two chapters (Chapters 5 and 6) are complementary; they address the issue of associated biodiversity and agricultural transformation in two case study tropical agroecosystems — rubber in Sumatra and coffee in northern Latin America. The *sisipan* system of regenerating a jungle rubber system in Indonesia focuses on local knowledge, as described in the chapter by Joshi and colleagues (Chapter 5). They discuss the way in which these local knowledge systems eventually affect biodiversity and its conservation. Similarly, Perfecto and Armbrecht discuss recent improvement of coffee production in northern Latin America and its effect on biodiversity. They also discuss current international political efforts at creating an economic system in which world consumers can aid in the preservation of biodiversity through sensible purchasing decisions, in the form of certified shade or biodiversity-friendly coffee.

The final three chapters in the collection (Chapters 7 through 9) treat more landscape-level phenomena. Geographer Rice, in Chapter 7, summarizes the distribution of tropical agroecosystems in the world and discusses the mechanisms that cause such distributions, raising concerns about current directions of political and economic change and how that will affect future distributions. In the next chapter, Foufopoulos and colleagues discuss the historically significant interplay of wildlife and domesticated animals, especially in Africa. They further present some disturbing questions about the future of both conservation of large ungulates and the problems posed by the latter for agricultural expansion. Finally, Wilson treats a subject rarely acknowledged in the ecological literature, that of the interaction of agriculture and human health. More than just a question of producing food for people, agroecosystems represent massive modifications of human habitats and consequently may have important effects on all aspects of human existence. Wilson documents how, in the past, diseases from malaria to Rift Valley fever have been intimately interactive with agroecosystem transformation.

REFERENCES

Angelsen, A. and D. Kaimowitz, *Agricultural Technologies and Tropical Deforestation*, CABI Publishing, CAB International, NY, 2001.

Carroll, C.R., Vandermeer, J.H., and Rosset, P., Eds., *Agroecology,* McGraw-Hill, New York, 1990.

Lappe, F.M., Collins, J., and Rosset, P., *World Hunger 12 Myths*, Institute for Food and Development Policy, San Francisco, 1998.

Levins, R., personal communication.

Loomis, R.S. and Connor, D.J., *Crop Ecology,* Cambridge University Press, Cambridge, 1992.

Morales, H. and Perfecto, I., Traditional knowledge and pest management in the Guatemalan highlands, *Agriculture and Human Values*, 17:49–63, 2000.

Schraeder, E., *Intercropping in a Traditional Production System in Nicaragua,* M.S. thesis, University of Michigan, 1999.

Swift, M.J. et al., Biodiversity and agroecosystem function, in *Biodiversity and Ecosystem Function. Global Diversity Assessment,* Mooney, H.A., et al., Eds., Cambridge University Press, London, 1996, pp. 433–443.

Vandermeer, J.H. and Perfecto, I., *Breakfast of Biodiversity,* Institute for Food and Development Policy, San Francisco, 1999.

CHAPTER 2

Plant–Plant Interactions in Tropical Agriculture

Luis García-Barrios

CONTENTS

0-8493-1581-6/03/$0.00+$1.50

INTRODUCTION

At the end of the last millennium, tropical agricultural systems extended over a surface of about 20 million square kilometers, occupied by nearly a billion people from 50 different African, Asian, and Latin American underdeveloped countries (Anders, 1990). Tropical agricultural systems (TASs) are found in an enormous variety of contrasting environmental conditions (MacArthur, 1976a; National Research Council, 1993): topography ranges from flat lowlands to very steep highlands; soils range from moderately fertile to very unfertile; temperature ranges from cool to hot; humidity ranges from extremely wet to very dry; primary vegetation ranges from tropical rainforests to semiarid shrub lands. The variety of TASs is even greater (ranging from shifting agriculture to highly technified agroindustrial plantations) due to the distinct social, cultural, and economic conditions and to the different intensities under which these environments and their resources are being used (Ruthenberg, 1976; Hougthon, 1994). Nevertheless, most of tropical agriculture shares the following relevant features:

- Their environmental conditions are frequently restrictive and fragile. Most tropical soils have low fertility due to water erosion, leaching, and acidification in more humid climates and to high temperatures and wind soil erosion in dryer ones (Bennema, 1977; Lal and Miller, 1990). Warm humid and subhumid regions are exposed to explosive pest populations, flooding, and crop storage and transportation problems (National Research Council, 1993), while semiarid regions suffer frequently from prolonged droughts (Okigbo, 1990).
- Most TASs are part of the livelihood of peasant smallholders who often confront severe socioeconomic restrictions on production (Okigbo, 1990). A significantly smaller proportion of TASs are part of large agroindustrial plantations, which in some cases can extend over vast portions of land in the subhumid and humid tropics (National Research Council, 1993).
- Many TASs are established in the so-called megadiversity regions of the world, and all TASs contain and foster significantly more biodiversity than their corresponding temperate counterparts (Perfecto et al., 1996; Collins and Qualset, 1999).
- Market-based economies and social pressure on land are rapidly driving TASs toward increasing levels of intensification, specialization, and simplification (MacArthur, 1976b; Hougthon, 1994), which produce short-term economic benefits to farmers, but only add to these systems' economic and ecological fragility in the long run (Vandermeer et al., 1998).

Traditional TASs were able to persist during millennia in spite of many restrictions. Land-use intensification and other expressions of social change are confronting them with enormous sustainability challenges: in the absence of significant soil conservation measures and costly exogenous inputs, agricultural productivity rapidly decreases and potential land degradation increases (Ruthenberg, 1976; Lal and Miller, 1990). Semiarid tropical regions have always had significantly higher restrictions than more humid ones (Stewart, 1990). Unfortunately, as population and land-use intensification increase, the natural productivity and potential land degradation are dangerously closing toward their lower ends (Figures 2.1–2.3).

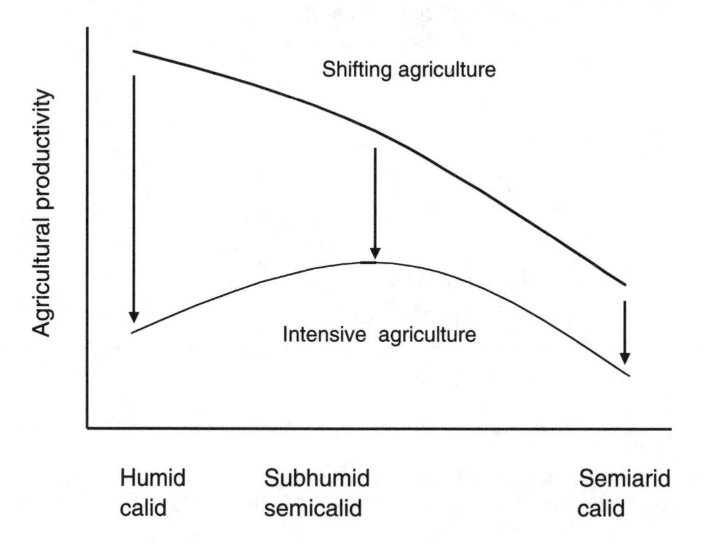

Figure 2.1 Schematic representation of the effects of climate and land-use intensification on agricultural productivity in the tropics, according to Ruthenberg (1976) and MacArthur (1976b).

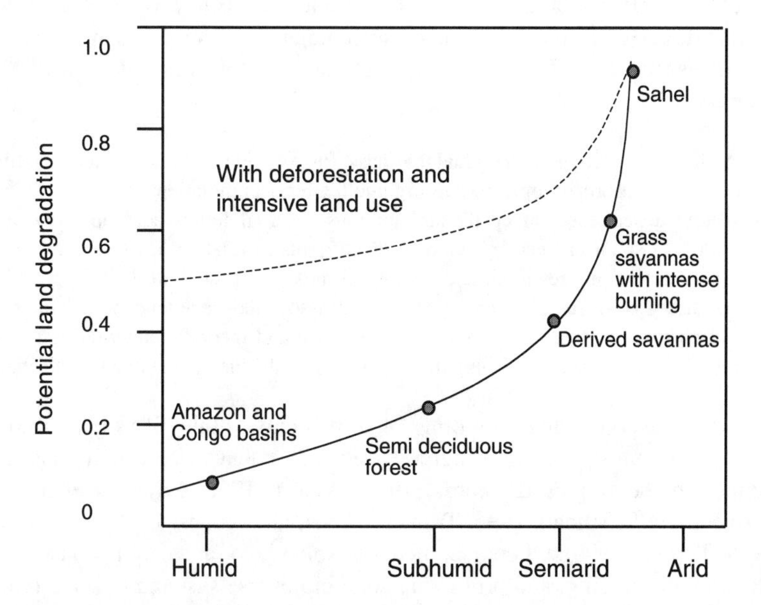

Figure 2.2 Schematic diagram of potential land degradation in the tropics, in relation to climatic aridity. (Modified from Lal and Miller, 1990, Figure 2.)

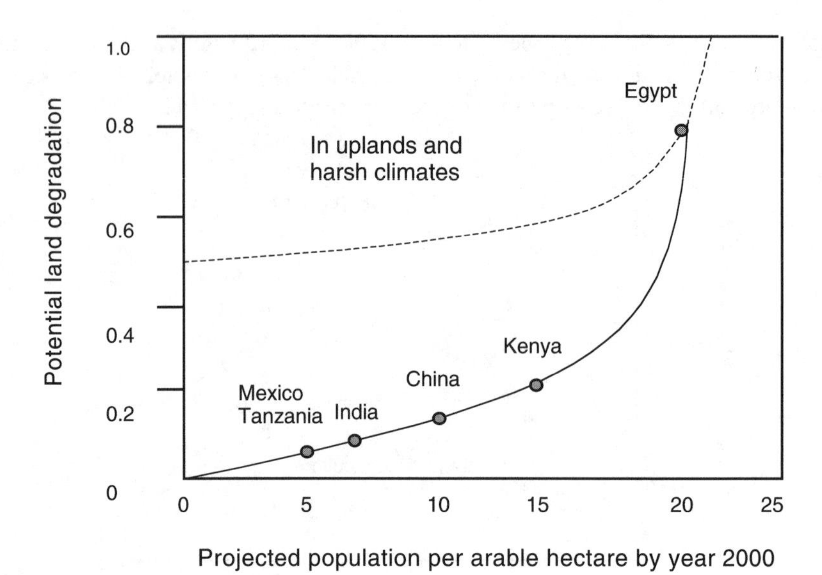

Figure 2.3 Schematic diagram of potential land degradation in the tropics, in relation to population per arable hectare. (Modified from Lal and Miller, 1990, Figure 1.)

Sustainability has become a strongly debated concept and a major issue in international development policies (Schaller, 1993; Demo et al., 1999). It has focused attention on the need to develop and promote the necessary social conditions and the ecologically sound technologies required for a sustainable agriculture in the tropics (National Research Council, 1993; Hatfield and Keeney, 1994; Buck, Lassoie, and Fernandes, 1999). In order to increase ecological sustainability in these fragile environments, the ideal TAS should have at least the following biologically based attributes:

- A high plant cover or residual biomass for efficient light and water capture, constant soil protection, and soil organic matter accumulation
- A low dependence on costly and noxious external inputs accomplished by a relatively small harvest, by removal of nutrients in relation to total biomass, by efficient nutrient recycling, and by natural means of pest control
- A variety of different types of crops and associated beneficial organisms as a means for increasing and diversifying produce and income, reducing the risk of total losses, and accomplishing the conditions established in the first two attributes

In short, preserving and promoting plant diversity within TASs is an important (and maybe the most available) ingredient in the endeavor for a more sustainable agriculture in the tropics (Edwards, 1990; Altieri, 1992; Edwards et al., 1993; Vandermeer, 1995; Tilman, 1996; Tilman, Wedin, and Knops, 1996; Vandermeer et al., 1998; Thrupp, 1998). Of course, means to solve labor and capital constraints for maintaining, developing, and promoting such biodiverse systems are also required.

The ecology, agronomy, and economy of multispecies systems have received increasing attention (Kass, 1978; Vandermeer, 1989). A number of advantages over monospecific crop systems have been repeatedly pointed out. In my opinion, such advantages are frequently overgeneralized in the sustainable agriculture outreach literature. It is necessary to stress that (1) not all potential crop mixtures can coexist, nor do they result in economically and ecologically sound systems; (2) their different benefits are not equally valued by all farmers, nor can they be maximized simultaneously; and (3) they demand more complex tasks and higher inputs of labor and of ecological knowledge than are sometimes available (García-Barrios et al., 2001).

Multispecies systems have developed historically as a continuous trial-and-error process through which specific groups of farmers have identified crops and other associated organisms that can be brought together to their advantage. Social and environmental production conditions change continuously and, with them, the viability of seemingly well-established agricultural systems. In order to develop and maintain agrodiversity in the face of economic, social, and environmental changes, which nowadays tend to advance at a speedier pace than the farmer's empirical exploration and adaptation capacity (García-Barrios and García-Barrios, 1992), it can be useful to support the farmer's effort with a more systematic and extensive theoretical and practical exploration, which can help evaluate current multispecies systems and design those appropriate to new circumstances. From an ecological perspective, the emphasis should be on developing and applying knowledge and skills that can help farmers to manipulate ecological interactions within these systems and foster those interactions that enhance crop productivity and lower risk while reducing external inputs and conserving soil, water, and biological resources.

During the last century, plant ecologists and agronomists have developed important ecological knowledge on plant interactions (i.e., interference, allelopathy, facilitation) (Harper, 1990). Mainstream agronomic research has focused on monocrop situations, and its interest in the details of ecological interactions has been marginal. Plant ecology research has been more concerned with the theory and details of such interactions and has engaged in studying far more diverse and complex plant communities. Agroecologists are making important efforts in bringing together the contributions of both disciplines in order to help understand, develop, and successfully manage both simple and complex multispecies agricultural systems (e.g., De Wit, 1960; Vandermeer, 1981, 1989; Firbank and Watkinson, 1990; Radosevich and Rousch, 1990; Gleissman, 1998). The purpose of this chapter is to contribute to this effort.

The following three sections briefly review the current ecological knowledge on plant interference, allelopathy, and facilitation. Interference has received much attention in crop research and is therefore presented more directly related to agricultural systems and in an analytical fashion. Allelopathy and facilitation are far more diverse interactions and have seldom been studied analytically (but see Vandermeer, 1989). They are treated on a more general basis, derived from the plant ecology literature, but the implications for TASs are discussed. The section titled "The Interplay of Plant Interactions in Environmental Gradients" analyzes the interplay of these plant interactions and how they are modified in productivity gradients. This topic has recently attracted the attention of plant ecologists and, in my opinion, has important implications for TASs when considering how plant–plant interactions can vary in the heterogeneous and contrasting soil and climatic conditions encountered in the tropics at the regional, local, and field levels, and how these interactions can change as a consequence of land-use intensification. The section titled "Some Consequences for Tropical Agriculture" examines the way positive and negative interactions come together in the major TASs and the possibilities of benefiting from these interactions through proper management. For the purpose of this discussion, TASs are classified according to (1) the permanence of a specific plant assemblage on a patch of land or, conversely, the frequency of land-use rotation; (2) the intensity of intercropping, meaning the number, type, and level of spatiotemporal concurrency of crops within the field; and (3) the percentage of tree canopy cover in the system. Finally, the section titled "Plant Interactions and System Design and Management" presents some concluding remarks.

INTERFERENCE

In order to grow and develop, all plants require solar radiation, water, nutrients, and space. As a plant grows, it continuously expands the above- and below-ground zone of influence from which it can actually or potentially acquire such resources. Interference occurs when two plants that have developed overlapping zones of influence reduce one or more of these resources to the point where the growth, survival, or reproductive performance of at least one of them is negatively affected (Begon, Harper, and Townsend, 1986). Interference interactions between growing neighbors constitute a dynamic process whereby both individuals continuously modify the other's above- and below-ground environment and respond to such modifications. Goldberg (1990) considers that a plant's competitive ability comprises the capacity to affect environmental resources (effect competitive ability) and the capacity to tolerate reduced environmental resources (response competitive ability). The effect on resources is related to uptake traits as well as to nonuptake processes that affect resources either positively or negatively, while response to resources is related to the balance between rates of resource uptake and loss at the individual or population level. (See Goldberg [1990] Table 2 for a useful description of these traits and processes.) The ways in which the net effect of one species on another is determined by their effect on and response to environmental resources are nontrivial and are just beginning to be understood (Goldberg, 1990).

Agricultural systems are commonly established at densities that imply highly competitive conditions. In multispecies agroecosystems where different crops, weeds, and trees grow together, the interplay between intraspecific and interspecific competitive abilities strongly influences what species will be able to coexist and what the per-species and per-stand yield will be. Understanding the mechanisms of competitive abilities for the sake of predicting community structure and productivity has proved elusive and controversial (Tilman, 1987; Grace, 1990). Further complications arise because competitive abilities above and below ground are context dependent, for they change in complex ways along resource gradients. Nevertheless, important progress has been made on the subject; see Vandermeer (1989), Grace (1990), and Holmgren, Scheffer, and Huston (1997) for further details.

I take a very general and phenomenological approach to interference by focusing mainly on how net intraspecific and interspecific interference can be evaluated and on their consequences for multicropping yields. I begin by looking at the effects of intraspecific interference at the stand level, both for the sake of analyzing its role in tropical monocrops and to better understand the interplay of both kinds of interference in multiple cropping systems.

Intraspecific Interference in Monocrops

The probability that an individual plant will be adversely affected by its conspecific neighbors increases the more their zones of influence overlap, as a consequence of growth or increased plant density. Competitive effects in a dense monospecific stand have important consequences for the population as a whole. They strongly influence its size distribution dynamics (Koyama and Kira, 1956; Gates, 1982; Hara, 1988), its self-thinning trajectory (Westoby, 1984), and the particular form taken by the yield–density relation (Willey and Heath, 1969; Vandermeer, 1984b).

Individual seedlings seldom grow at the same pace within a monospecific stand due to small differences in genotype, germination time, microenvironmental conditions, or tissue loss to herbivores and pathogens. These initial differences are further enhanced nonlinearly by competition, more so when it is intense and asymmetric (Thomas and Weiner, 1989; Weiner, 1990). This leads the approximately normal seedling size distribution (Figure 2.4a) to become increasingly skewed to the left as individuals grow (Figure 2.4b). Eventually, the smallest individuals, most strongly affected by interference, die out. Thus, if interference is sufficiently intense, density is reduced as the average plant in the population grows in size, and eventually

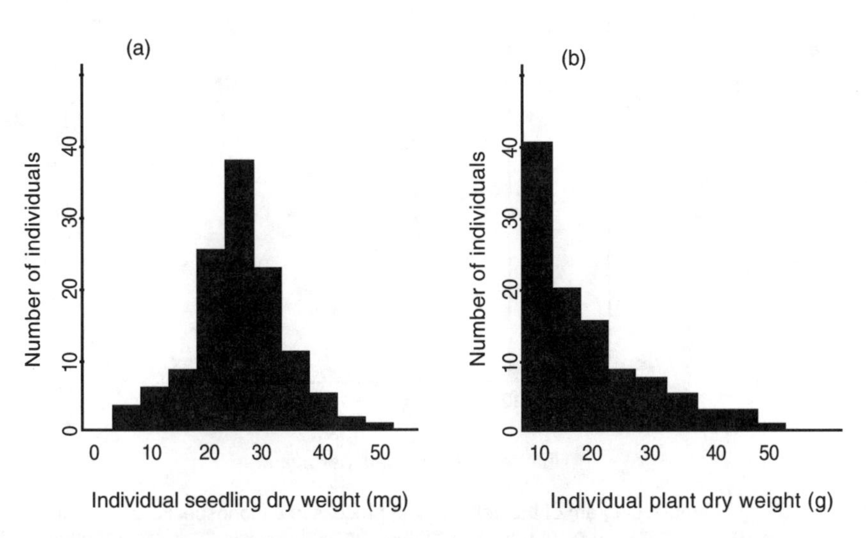

Figure 2.4 Schematic representation of change in size distribution skewness in a monospecific plant stand. Subtle differences in time and size of birth as well as unequal intrinsic growth rates are further exaggerated through interference: (a) early stage; (b) late stage. (Modified from Begon and Mortimer, 1986, Figure 2.10.)

stabilizes at a value that is specific to the particular environment and species. For any initial arbitrary density, the biomass of the average plant in the population grows to a critical point beyond which further increase can only be achieved with a concomitant loss of individuals (Begon and Mortimer, 1986) (Figure 2.5). The so-called self-thinning rule (Westoby, 1984) states that in an overcrowded situation, the number of individuals in the stand must be reduced tenfold in order for a survivor to increase its biomass a hundredfold.

The most obvious effects of intraspecific interference are reduced growth rate, final biomass, and seed set weight of the average plant. These per-plant variables normally are descending geometric functions of sowing density. When the population as a whole is considered, above-ground plant biomass per unit area is commonly an asymptotic function of sowing density, while seed yield per unit area is either asymptotic or quadratic (Willey and Heath, 1969; García-Barrios and Kohashi, 1994; Figure 2.6). Asymptotic behavior occurs when reduction in per-plant growth is exactly compensated for by the increase in plant number, which leads to constant final yield, a condition most common in species with plastic, indeterminate growth. Quadratic behavior occurs when increased density disproportionately reduces seed setting or seed weight and when severe self-thinning cannot be compensated for by the remaining population. These conditions are mostly found in species with less plastic, determinate growth.

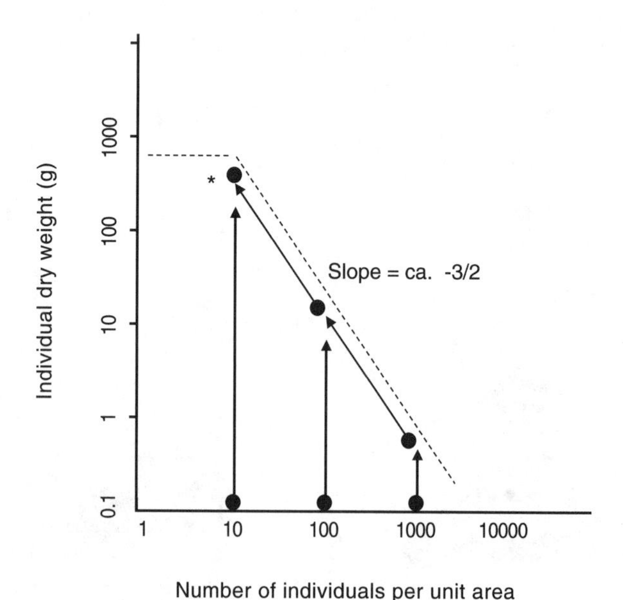

Figure 2.5 Schematic diagram of the self-thinning process in a monospecific plant stand: the relation between plant density and mean individual's weight. Maximum individual weight is marked with an asterisk. (Modified from Begon and Mortimer, 1986, Figure 2.13b.)

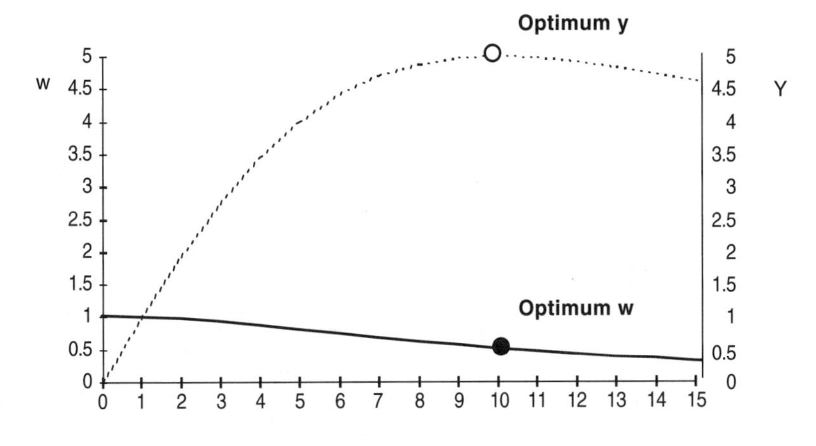

Figure 2.6 Schematic diagram of the parabolic relation between plant density and yield. *w* means average yield per plant and *Y* means yield per unit area. In the example, optimum density = 10; optimum *w* = 0.5; maximum *Y* = 5.

Evaluating Intraspecific and Interspecific Interference Effects in Two-Species Stands

In multiple species stands, both intraspecific and interspecific interferences are encountered simultaneously. Comparing the intensities of intraspecific and interspecific interference helps to explain plant species coexistence, plant mixture overyielding, and weed–crop interactions. Consider the case where two monocrops (species A and B) are sown in separate unit-area plots, each at its optimum density (i.e., the density that produces maximum per-unit-area yield). In such conditions, each species' population uses resources as efficiently as it can. Then consider a substitutive intercrop where 50% of plants in the B monocrop are substituted for by species A plants. When comparing the latter species' per-plant yields in monocrop and intercrop, six basic scenarios can result (Figure 2.7). Condition 3 in the figure is an interesting reference case, where A's per-plant yield remains the same as in the monocrop. This suggests that per-individual interspecific and intraspecific interference should be equal (i.e., A and B individuals are competitively equivalent). In other terms, although intraspecific interference is obviously reduced in the intercrop due to substitution, it is exactly compensated for by interspecific interference. As shown in Figure 2.7a, interspecific interference can also be greater or lower than intraspecific interference. It can also be zero (no interspecific interference), or even negative if net interspecific facilitation occurs. The consequences on A's per-unit-area yield are shown in Figure 2.7b.

A similar analysis is presented in Figure 2.8 for an additive intercrop in which 50% of B's optimum monocrop population is added to the full A monocrop. Again, condition 3 is the competitive equivalence case. Both analyses can also be done for species B. I will now briefly address the consequences of some of these possible outcomes for species coexistence and land-use efficiency in an intercrop.

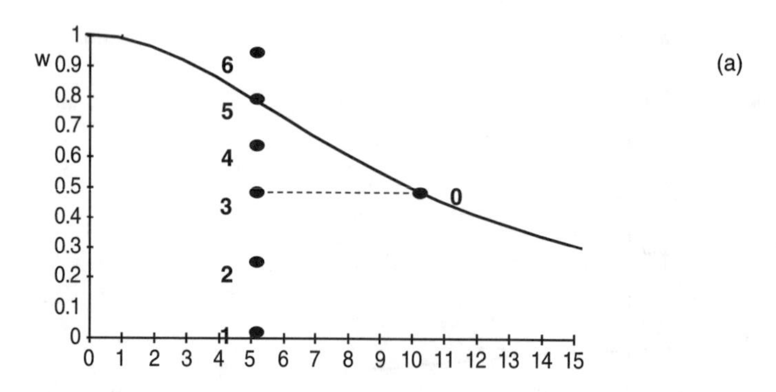

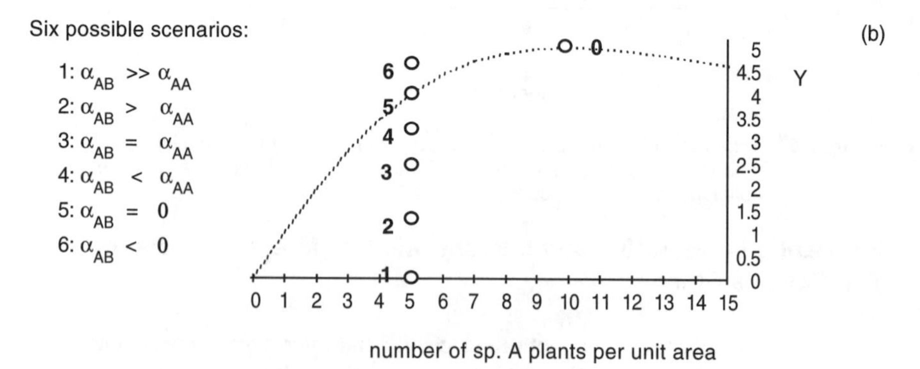

Figure 2.7 Identifying the relative intensity of interspecific interference in a 50%A:50% substitutive intercrop design: (a) representation on a per-plant basis; (b) representation on a per-unit-area basis. Six possible scenarios are depicted in each figure. α_{AA} = intraspecific interference between two A plants. α_{AB} = interspecific interference exerted by plant B on plant A. Condition 3 is a reference case, where per-individual interspecific and intraspecific interferences are equal ($\alpha_{AA} = \alpha_{AB}$). As in Figure 2.6, optimum density = 10; optimum $w = 0.5$; maximum $Y = 5$. See text for further details.

Interference, Coexistence, and Overyielding in a Two-Species Stand: The Competitive Production Principle Revisited

For a multiple crop system to be viable, its component species must be able to coexist, and the system must have advantages over its competitors (the corresponding monocrops). The land equivalent ratio (LER) is the most commonly used intercrop performance index (Willey, 1979). LER is defined as

(A's intercrop yield/A's maximum monocrop yield) +
(B's intercrop yield/B's maximum monocrop yield)

The use of this criterion assumes that the farmer is interested in both crops, and it defines how many monocrop surface units yield the same as one intercrop surface unit. An LER greater than 1.0 implies both coexistence and overyielding, while an LER less than 1.0 can still mean the former, but not the latter.

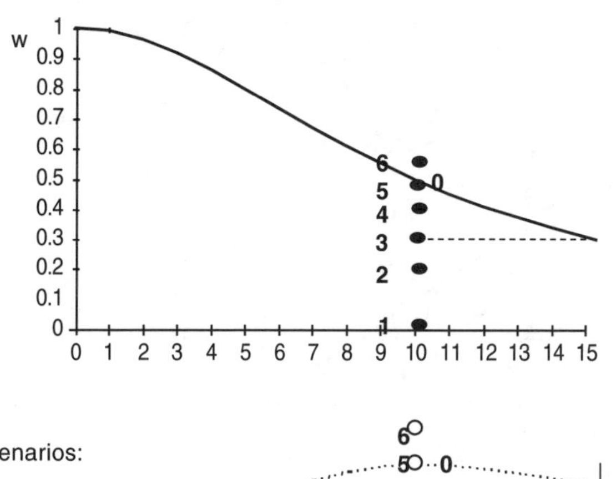

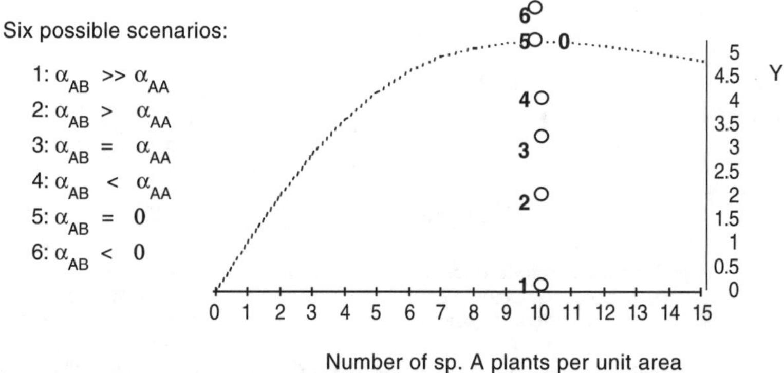

Figure 2.8 Identifying the relative intensity of interspecific interference in a 100%A + 50%B additive intercrop design: (a) representation on a per-plant basis; (b) representation on a per-unit-area basis. Six possible scenarios are depicted in each figure. α_{AA} = intraspecific interference between two A plants. α_{AB} = interspecific interference exerted by plant B on plant A. Condition 3 is a reference case, where per-individual interspecific and intraspecific interferences are equal ($\alpha_{AA} = \alpha_{AB}$). As in Figure 2.6, optimum density = 10; optimum $w = 0.5$; maximum $Y = 5$. See text for further details.

Figure 2.9 illustrates how the interplay of interferences modifies LER, both qualitatively and quantitatively, for a 50%:50% substitutive and a 100% + 100% additive intercrop. In the substitutive case, a modest to high intercropping advantage (LER > 1.0) occurs when the reduction of intraspecific interference due to substitution is not fully compensated by interspecific interference. This situation is to be expected if A and B exploit environmental supplies differently, such that resources that limit yield are available in greater quantities to the intercrop than to the pure stands. In such a case, the competitive production principle operates (Vandermeer, 1989). Additive intercrops can also render LER greater than 1.0 if interspecific interference is lower than the intraspecific interference experienced by full monocrop species.

LER less than 1.0 occurs when one or both species suffer from more interspecific than intraspecific interference. The most common case is one species exerting an ever-increasing interference on the other, while the latter reduces it interference on the former. Such dominance-reduction relations between individuals can eventually lead the weaker competitor to exclusion. In substitutive intercrops, the dominant

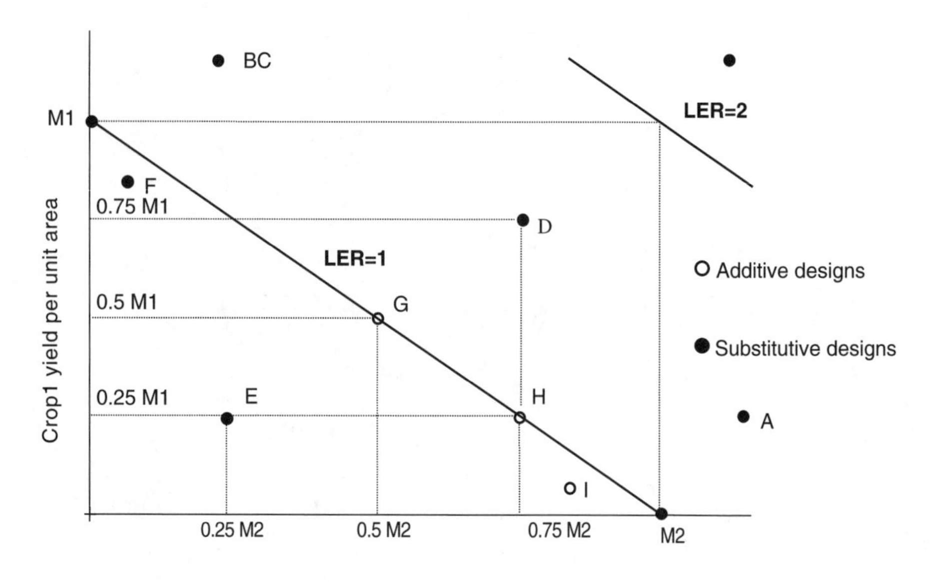

Figure 2.9 The intensity of interference and facilitation in a diculture define its position in a diculture evaluation plane. Points A–F correspond to 1 + 1 additive designs; points G–I to 1:1 substitutive designs. When monocrops 1 and 2 (at their respective optimum densities M1 and M2) are added, some possible outcomes are (A) 1 facilitates 2 while 2 strongly interferes 1; (B) 2 facilitates crop 1 while 1 strongly interferes 2; (C) 1 and 2 facilitate each other; (D) both crops interfere each other weakly; (E) both crops interfere each other strongly; (F) 1 strongly interferes 2. When crops 1 and 2 are added (each at half its optimum monocrop density), some possible outcomes are as follows: (G) Both species maintain their per-plant monocrop yield because, although intraspecific interference has been strongly reduced, it is compensated exactly by interspecific interference. (H) Species 1 individuals are strongly interfered by species 2 and lose 50% of their weight in spite of the fact that intraspecific interference has been substantially reduced. Species 2 individuals gain weight in the same proportion as lost by species 1 individuals. This dominance-suppression process with perfect compensatory growth produces a LER = 1.0 as in (G), although the ecological situation is quite different. (I) As in (H), but species 2's per-plant weight gain does not fully compensate species 1's per-plant loss.

individuals benefit from a certain degree of compensatory growth due to lower overall interference. A neutral result (i.e., LER = 1.0) occurs either when intraspecific and interspecific interferences are identical within each species (Vandermeer, 1995) or when dominance reduction occurs, but with perfect compensatory growth by the dominant species (Trenbath, 1974; García-Barrios, 1998).

Until recently, agroecologists accepted niche differentiation theory (Gause, 1934; MacArthur and Levins, 1967) as the most important explanation of plant coexistence and stressed the fact that it predicted a high frequency of intercropping conditions where the competitive production principle should hold (e.g., Vandermeer, 1989). Unfortunately, the contrary seems to be the case (Vandermeer, 1995; García-Barrios, 1998). In the past two decades, ecologists have recognized that weak interspecific competition is only one of many conditions — and perhaps not the most common

— that can explain species coexistence in a natural plant community (e.g., Hubbell and Foster, 1986; Silvertown and Law, 1987; Fowler, 1990) although the topic is still controversial (e.g., Tilman, 1987; Huston and DeAngelis, 1994; Grace, 1995).

An LER greater than 1.0 is certainly the case in many legume–cereal dicultures where light and nitrogen source partitions are possible (Vandermeer, 1989). Yet it is becoming clear that (1) most plants share the same niche (Silvertown and Law, 1987); (2) *ceteris paribus*, intraspecific and interspecific interferences are roughly equal *on a per-gram basis* (Goldberg and Barton, 1992); (3) individual size differences explain a major part of unequal effect competitive abilities (Goldberg and Werner, 1983; García-Barrios, 1998); and (4) consequently, the case to be most commonly expected is an LER of about 1.0, either due to individuals being competitively equivalent (Vandermeer, 1995) or — more commonly — due to dominance reduction with compensatory growth (Trenbath, 1974; García-Barrios, 1998). Fortunately, in some cases this unfavorable result can be potentially changed into an LER greater than 1.0 situation if dominance can be reduced or even reverted through appropriate sowing and harvesting time differences between species within the same growing season. This has been demonstrated both empirically (Fukai and Trenbath, 1993) and through computer simulation (García-Barrios, 1998).

A Word on Interference in Weed–Crop Systems

Analyzing and manipulating interference is also in order when considering weed–monocrop systems. These have been frequently studied as additive intercrops. The relevant questions to be answered in terms of interference are (1) how strong is the per-individual interference (or per-unit-leaf-area effect; Kropff and Lotz, 1992) of the weed over the crop, and (2) how much can interference be reduced by increasing crop density or hastening crop emergence? (Zimdahl, 1988; Kropff, Weaver, and Smits, 1992; Liebman and Gallandt, 1997). Figure 2.10 shows the typical effect of relative weed density on relative crop yield and how the relative timing of crop emergence can either revert or aggravate this effect. Weed–intercrop interactions have been less thoroughly studied. The simplest approach considers only additive intercrops and their negative effect on weeds through a higher global crop density (Liebman and Stavers, 2000). A more general and complex approach considers a three-party system where indirect interference relations produce a facilitative relation between crops (Vandermeer, 1989).

Studying Competitive Interactions in Many Species Stands

Many relevant intercropping situations comprise two or three crops whose major interactions can be analyzed as explained previously. Nevertheless, it does not seem sound to empirically approach all possible interactions at the species level in a highly diverse agricultural plant community. Statistical analysis of yield performance for three or more crops is a cumbersome task (e.g., Federer, 1999), and the number of bilateral interactions to be considered grows geometrically with species richness, rendering analysis and data gathering impractical (Tilman, 1990). Following Goldberg and Werner (1983) and Tilman's (1990, 1996) work with natural communities, I would

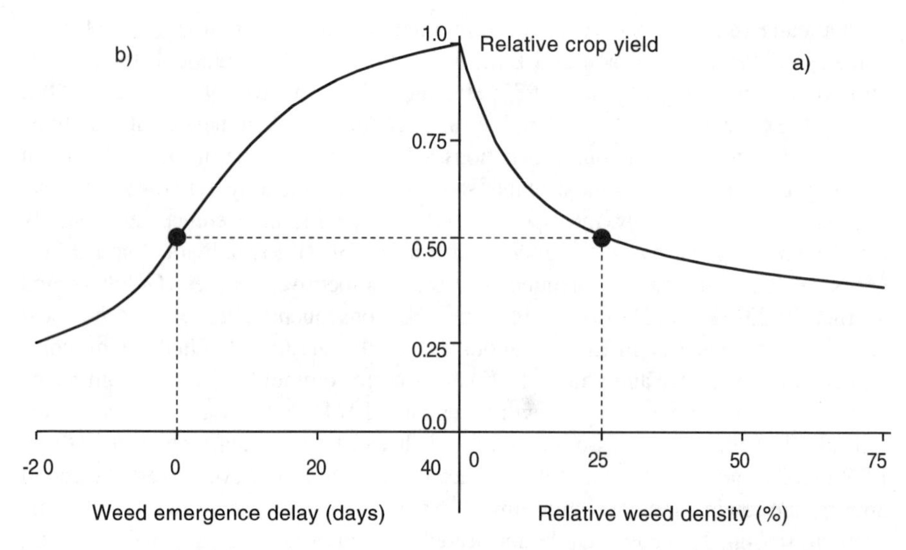

Figure 2.10 The relative timing of weed emergence (WED) can either revert or aggravate the negative effect of relative weed density (RWD) over relative crop yield (RCY). Relative crop yield = Uninfested Y_{max}/Weed infested Y; Relative weed density (%) = $(100N_w)/(N_w + N_c)$ where N_c = number of plant crops and N_w = number of weed plants. Weed emergence delay = days after the crop has emerged. (a) RCY response to RWD when WED = 0; (b) RCY response to WED when RWD = 25%. Within reasonable ranges, the effect of RWD can be modified by manipulating WED, and vice versa. The qualitative form of function (a) is based on Radosevich (1988) and (b) on Kropff, Weaver, and Smits (1992); the relation between RWD and WED is inferred from Liebman and Gallandt (1997).

venture at least three alternative approaches in order to understand the performance of a plant species in a diverse plant community: (1) species with similar ecological attributes could be grouped together, and intraguild and interguild interactions could be considered; (2) a target species' performance could be analyzed as a function of its plant neighborhood's diversity, regardless of the species involved; and (3) functions and physiological processes, rather than species, could be considered mechanistically.

ALLELOPATHY

Allelopathy, in its broadest sense, encompasses all types of inhibitory or stimulatory chemical interactions among plants and between plants and microorganisms. Several hundred different nonnutritional organic compounds released from plants and microbes are known to affect the growth, development, behavior, and distribution of these and other organisms in natural communities and agroecosystems. Recent research developments have stressed the importance of allelopathy and provided new insights into the complexity of interactions that occur in natural and agricultural communities (Einhellig, 1995).

Plants can chemically influence other plants — of the same or different species — in a direct and active manner, or passively through indirect processes that are sometimes very complex and have proved difficult to elucidate (Figure 2.11).

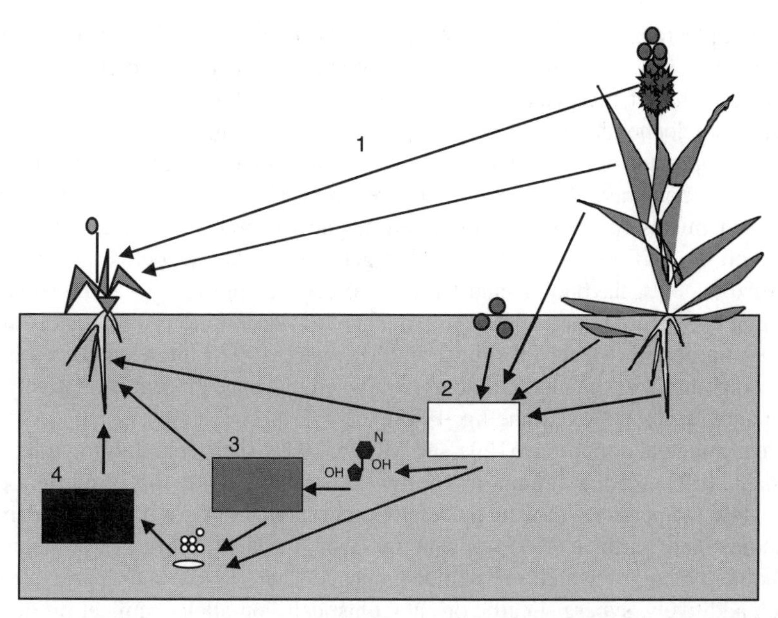

Figure 2.11 The diverse pathways of allelopathic compound formation. Allelopathic compounds that act upon a target plant can derive directly or indirectly form another plant's tissues: (1) Chemicals evaporated or exuded from aerial plant parts can reach the target plant directly. (2) Chemicals derived form plant debris and exudates can be carried in the soil solution without further transformation, or (3) they can react with organic and inorganic soil matrix compounds, or (4) they can be metabolized by soil microorganisms before acquiring their allelopathic attributes. The possible combinations of plant precursor molecules and of soil and microorganism mediated chemical reactions are enormous and can conform extremely complex pathways that render myriad different potentially allelopathic compounds.

Allelochemicals or their precursors volatilize from plant surfaces or are released as leaf and seed leachates and root exudates (Einhellig, 1995; Zimdahl, 1993). Allelopathic compounds derived from plants can also form passively, as byproducts of tissue decay and decomposition. Compounds released by living tissues or plant residues can show immediate allelopathic activity or acquire it after being transformed by microorganisms in the soil matrix. Plant-derived allelochemicals can act upon other plants, either directly or by affecting organisms that interact with them.

With a few exceptions, allelochemical agents produced by higher plants are secondary compounds that arise either from the acetate or shikimate pathway, or their chemical skeletons come from a combination of these two origins (Einhellig, 1995). Shikmate is an organic, aromatic compound and a precursor for aromatic aminoacids. Most are thought to have no central metabolic function in the plant producing them, but to have been selected for by their indirect serendipitous consequences on fitness. However, a few have structural or physiological functions within the producing plant (Hedin, 1977). Whittaker and Feeny (1971) classify plant allelochemicals into five groups: phenylpropanes, acetogenins, terpenoids, steroids, and alkaloids. Rice (1984) designates 14 categories of allelochemicals, plus a miscellaneous group.

Often, the immediate source of a compound involved in allelopathy is obscure, especially if contacted through the soil medium. Further, the same compound may have multiple roles, and plant response may be elicited by a group of different compounds, acting by additive, synergistic, or antagonistic means, depending on the relative amounts of each compound (Gerig and Blum, 1991).

Some allelochemicals can be transported across relatively large distances as volatile compounds or soil water solutes, while others are circumscribed to the root zones or the spot where they are formed or released. The range for biological activity of allelochemicals covers several orders of magnitude, and bioactive concentrations depend on the particular compound and target species. Many coumarins, cinnamic and benzoic acids, flavonoids, and terpenes affect seedling growth at thresholds of inhibition between 100 and 1000 μmol, but active concentrations as low as 10 μmol have been reported (Macías, Galindo, and Massanet, 1992). Interestingly, an inhibitory compound will often stimulate growth when its concentration is relatively low (Chou and Patrick, 1976; Einhellig, 1995).

Environmental conditions (Hale and Moore, 1979) such as high ultraviolet light (Zimdahl, 1993), strong gamma irradiation (Alsaadawi et al., 1985), nutrient deficiency, low temperature, moisture stress (Gershenzon, 1984), and predator damage (Sembdner and Parthier, 1993) appear to favor a general increase of secondary metabolites, many of which are allelochemicals. These biotic and abiotic stresses can act additively, synergistically, or antagonistically on allelochemical production and action (Einhellig, 1995).

The mechanisms of action of allelochemicals are many and as yet are not well understood. At present it is known that coumarins and phenolic compounds interfere to some extent with many vital plant processes, including cell division, mineral uptake, stomatal function, water balance, respiration, photosynthesis, protein and chlorophyll synthesis, and phytohormone activity. Membrane perturbation may be a starting point for the multiple action of these compounds (Einhellig, 1995). Plant inhibition is often indirect, through suppression of fungal root colonization (Rose, 1983) and poor *Rhizobium* spp. nodulation or from inhibition of free-living nitrogen-fixing bacteria and blue-green algae as well as other microbes that are critical for the nitrogen cycle (Einhellig, 1995). In some cases, susceptibility to pathogens is increased by allelochemicals released from decomposing plant residues (Hartung and Stephens, 1983).

In natural communities, allelopathic effects ultimately influence vegetational associations and patterns (Muller, 1969), secondary plant succession (Rice, 1984), exotic plant invasion, and other community processes (Einhellig, 1995). Allelopathy in agricultural fields, pasture lands, and agroforestry systems is also important: crops, weeds, or microorganisms can be either the source or the target of allelochemical compounds. The most common consequences of allelopathy in agroecosystems are crop autotoxicity, difficulties for intercropping, and weed infestation. However, actual and potential benefits of allelopathy for biological pest control are also worth considering (Zimdahl, 1993).

FACILITATION

The positive effect of plants on the establishment or growth of other plants has long been recognized as an important driving force in structuring plant communities (e.g., Kropotkin, 1902; Clements, Weaver, and Hanson, 1926). Even so, for over half a century, competition received far more attention in ecological research (see reviews by Connell and Slatyer, 1977; Keddy, 1989; and Goldberg and Barton, 1992). Recently, however, there has been renewed interest in the topic of beneficial interactions (Vandermeer, 1984a; Boucher, 1985; Hunter and Aarsen, 1988; Goldberg, 1990; Callaway and Walker, 1997), perhaps as a consequence of a waning faith in the importance of interspecific competition (Schoener, 1982). From an evolutionary perspective, traditional theory depicts survival of the fittest as a difficult existence based on danger, conflict, and strife. But another view is emerging of a more synergistic organization in which ecosystems on the whole provide hospitable conditions for life. In this view, the world is populated by organisms mutually adapted and beneficial by virtue of their direct and indirect interactions (Fath and Patten, 1998). Such synergistic networks might well have developed and could actually be operating in many biodiverse agroecosystems.

A neighboring plant may benefit others directly by improving microclimate, providing physical support, and ameliorating soil conditions and plant nutrition. Indirect benefits may result from reducing the impact of competitors, distracting or deterring predators and parasites, encouraging beneficial rhizosphere components, and attracting pollinators or dispersal agents. I shall now consider these forms of facilitation, following the model Hunter and Aarsen (1988) have used to organize them. Although most of the cited research has been done in nonagricultural plant communities, it readily applies to TASs.

Direct Facilitation

In very dry environments, certain plant seedlings require the relatively humid conditions found beneath the canopy of other species in order to grow to a stage where they can tolerate high soil and air temperatures (Bertness and Callaway, 1994; Briones, Montaña, and Ezcurra, 1994). However, the so-called nursing syndrome is not exclusive of arid conditions. In mesic and humid environments, forest understory plants can tolerate and even require a certain amount of shade provided by canopy species (Ramírez, Gonzáles, and García-Moya, 1996). Even in waterlogged or very humid soil environments, some plants have high transpiration rates that allow them to lower the water table, improving soil aeration for themselves and neighboring species (Berendse and Aerts, 1984).

Some plants act as windbreaks, preventing nearby plants from being overthrown. Others offer support to lianas and vines, thus allowing them to economize on expensive support structures. In turn, by temporarily capturing falling plant debris, these can slow down residue decomposition, nutrient mineralization, and leaching (Hunter and Aarsen, 1988).

Plants stabilize loose surfaces and improve litter accumulation, soil structure, cation exchange capacity, and water holding ability. Some plants produce stimulatory allelopathic effects in others. Many legumes associate with microorganisms that provide them with atmospheric nitrogen; this resource is eventually incorporated into the soil through root and shoot residues and made available to other plants. Nutrients are also shared directly by some plants with their conspecifics through naturally occurring root grafts; interspecific connections are possible although rare. Mycorrhizal fungi exhibit low host specificity, and the same network of hyphae may join a large number of plants of different species (Finlay and Read, 1986). Transfer can be direct or nutrients can be leached by a plant, taken up from the soil pool by fungi, and transferred to another plant.

Indirect Facilitation

"Your enemy's enemy is your friend" is a precept that could be operating in communities where plant A would be outcompeted by B in the absence of plant C, which nevertheless does not strongly suppress plant A. This form of beneficial interaction has been outlined theoretically (Vandermeer, 1989), but experimental evidence is still scarce although encouraging (Haines, Haines, and White, 1978; Pennings and Callaway, 1996; Levine, 1999). Miller (1994) found that direct effects of five species in an old-field community were generally competitive, while indirect effects were generally positive (although some indirect effects were also negative). In several cases, the magnitude of the indirect positive effect was greater than that of the direct negative effect, resulting in a facilitative effect overall. Weed control in multiple crop systems can also be seen as indirect facilitation (Vandermeer, 1989; Liebman and Stavers, 2000).

Lower levels or higher stability of herbivorous insect populations is more common in diverse vegetation than in monospecific stands. This may be due to higher predator numbers (Dempster, 1969), altered wind flow and shading (Risch, 1981), chemical signals interfering with host location mechanisms (Tahvanien and Root, 1972), alternative or decoy hosts (Atsatt and O'Dowd, 1976), or lower resource concentration (Root, 1973). Some spiny or unpalatable plants can also protect their close neighbors from vertebrate herbivores (Hunter and Aarsen, 1988). Referring specifically to agroecosystems, Vandermeer (1989) has revised the various hypotheses (Aiyer, 1949; Root, 1973; Trenbath, 1976) on the mechanisms that lead to fewer pests in intercrops. He reduced them to the following three. A second species can (1) disrupt the ability of a pest to locate and attack its proper host efficiently, largely applied to specialist herbivores; (2) serve as a decoy and distracter for pests that would normally damage the principle species, largely applicable to generalist herbivores; (3) attract, for whatever reason, more predators and parasites of the pest than the monoculture.

Plants exude and leach organic compounds that enhance growth, promoting bacterial and mycorrhizal activity in the rhizosphere of other plants (Hunter and Aarsen, 1988).

Some plants may benefit from proximity to species that are especially successful at attracting pollinators. Under certain density conditions there are trade-offs in developing together in space and time because the beneficiary might lose too much pollen to the other species (pollen competition) and might have to compete with the latter for other resources (in Hunter and Aarsen, 1988). Fruit dispersal also seems to be more successful when fruiting individuals occur in conspecific or polyspecific clumps. However, in some conditions, sequential flowering and fruiting are convenient and even necessary in order to maintain pollinators and fruit dispersers year-round. (Hunter and Aarsen, 1988.)

To conclude, it should be stressed that ecologists are increasingly paying attention to facilitation and finding it to be common across many different environments. Hopefully, this will promote research on the topic in agricultural systems, beyond the most obvious cases of pest control and *Rhizobium*–legume crop symbiosis. Positive plant interactions might prove to be particularly common and relevant in agricultural communities developing under high physical stress (e.g., semiarid tropics) or with high herbivore and parasite pressure (e.g., subhumid tropics) (Bertness and Callaway, 1994).

THE INTERPLAY OF PLANT INTERACTIONS IN ENVIRONMENTAL GRADIENTS: SOME CONSEQUENCES FOR TROPICAL AGRICULTURE

My understanding of how interference, allelopathy, and facilitation occur and affect plant community structure and yield is largely based on studies in which each has been isolated and analyzed separately. This has been my approach in previous pages. But positive and negative plant interactions do not act in isolation, and by occurring together within the same plant community, and even between the same individuals, they may produce complex and variable effects, which are further complicated by the fact that they can be modified by environmental changes (Callaway and Walker, 1997). For example, production of allelochemicals is enhanced by mineral deficiency and drought stress. Therefore, a harsh environment and strong competition for limited soil resources can increase allelopathic interactions. On the other hand, these can modify competitive relations. For example, a plant can change the environment to its competitive advantage by subtle means such as changes in the nitrogen relationships caused by the release of specific inhibitors of nitrogen fixation or nitrification (Zimdahl, 1993).

Another example can be readily found in interference–facilitation interactions. At common cropping densities, plants will most certainly compete, whether or not facilitation is operative. When positive and negative interactions are present — and both are strongly affected by the benefactor plant's density — one can expect the net effect to change as plants are brought closer together. Such a situation is to be expected, for example (1) in natural communities (Holmgren, Scheffer, and Huston, 1997) and cropping systems (García-Barrios, 1998) where nurse versus competitive effects of the benefactor plant are present; (2) in alley cropping where there is a trade-off between the contribution of tree foliage to soil fertility and its light depletion effects (Ong, 1994; Vandermeer, 1998), and (3) in intercrops, where the smaller crop hosts an insect pest and the taller one acts as a deterrent or harbors the pest's natural enemies (Figure 2.12; Vandermeer, submitted). The net effect of both interactions will either be net facilitation, no effect, or net interference. In some circumstances, a switch from a positive to a negative net result can be expected along the density gradient (Figure 2.13). An optimum benefactor density can be expected at which net facilitation is maximum or — more commonly — net interference is minimum but this optimum density can be expected to change if the environment is modified (Vandermeer, 1989).

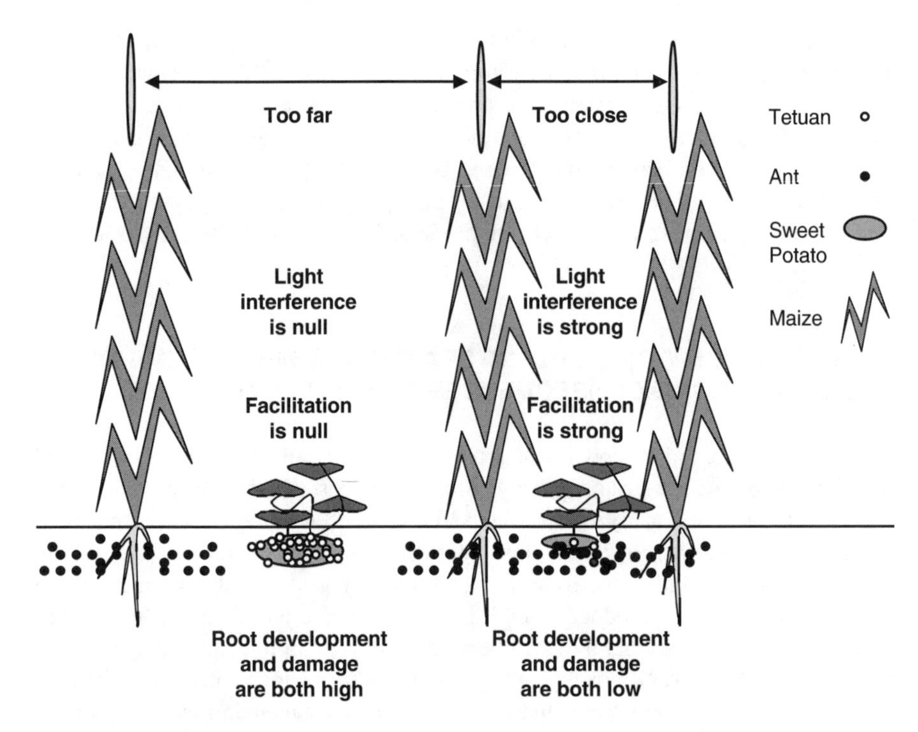

Figure 2.12 Maize (*Zea mays*) both interferes and facilitates sweet potato (*Ipomoea batatas*) (Castiñeiras et al., 1982). Maize casts strong shade on sweet potato, but its root zone harbors the lion ant (*Pheidole megacephala*), which is a natural enemy of the Tetuan (*Cylas formicarius elegantulus*), an important sweet potato root plague. The net effect of these antagonic interactions depends, among other things, on plant spacing. The figure depicts two suboptimal plant spacings that are conducive to low sweet potato net production. See Vandermeer (submitted) for an analytical model to determine optimum plant spacing.

The interplay of negative and positive effects is nonlinear and trade-offs are seldom a zero sum game. For example, removing a plant species that competes aggressively with a crop but also harbors pollinators might improve the latter's vegetative growth but preclude its reproduction. Likewise, when an allelopathic weed severely reduces a crop's growth, this does not necessarily mean that weed biomass growth will exactly compensate crop biomass loss.

The net outcome of plant interactions in TASs, both in terms of coexistence and yield, depends on a host of circumstances (e.g., species involved and their effect on and response to above- and below-ground interactions, plant age, density, environmental conditions). Therefore, it is impossible to analyze the myriad of possible outcomes to be found in tropical agriculture. In the search for useful generalizations, it is important to bear in mind that in tropical agriculture we find conditions ranging from (1) very limited to very high soil moisture availability; (2) mild to very high solar radiation and temperatures; (3) unfertile to fertile soils, and low to high nutrient availability. These gradients in resource availability (and concomitant plant stand productivity) are both natural and a result of the positive and negative effects of

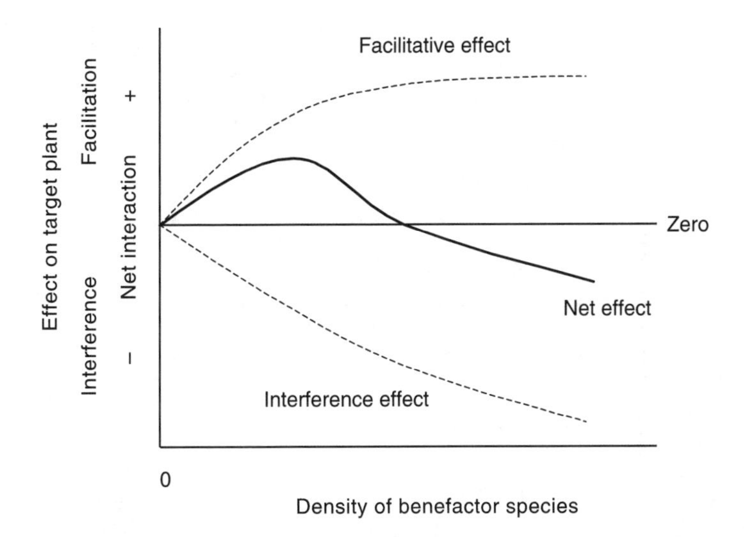

Figure 2.13 A benefactor plant will commonly also have negative effects on a target plant due to interference. The net effect of positive and negative will either be net facilitation, no effect, or net interference. In some circumstances, a switch from a positive to a negative net result can be expected along the benefactor's density gradient. An optimum benefactor density can be expected at which net facilitation is maximum or — more commonly — net interference is minimum. (Modified from Vandermeer, 1989, Figure 4.5.)

agricultural intensification. Such gradients occur on a geographical scale, but sometimes also within a household or field, or in the same place over years. In the face of such tremendous variability, it is important to identify, if possible, some general trends that we can expect in plant interactions and their consequences when moving along one or more of these gradients.

In past decades, plant ecologists have been intensively studying interference and facilitation along resource (productivity) gradients in natural plant communities. To date, empirical results are ambiguous and explanatory theories are controversial and mostly at the hypothesis level. Monocrop yields (e.g., Wallace, 1990) and, to a lesser extent, intercrop yields (e.g., Rao and Willey, 1980) have been studied in nutrient and soil moisture gradients. Results are also ambiguous and have seldom been explained in terms of plant interactions (but see Vandermeer, 1989; Santiago-Vera, García-Barrios, and Santiago, submitted; Santiago-Lastra et al., submitted).

Interference in Productivity Gradients

In the plant ecology literature, productivity gradients are defined in terms of soil resource availability and standing plant biomass. Interspecific competitive intensity is usually defined in relative terms as the difference in performance between the average individual in a monospecific and a mixed stand, divided by its performance in the monospecific stand (Grace, 1993).

Two conceptual models are most commonly used to predict changes in relative competitive intensity (RCI) with increasing environmental productivity (Figure 2.14).

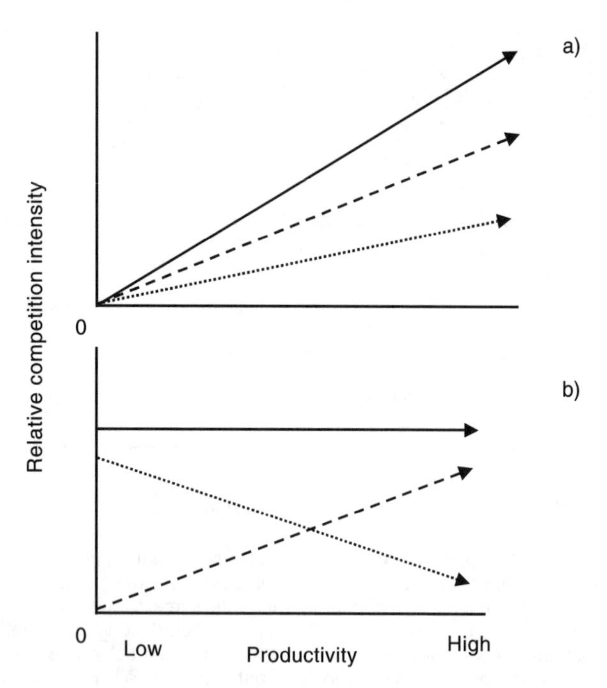

Figure 2.14 The two most common hypotheses regarding the response of above- and below-ground relative competitive intensities to increasing productivity gradients: (a) both RCIs and net RCI increase (Donald, 1958; Grime, 1977; Keddy, 1958); (b) above-ground RCI increases while below-ground RCI decreases, such that net RCI is fairly constant (Newman, 1973; Tilman, 1988; Wilson and Tilman, 1995).

The first is that RCI for both light and soil resources increases with productivity (Donald, 1958; Grime, 1977; Keddy, 1989). The second is that RCI for light increases with productivity but RCI for below-ground resources declines (Newman, 1973; Tilman, 1988; Wilson and Tilman, 1988). In the second model, total competition intensity may not increase with productivity if below- and above-ground are negatively correlated and are of similar intensities along resource gradients (Peltzer, Wilson, and Gerry, 1998).

In the past few years, many studies have addressed these hypotheses. Theoretical discussion has been highly controversial, and empirical results bearing on this question are variable and contradictory. There have been recent theoretical efforts trying to explain and reconcile these contradictory results (e.g., Goldberg and Novoplansky, 1997; Stevens, Henry, and Carson, 1999) as well as comprehensive metaanalyses to identify dominant trends in the available data (Goldberg et al., 1999). No clear-cut answer has yet emerged, and some authors urge the need for new hypotheses and methods (Grace, 1995; Cahill, 1999). Five possible reasons for inconsistent and complex results are as follows. First, the level of diversity and the presence or absence of species turnover can affect results (Peltzer, Wilson, and Gerry, 1998). Second, in all studies, it has been assumed that above-ground competition does not influence a plant's ability to compete below ground, and vice versa. Yet, independence (additivity) versus interaction between above- and below-ground RCI is itself a function of resource availability (Cahill, 1999). Third, subtle differences in the metrics used to calculate competitive intensity have important consequences on observed trends (Grace, 1995). Fourth, both hypothetical situations can be found in the same experiment, depending on the range of soil fertility explored and the density of competitors (Miller, 1996). Finally, both hypotheses are special cases that apply under different types of resource dynamics and different types of interactions between the growth and survival components of fitness (Goldberg and Novoplansky, 1997).

In spite of the complexities and controversies of the topic, a few simple ideas that are widely accepted, or constitute reasonable hypotheses, can prove useful for a discussion of TASs and are outlined below.

As soil resources (minerals, metabolites, water) and standing biomass increase, light available per unit biomass is reduced (Goldberg, 1990).

Competition for soil resources is mitigated by the fact that individual depletion zones tend to overlap less than could be expected for a given density. This is because plants tend to avoid excessive interspecific root competition and, under certain circumstances, can develop vertically stratified root systems leading to complementarity in the use of soil resources (Schroth, 1999). Light competition is more global and zones of influence are more extended. Above-ground vertical stratification has the opposite effect of below-ground stratification, that is, light preemption by the taller plants (Huston and DeAngelis, 1994).

Depletion zones and soil resource uptake tend to be proportional to each competitor's biomass (symmetric interference) (Weiner, Wright, and Castro, 1997). The amount of light appropriated by the larger plant competitor tends to be more than proportional to its biomass, while interference exerted by the smaller plant is less than proportional to its size and frequently negligible (asymmetric interference) (Weiner, 1990).

Asymmetric interference constitutes a positive feedback loop that enhances the slightest initial differences in per-unit biomass light availability between competing individuals (García-Barrios, 1998; Cahill, 1999).

Light and soil resource availability per capita in increasingly productive environments differs hypothetically in various ways. Light availability decreases in the presence of neighbors and differs markedly between canopy and subcanopy individuals (Figure 2.15). Soil resource availability increases (by definition), is relatively less affected by neighbors, and differs less between unequally sized individuals (Figure 2.16). It is reasonable to expect that in low productivity environments, interference will occur mostly below ground and will be symmetric, while in highly productive environments it will occur mostly above ground and will be asymmetric (Figure 2.17).

Below-ground competition will not affect above-ground competition if (1) plant biomass is sparse and unaffected by light competition, (2) the canopy is closed but the plants of interest are perennially in the understory and are adapted to shade tolerance, (3) the dominant plant's biomass is so high that initially lower plants never get a chance to compete for the canopy (Cahill, 1990). If, on the contrary, a plant is affected by below-ground interference and this impairs its ability to avoid being overtopped by neighbors, one can expect a synergistic increase in both above- and below-ground interference. Therefore, it is reasonable to expect in many cases

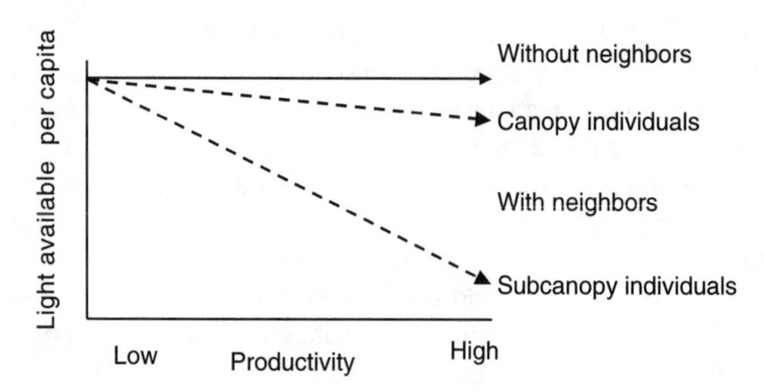

Figure 2.15 Schematic diagram of light availability per plant in increasingly productive environments. Canopy and subcanopy target plants are considered, with and without neighbors.

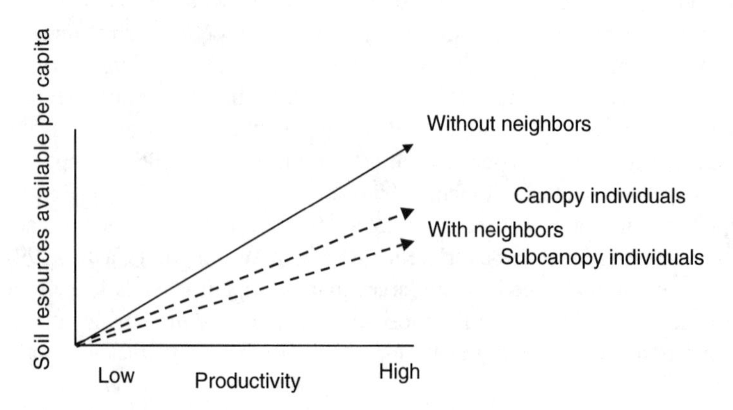

Figure 2.16 Schematic diagram of soil resource availability per plant in increasingly productive environments. Canopy and subcanopy target plants are considered, with and without neighbors.

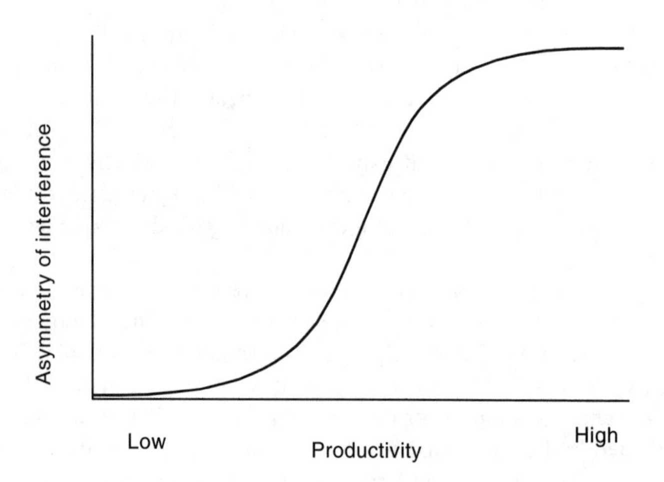

Figure 2.17 Overall asymmetry of plant interference as affected by environmental productivity.

low interaction between above- and below-ground interference at the ends of a water or nutrient gradient, and strong interaction in the middle. Two hypothetical examples of this situation are presented in Figure 2.18.

For canopy-dominant species and subcanopy shade-tolerant species, one can expect below-ground competition either to decrease or to remain constant along the productivity gradient, with no significant increase in above-ground competition. In tropical agroecosystems, decreasing below-ground competition should hold for a tall stature monocrop moderately infested with weeds. Decreasing or constant below-ground competition would be expected in variably sized multiple crops where one species clearly dominates the canopy and subcanopy species are shade tolerant. A clear example of this latter situation is the *Inga* spp. tree shaded coffee plantation.

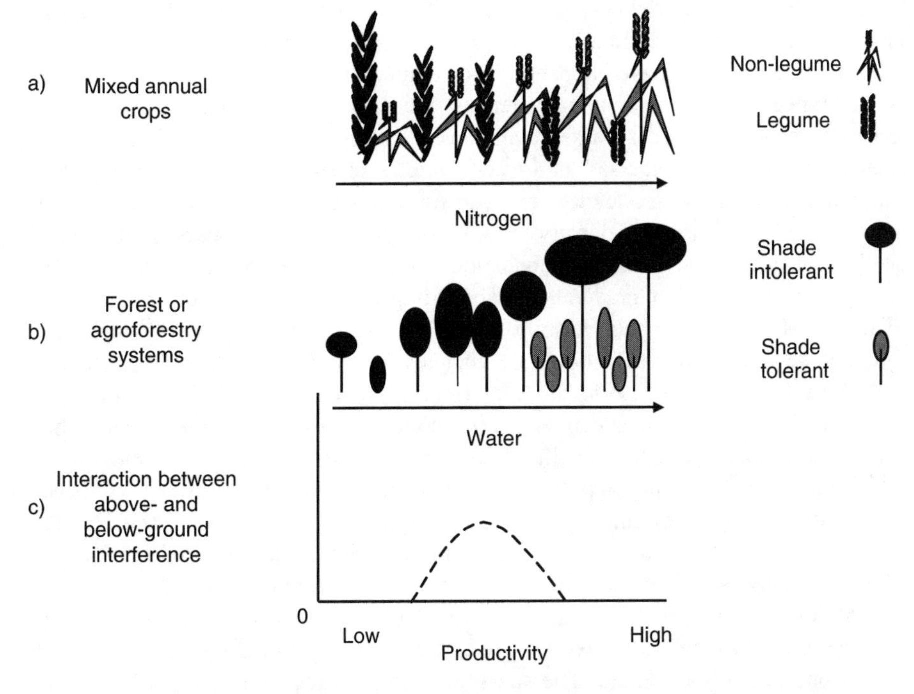

Figure 2.18 Interaction between above- and below-ground interference can change in a strongly nonlinear fashion along a productivity gradient. Consider the following hypothetical examples, (a) As water becomes more available, plant density, plant cover, and vertical stratification tend to increase, both in natural plant communities and in agroforestry systems. At intermediate water levels, where two tree individuals (or species) have roughly equal possibilities of occupying the canopy, below- and above-ground competitive status affect each other. (b) As nitrogen becomes more available, a canopy dominance shift can occur between a nitrogen-fixing legume and a nonlegume annual crop. At the ends of the gradient, clear canopy dominance by either one of them precludes above- and below-ground interference interaction. In the middle of the gradient, both can potentially dominate the canopy and, as in (a), interaction is expected. For both (a) and (b), a humped interaction curve, depicted in (c), would be expected.

Whether weed infested, short sized monocrops, even sized multiple annuals, perennial crops, or alley crops, most tropical agroecosytems have shade-intolerant species strongly competing for the canopy. In such cases, increasing productivity will enhance positive feedback, which can drive a plant either to grow further and to dominate the canopy or to be overtopped and reduced by others. The outcome for a given species or individual is highly sensitive to factors such as density, sowing time differences, microhabitat, and genotype. Canopy-dominant species should not experience significant change in above-ground relative competitive intensity along productivity gradients, while shade-intolerant species left behind in the subcanopy should be very sensitive.

Noncultivated plant diversity in an agroecosystem (e.g., weeds and other successional flora) can also be influenced by productivity gradients. In natural plant communities, species richness declines (either monotonically or with a hump) as soil fertility increases (Figure 2.19) (Abrams, 1995). Some authors (e.g., Newman, 1973) consider this to have happened when competitively stronger species suppress weaker species and exclude them as fertility increases. Others (e.g., Oksanen, 1996) have recently suggested an assemblage-level thinning hypothesis, which proposes that individuals of all species tend to become larger as fertility increases and that individuals of all species tend to exclude subordinate individuals of their own or other species. Because total density declines, samples of finite numbers of individuals will result in fewer species by chance alone. In other words, plant species richness along many productivity gradients may be strongly influenced by total stem density. Thus, differences in competitive ability among species, although generally important, are not necessary to create dramatic changes in species richness along fertility gradients (Vandermeer, 1996; Stevens, Henry, and Carson, 1999). The complex interplay of intraspecific and interspecific competition can give rise to situations predicted both by the interspecific competitive exclusion hypothesis and by the assemblage-level thinning hypothesis. The second one could prevail in very dense, highly diverse plant communities. Weed (and other early successional species) floristic richness in tropical agroecosystems might be influenced by productivity gradients, although management might mask or entirely eliminate this effect.

Multiple crop coexistence and yield advantage are also affected by increased soil resources (Figure 2.20). No unique and consistent trend has been established, nor is one likely to be found. The sparse literature on the subject suggests that, as nitrogen is increased, the land equivalent ratio (LER) decreases (Rao and Willey, 1980), follows a U-shaped curve (Oelsligle and Pinchinat, pers. comm., 1975), or remains fairly constant (Kavamahanga, Bishnoi, and Aman, 1995). As soil moisture is increased, LER either shows no statistically significant variation (Rao and Willey, 1980) or is reduced (Kass, 1978; Natarajan and Willey, 1986; Santiago-Lastra et al., submitted; Santiago-Vera, García-Barrios, and Santiago, submitted). Of special interest for tropical rain-fed agriculture is the case where an increase in soil moisture can shift the LER > 1.0 situation into an LER <1.0 situation, in which the intercrop advantage over monocrops is lost and crop coexistence is ecologically and economically discouraged. This possibility has not been reported earlier but is supported by preliminary data (García-Barrios, 2000) and is currently under trial.

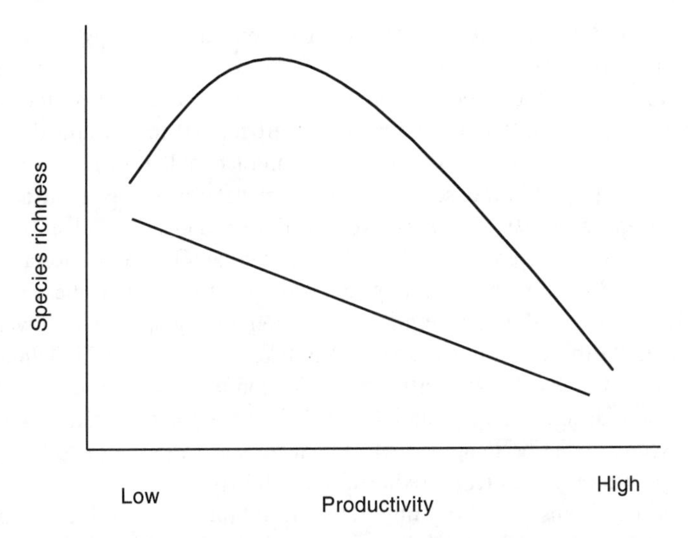

Figure 2.19 In natural plant communities, species richness declines (either monotonically or with a hump) as soil fertility increases. (Based on Abrams, 1995.)

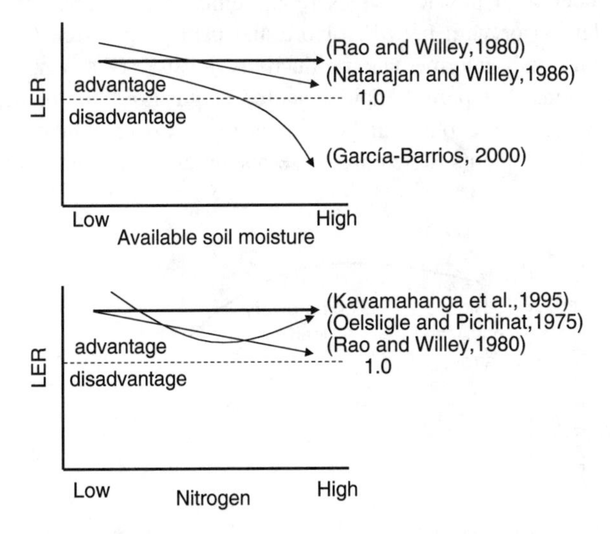

Figure 2.20 Schematic diagram of intercropping land equivalent ratio (LER) response to available soil moisture and nitrogen gradients, as reported in different sets of experiments. If LER is greater than 1.0, the intercrop uses land more efficiently than the corresponding monocrops.

Plant competitive abilities and stand composition are expected to change in a productivity gradient (Tilman, 1984; Grace, 1990). For example, in some low-nitrogen grass–legume swards, the legume can outcompete the grass due to its N-fixation symbiosis with *Rhizobium* spp. As nitrogen accumulates in the soil through legume debris, the grass eventually recovers and outcompetes the legume through shading, until nitrogen is lost again through herbivory and leaching and the legume is again able to recover an important status in the community. An interesting dominance oscillation is established, and both species tend to coexist over the long term (Hunter and Aarsen, 1988). On the contrary, if nitrogen fertilizer is added repeatedly, the legume will eventually be excluded (Stern and Donald, 1962).

Most tropical agricultural soil conditions are heterogeneous even at the field level (Vandermeer, 1989). Farmers have coped with and taken advantage of this situation by sowing a different crop in each microhabitat (García-Barrios and García-Barrios, 1992). Noncultivated plants might also coexist better in these conditions that preclude absolute dominance and suppression among species by favoring different species at different points in space (Whittaker and Levin, 1977; Tilman, 1982; Huston and DeAngelis, 1994). Fertilizer application tends to homogenize soil conditions and to reduce the need of and the conditions for plant diversity (Hall, 1995). A similar scenario can be imagined when year-to-year variation in soil moisture in rain-fed agriculture is drastically reduced through irrigation.

Competition seems to be very important toward the more productive end of the environmental gradient, while facilitation could be more important under harsh conditions (Figure 2.21) (Bertness and Callaway, 1994). Positive plant interactions might prove to be particularly common and relevant in agricultural communities developing under high physical stress (e.g., semiarid tropical systems with low productivity levels) or with high herbivore and parasite pressure (e.g., subhumid tropical systems with intermediate productivity levels) (Oksanen, 1990). Under physical conditions that permit rapid resource acquisition, competition could be intense. However, severe physical conditions (e.g., extreme heat, salinity) may restrict the ability of plants to acquire these resources. Any amelioration of severe

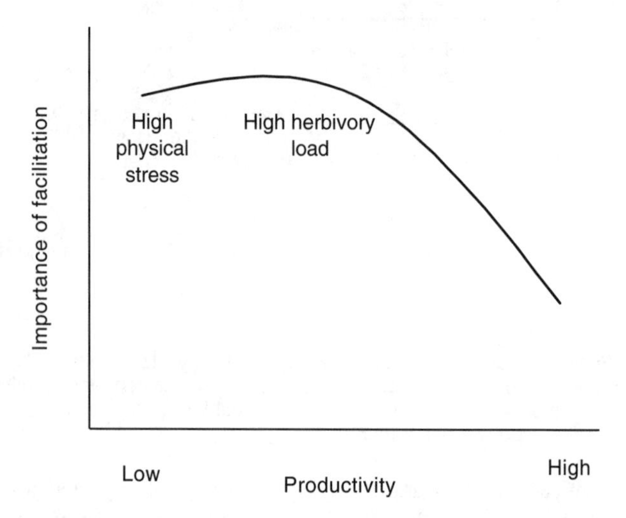

Figure 2.21 Positive plant interactions might prove to be particularly common and relevant in agricultural communities developing under high physical stress (e.g., semiarid tropical systems with low productivity levels) or with high herbivore and parasite pressure (e.g., subhumid tropical systems with intermediate productivity levels).

stress by neighbors may be likely to favor growth to a greater extent than competition for resources is likely to restrict growth. For example, light limitation outweighs moisture limitation in mesic habitats, thus negating the possible favorable effects of slight increases in moisture under plant canopies. In contrast, moisture limitation is more important in xeric habitats than light limitation. Thus, the slight decreases in understory light are outweighed by proportionally high increases in soil moisture (Holmgren, Scheffer, and Huston, 1997).

In practice, there will always be a complex gradient of various resources and disturbances, from the low-productivity open stand to the high-productivity closed canopy. Over this gradient, some factors (e.g., nutrients, water) might change for the better, whereas others (e.g., light, allelopathic exudates) might change for the worse. The net effect of these correlated changes will depend on the combined response of plants to all factors involved. If a factor that improves under the canopy happens to be the only limiting factor, that is, if the effect of the simultaneous change in other factors is nil, the net effect of the canopy will obviously be facilitative. However, in reality, the matter will usually be more complicated. Rarely will the impact of one environmental factor be independent of the value of others (Holmgren, Scheffer, and Huston, 1997).

PLANT INTERACTIONS AND SYSTEM DESIGN AND MANAGEMENT

In the tropics, a typical agricultural system will have annual crops, trees, bushes, weeds, arthropods, microoganisms, livestock, and other associated fauna, all interacting both positively and negatively with each other. Some of these species are deliberately introduced and displaced in space and time by the farmers, while others are wild species that are either fought, tolerated, or fostered as semidomesticated organisms. Ecological interactions in this species ensemble have important consequences on the system's overall performance. The way the system is designed and managed both responds to and affects such interactions.

A myriad of different agricultural systems can be identified in the tropics on the basis of their composition, design, and management. This is a consequence of the diversity of biophysical environments, cultures, production purposes, land-use intensities, and resource availabilities encountered. It would be impossible to consider all the ways in which system design and management relate to plant interactions, even in the major tropical systems. Rather, I will attempt a more general analysis and will focus on three design and management factors that relate to different aspects of the system's functional and structural diversity in space and time and that can be important when trying to understand and optimize plant interactions for agricultural sustainability purposes. These factors are:

- The permanence of a specific crop plant assemblage on a patch of land or, conversely, the frequency of land-use rotation
- The intensity of intercropping, meaning the number, type, and level of spatiotemporal concurrency of crops within the field
- The percentage of tree canopy cover in the field

Each of these three factors can be divided into somewhat arbitrary levels, with the gradient displayed on an axis (Figure 2.22). TASs can then be classified and placed accordingly in the resulting three-dimensional space. Figures 2.23 and 2.24 show some major TASs, classified according to these criteria.

Frequency of Land-Use Rotation

Most TASs derive originally from forest clearing (Hougthon, 1994). Under very low social pressure on land, shifting agriculture constitutes an adequate long-term rotation system where mature forest alternates with highly diverse, very productive, ephemeral agricultural fields. As land use is intensified, fallow gives place to rotations or intensive intercropping systems where short-term rotations are common. Further intensification and specialization can eventually eliminate intercropping and crop rotation. Although there is generally a loss of species diversity and of tree components along the intensification process, there are many permanent TASs (both traditional and modern) that conserve these attributes, albeit to a lesser degree (Figure 2.23).

Slashing and burning of primary or secondary forest reduce native seed banks and produce an important flush of nutrients available for crop growth. Where shifting agricultural cycles are still relatively long (10–20 years), during the first year a diversity of crops is combined in the different soil microhabitats encountered by the

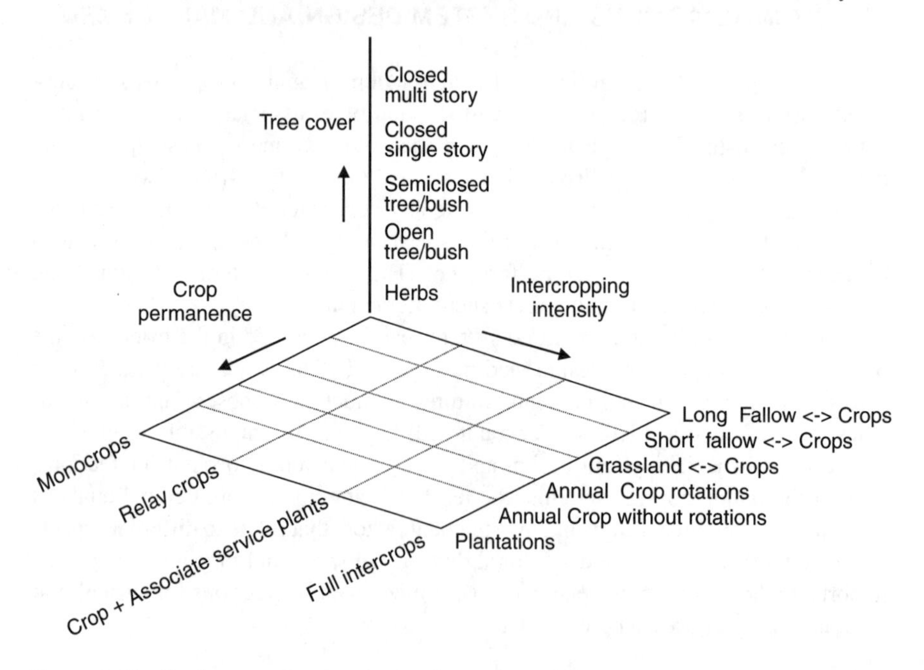

Figure 2.22 For the purpose of understanding and managing plant interactions, it is useful to classify major tropical agricultural systems according to their position on the following three axes: *X:* crop permanence on a patch of land, *Y:* intensity of intercropping, and *Z:* percentage of tree canopy cover. See text for further details, and Figures 2.23 and 2.24 for examples, which the reader can easily place in this figure.

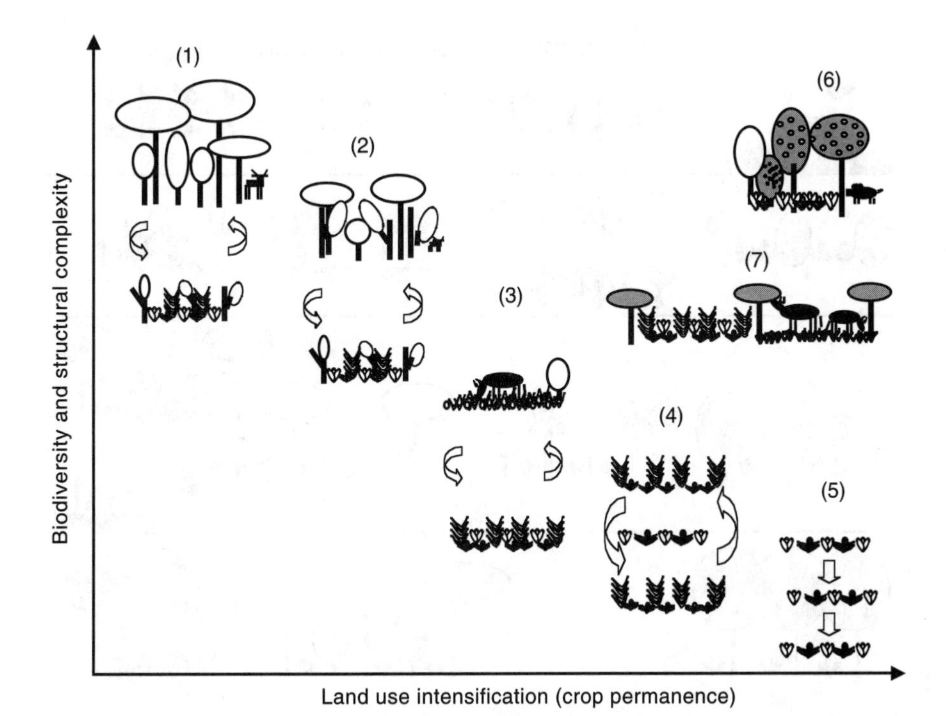

Figure 2.23 As land use is intensified and rotation cycles are shortened or eliminated, the biodiversity and structural complexity of tropical agricultural systems tend to diminish (icons 1–5). Nevertheless, some intensive systems (both traditional (6) and more recently developed (7)) maintain these characteristics, albeit to a lesser degree. (1) Long fallow systems; (2) short fallow systems; (3) ley systems, grass and small bushes in rotation with annual crops; (4) annual crop rotations; (5) annual monocrops without rotation, and plantations; (6) mature multistory home gardens; (7) multipurpose alley cropping.

farmer. Crops interact mainly among themselves, and with low-density weed populations and small tree stump resprouts that commonly facilitate shade-tolerant species and vine crops (Alemán, 1985). After the first year, production occurs in the midst of increasing nutrient depletion, of weed population buildup, and of the successional process typical of the cleared vegetation stand (Mariaca et al., 1995). The farmer does not fight against these processes but abandons the field after 2 or 3 years to a new fallow period.

Unfortunately, fallow periods are being severely reduced throughout the tropics to 4 to 6 years or less (Houghton, 1994). In such cases, the tree and bush regrowth component decreases, weed interference and allelopathy grow very rapidly to become a major problem, and crop diversity and productivity tend to diminish. (Ramakrishnan, 1988; Liebman and Stavers, 2000). Efforts are being made — still at an experimental and small-scale level — to improve the short fallow phase in order to reduce its weed components and to build up fertility more rapidly. Improvements include introducing legume cover crops and rapid-growing legume trees (e.g., *Sesbania sesban, Gliricidia sepium, Leucaena leucocephala*; Schroth, 1999) during

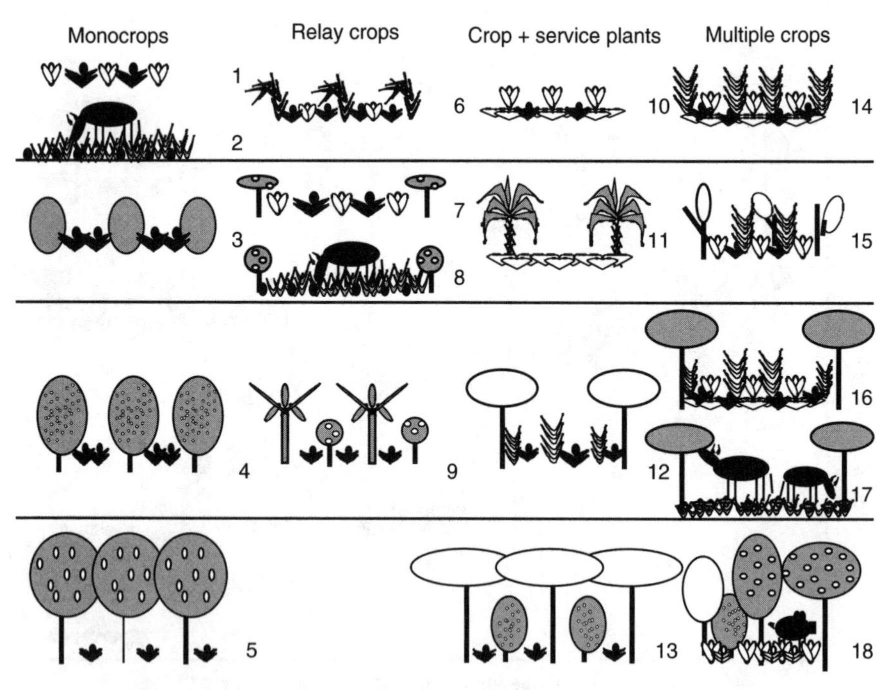

Figure 2.24 Schematic representation of some major tropical agricultural systems, classified according to intercropping intensity (rows) and percentage of tree cover in the field (columns). One or more examples of each system are presented. *Monocrops*: (1) sugar cane, corn, millet, rice, yucca; (2) cultivated grasslands; (3) tea plantation; (4) unshaded coffee plantation, citric orchard; (5) rubber plantation, mango plantation. *Relay crops*: (6) pole bean–mature maize, potato–young maize; (7) (taungya) vegetables or staple crops in young fruit tree plantations; (8) (taungya) cultivated grasses in young fruit tree plantations; (9) bananas in young mango plantation. *Crop + service plants*: (10) yucca or corn + legume cover crop; (11) (taungya) young oil palm plantation + legume cover crop; (12) (alley cropping) corn + green manure legume tree; (13) monospecific tree shade coffee plantation. *Full intercrops*: (14) maize–sorghum, rice–soybean, millet–groundnuts, maize–cassava, tomato–watermelon; (15) (milpa) maize–bean–squash + tree stump regeneration; (16) (alley cropping) maize–bean with multipurpose legume trees; (17) mixed pastures with multipurpose legume trees; (18) mature multistory home gardens, mixed multipurpose tree plantations with coffee and cacao.

the fallow period, which are not burned but incorporated as green manure or dead mulch in the subsequent cropping phase, both for further weed control and for slower nutrient release to avoid runoff and leaching (Rejintjes, Haverkort, and Waters-Bayer, 1992; Thurston, 1994; Rao, Nair, and Ong, 1998).

Where fallow periods have disappeared, short-term rotations still have important effects on plant interactions (most of them favorable to the farmer). Rotating crops, subject to dissimilar management practices and with contrasting phenology, physiology, and competitive characteristics, can sometimes stop the buildup of weed, insect, and pathogen populations and prevent allelopathic autotoxicity (García-Espinosa, Quiroga, and Granados, 1994; Liebman and Gallandt, 1997). Cereals rotated with grain legume crops can strongly outyield continuous cereal cropping (up to

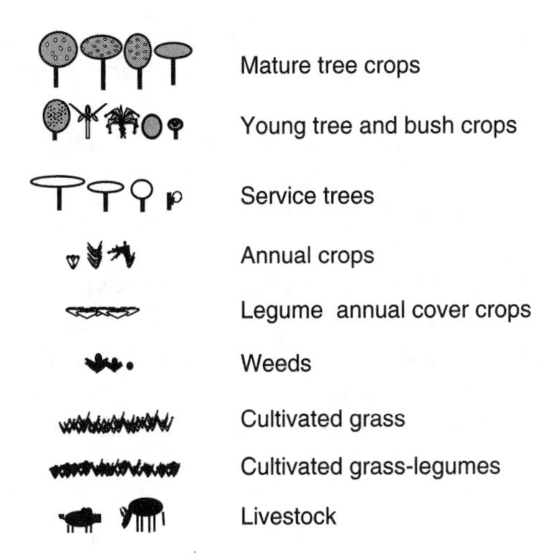

Figure 2.24 (Continued).

60%), more so if exogenous nitrogen input is low (Francis and Clegg, 1990). Tropical grass (*Pennisetum purpureum, panicum maximum*) (Rejintjes, Haverkort, and Waters-Bayer, 1992) and legume cover crops (e.g., *Canavalia ensiformis, Stizolobium* spp., *Mucuna pruriens*) in rotation with major crops remain as dead mulches and produce allelopathy, shading, cooling, and obstruction, which deter weed establishment (Mohler, 1991; Bunch, 1994; Liebman and Davis, 2000). Nevertheless, some crop rotations have their drawbacks: dead mulches can attract noxious fauna, and allelopathic residues from previous crops can affect later ones, so compatibility must be considered.

Intercropping Intensity

Except for weed-free monocrops (an uncommon condition in the tropics), all tropical agricultural fields include a diversity of plant species and can be loosely considered as intercrops. Some of their plant components are perceived as clearly noxious and others as clearly beneficial. Yet, most have both positive and negative effects on the system's sustainability. When designing and managing a crop field, a tropical farmer might focus on short-term benefits (e.g., net income this year), medium-term benefits (e.g., economic and environmental risk reduction), or long-term benefits (e.g., soil conservation and natural pest population control). These different criteria are not always compatible and, at the present time, short-term benefits tend to dominate over long-term ones (García-Barrios and García-Barrios, 1992). We can group plant species into the following types, according to how the farmer perceives their benefits: (1) highly valued crops (e.g., major staple crops, cash crops), (2) secondary crops (e.g., produced for sporadic local consumption), (3) nonharvestable service plants (e.g., cover crops and tree foliage used as green manures), and (4) nonbeneficial weeds.

In tropical agroecosystems, two or more of these plant types can be found together in the same field and can co-occur either partially or totally in space or time. We find it useful to define four categories of intercropping intensity according to the plant types and the level of concurrence involved: (1) monocrops, (2) relay crops, (3) crops associated with service plants, and (4) full intercrops. Weeds are an additional component in all four groups. As can be expected, when intercropping intensity increases, more opportunities for positive interactions occur, but more trade-offs have to be considered in the case of management practices that affect plant species in opposite ways.

In monocrops, intraspecific interference is the most obvious interaction. It defines optimum density (Willey and Heath, 1969), as well as crowding effects on individual plant variation, which can be important when produce homogeneity is an important issue (Benjamin, 1992). Autotoxicity due to allelopathy has been found for some major crops such as sunflower (*Heliantus annuus*), sorghum (*Sorghum bicolor*), and, to a lesser extent, rice (*Oryza sativa*), wheat (*Triticum aestivum*), alfalfa (*Medicago sativa*), and other perennial legumes (Einhellig, 1995). Net facilitation in early growth stages due to microclimatic amelioration has also been reported (García-Barrios, 1998).

Although formally monospecific stands, tropical monocrops coexist in the field with a dozen to more than a hundred different weed species, which can reduce crop growth and yields from 10% to 25% (Liebman and Gallandt, 1997) through interference and allelopathy (Putnam and Weston, 1986; Tongma et al., 1998). Weed effects depend on their density, mass, leaf area, species composition, time of emergence, and period of association with the crop (Kropff and Lotz, 1992). Weeds can sometimes be beneficial; many are tolerated as secondary crops and service plants in peasant economies (García-Barrios and García-Barrios, 1992). In some circumstances, rapid-growing weeds can reduce barren soil erosion and protect the crop against desiccation during early crop growth (Altieri, 1988; Berkowitz, 1988), but an obvious trade-off is established when further weed growth can produce nutrient and water depletion. Weeds can either reduce or promote polyphagous herbivores and their natural enemies. Current understanding of these contrasting effects is quite limited (Andow, 1988).

Dense monocrops reduce weed performance significantly and, when a range of optimal crop densities exists, very high monocrop populations constitute a natural means of weed control (Weiner, Griepentrog, and Kristensen, 2000). Nevertheless, because most weeds exhibit immense phenotypic plasticity, and seed production per capita can vary by at least two orders of magnitude when the individual is confronted by intense crop interference, they can persist at a minimum level (Cussans, 1995), and eventually reconstruct their populations, assisted by efficient dispersal mechanisms. Crop allelopathic effects on weeds have been demonstrated but might not be as common as the reverse situation (Einhellig, 1995). Crops can also affect weeds in a more indirect fashion — many weed-feeding insects, weed pathogens, and weed seed predators can thrive in the crop stand or in its residues (House and Brust, 1989; Kremer, 1993). Optimum management strategies exploiting these interactions have yet to be developed (Wyse, 1992; Liebman and Gallandt, 1997).

Although the former considerations of weed–crop interactions apply to all levels of intercropping intensity, it is important to stress that weeds have become particularly intractable in monocrops. Weeds evolve, under selection pressures, to resist herbicides, to mimic crops, to readily synchronize with crop phenology, and to complete their life cycles between tillage and other elimination episodes. Ecological weed management requires subjecting weeds to multiple, temporarily variable stresses that disrupt selection. In other words, many small hammers are needed instead of one big hammer (Liebman and Stavers, 2000). There are limits to such management shifts in strict monocropping and better opportunities in more temporarily and spatially diverse systems

Relay Crops

Semiconcurrent or relay cropping is a modality of intensive monocrop rotation, mostly found where agriculture can be practiced year-round, which shares some characteristics with full intercropping. In the most common cases, crops coexist in the field during a relatively small part of their life cycle. A first modality occurs when a crop is almost ripe and has lost an important part of its foliage, so a second crop is planted in order to take advantage of the incoming light, the favorable microclimate, the water stored in the soil, and residual fertilizers. Typical examples are winter beans climbing on ripe corn stalks in milpa fields (Alemán, 1985) and cover legumes sown into a major crop near harvest time in order to produce a dead mulch for the next cycle (Rejintjes, Haverkort, and Waters-Bayer, 1992; Melara and Del Río, 1994). A second modality occurs when a rapid-growing species is planted during the initial stages of a slow-growing crop in order to take advantage of available space and resources and to protect barren soil from erosion and weed invasion. Early maturing vegetables associated with slower growing cereals and tree-based "Taungya systems" (discussed later in this chapter) are examples of this modality. In the second case, there is a trade-off between exploiting available space with a second crop and delaying the growth of the slower species, which requires fine-tuning of spatiotemporal design. Concurrence can extend for larger periods in some relay crops; a judicious selection of plant arrangements and of sowing and harvesting times for each crop can help avoid suppression of the less competitive species and obtain intercropping advantages (Fukai and Trenbath, 1993; García-Barrios, 1998).

Crops Associated with Service Plants

Natural plant communities have traditionally provided tropical agriculture with various ecological services such as pest control, soil protection, and fertility buildup. Since many of these communities have been dislodged by intensification processes, efforts are being made to promote adequate natural substitutes for these services. More than 30 herbaceous legumes and grasses and a number of secondary growth, nitrogen-fixing tree species (Rejintjes, Haverkort, and Waters-Bayer, 1992) are being domesticated and associated with major crops to serve as cover crops and green manures. Herbaceous service plants are most commonly intercropped additively while trees are intercropped substitutively. If these systems are improperly managed, positive interactions can be exceeded by negative ones, which explains why many farmers are slow to adopt them (Velásquez et al., 1999). Crop production can be severely reduced if service species attract pests (Melara and Del Río, 1994), have allelopathic effects (Einhellig, 1995), or strongly compete with small- and medium-sized crops (Velásquez et al., 1999). In the semiarid tropics there is generally a strong need to incorporate organic matter into the soil but, unfortunately, service plants will compete strongly for water with crops, so they have to be sown at very low densities or avoided altogether until better water conservation strategies are developed (Ruthenberg, 1976; Rejintjes, Haverkort, and Waters-Bayer, 1992).

Full Intercrops

More than 50 cereal–legume, cereal–cereal, vegetable–vegetable, and tuber–legume combinations of two to five annual crops are very commonly associated in the tropics (Kass, 1978; Vandermeer, 1989), and multistory tree-based systems can comprise as many as 200 plant species in a single home garden (Soemarwoto and Soemarwoto, 1984). In all but the simplest intercropping systems, every type of plant interaction discussed previously is to be expected. The interplay of intraspecific and interspecific interference, allelopathy, and facilitation depends on the environment, on the species involved, and on their spatiotemporal arrangement and management. It is extremely difficult to elucidate the synergisms, antagonisms, and relative contributions of the different interactions in these systems. Nevertheless, comprehensive theoretical frameworks have started to develop (e.g., Vandermeer, 1989), and some recent and very interesting experimental work has been undertaken with relatively simple intercrops (e.g., Kluson, 1995). A more phenomenological approach focuses on the overall consequences of such interactions by comparing the performance of variants of the same crop association with that of its monocrops (e.g., Willey, 1979). A number of multicropping advantages over monocrops (Francis, 1990) were pointed out at the beginning of the chapter. Note that these advantages (1) do not occur systematically in all crop associations, (2) are not equally valued by all farmers, (3) are modified by changes in intercrop design and management, and (4) can be optimized within constraints. When confronted with the possibility of planting a set of crops either together or separately, a farmer will explicitly or implicitly evaluate these choices in terms of land-use efficiency, absolute yield, economical profit, input replacement benefits, yield variability, and other criteria (Vandermeer, 1989; García-Barrios, 1998). A single intercropping design and management strategy for a set of crops will not normally satisfy or optimize simultaneously all desired criteria; so in selecting the most appropriate, trade-offs will be encountered between the different benefits that can result from interactions in diversified plant communities.

As to weeds in full intercropping systems, an important issue is whether weeds are better suppressed by increasing crop diversity per se or by its combined effect with increasing crop density. Studies addressing this issue are still scant and controversial (Liebman and Stavers, 2000). Although it is clear that total intercrop density reduces weed biomass as well as dominance by any single weed species, it is still difficult to predict which management options will maximize the strength of intercrop interference against weeds while reducing interference between crop components and maximizing their yield (Liebman and Stavers, 2000).

The Percentage of Tree Canopy Cover in the Field

Trees can simultaneously have strong positive and negative effects on agroecosystem structure and function and on its economic performance (Sharrow, 1999; Nissen, Midmore, and Keeler, 2001). Trees can ideally exert low below-ground interference on more superficially rooted crops and have major positive effects on agroecosystems, conditions that are not easily matched by other groups of plants.

Trees produce fuel, construction material, fodder, staple crops, and cash crops. They improve the fertility and physical structure of soils, reduce erosion, host nitrogen-fixing bacteria and mycorrhizal fungi, add organic matter, and slow down its decomposition rate. Trees reduce weed populations and change weed floristic composition. They provide shade and protection against wind and river overflow for certain crops and animals (Rao, Nair, and Ong, 1998). Tree-dominated systems (Perfecto et al., 1996) and strips of buffer vegetation around fields (Gliessman, 1998) can serve as shelter for arthropods, birds, and small mammals. On the other hand, as trees increase in density, size, and ability to capture resources, they become less susceptible to interference by annual and perennial herbaceous crops and weeds and can even suppress the latter's growth, thus reducing the yields of smaller crops if improperly selected and managed.

Trees can be found in systems with all levels of intercropping intensities, and can dominate the field at different levels. I review here some specifics of plant interactions in tree-based systems not considered in the previous section.

Tropical tree monocrop plantations are relatively modern systems. Their canopies can be relatively open (e.g., unshaded coffee, tea), semiclosed (e.g., citric orchards) or very closed (e.g., rubber, mango). Tree canopies do not overlap in monocrop plantations, so light interference is rarely a problem. Tree roots tend to be concentrated beneath their shoots (Atkinson, 1980) but can extend further in semiarid climates (Rao, Nair, and Ong, 1998). Therefore, we can expect below-ground interference between trees to be stronger, and available soil resource uptake, to be more complete in closed canopy systems. This, together with light preemption, leaves little space for herbaceous crops and weed development and produces barren soils. Consequently, wind and raindrop splash erosion can sometimes be a problem.

Even in closed canopy monocrops, during plantation establishment, there are abundant light and soil resources for annual crops, grasses, or cover crop production while the trees grow. This strategy, known as the Taungya system, allows for very successful relay cropping, but again it requires care to prevent temporary components from slowing tree growth excessively (Vandermeer, 1989; Bunch, 1994). Recent explorations consider ecological, social, and economic trade-offs in these systems and define the conditions under which Taungya system results are viable and beneficial for farmers (e.g., Nissen, Midmore, and Keeler, 2001).

In some mature plantations, optimum tree distance does not produce a closed canopy, and cover crops are recommended between tree rows (Liebman and Stavers, 2000). Trees themselves become service plants, for example when used as green manure in alley cropping systems or as shade in coffee and cacao plantations. The latter aromatic crops are normally shade tolerant and root interference is the major issue. Nevertheless, in all cases, trees should be pruned periodically and cuttings should be incorporated as dead mulch. The enhanced light condition and soil cover favor crop growth while weeds are still controlled. Most alley cropping research and development has concentrated on leguminous species whose main or only function is to improve nutrient cycling and soil fertility. This makes sense to farmers only if the annual crops outyield the sole crop in the short or long term. Relatively uncompetitive service tree species are desirable in this case, but these commonly are not very efficient in pumping up leached nutrients. More aggressive species can be used as long as shoot or root prunings are properly done (Schroth, 1999).

Alley cropping systems can be considered full intercrops when tree cuttings are used not only as green manure but also for other purposes (e.g., fodder, fuel wood). Tree selection should not be based only on consumption preferences but should consider ecological compatibility with associate crops. For example, different tree root characteristics are necessary for alley cropping with annual cereal crops, grasses, or nitrogen-fixing cover crops (Schroth, 1999). Tree root systems are already developed when the crop is still establishing its own roots, so tree roots should explore lower soil layers and not more superficial ones. Grasses have deep root systems and are normally there when the tree is established, so the trees should rapidly reach and explore even deeper strata. On the contrary, when trees are associated with legume cover crops, superficial tree-root systems are desirable in order to take better advantage of nitrogen fixed by *Rhizobium* and liberated by the cover crop's debris.

Wrong tree selection and lack of pruning can reduce by as much as 50% the potential annual crop yield obtainable in alley cropping systems (Rao, Nair, and Ong, 1998). Annual crop response to interaction with trees can change strongly along productivity gradients and differs according to crop species (Figure 2.25).

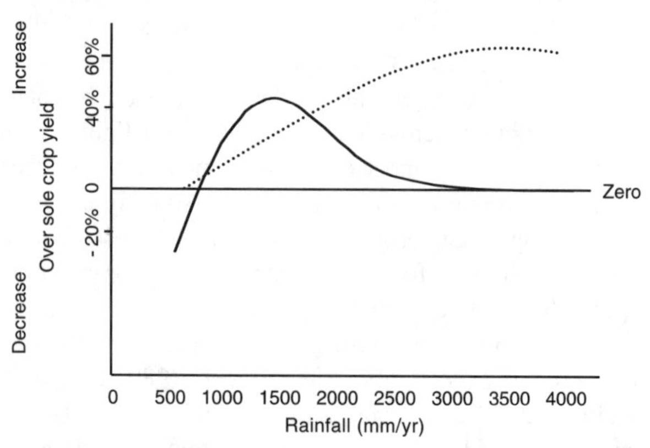

Figure 2.25 In alley cropping, relative benefit for the annual crop changes according to annual rainfall and crop species. A schematic representation of tendencies derived from 29 experiments throughout the tropics. (Modified from Rao, Nair, and Ong, 1998. Figure 3.)

Therefore, selecting in each environment the proper annual crops, tree species, and densities, and the timing, amount, and distribution of prunings among its various uses is critical to optimize trade-offs between tree interference and facilitation, so as to achieve the desired balance between tree growth, annual crop production, tree crop production, weed suppression, and soil improvement (Vandermeer, 1998; Rao, Nair, and Ong, 1998). Analytical tools such as those proposed by Vandermeer (1998) and Blair (1998) can be useful for this purpose. Unfortunately, alley cropping management can demand a great deal of experience and hand labor. Under current social conditions, the latter commonly constitutes a severe bottleneck, which partially explains the very slow adoption of alley cropping by tropical farmers (Giampetro, 1997; M. Liebman, pers. comm., 1999).

Mixed canopy coffee and cacao plantations and mature multistory home gardens are the most diverse and complex intercropping systems in the tropics and the world (Landauer and Brazil, 1990). Diet diversity, social life and recreation (Estrada, Bello, and Serralta, 1998), natural weed control (Christanty et al., 1986), and, in some cases, wildlife conservation (Soto-Pinto et al., 2000) are their most ubiquitous benefits. Plant interactions in these systems, and their consequences on yields, might be too complex to tackle mechanistically. Nevertheless, some studies have revealed, for example, that in mixed-canopy coffee plantations, tree interference (Nestel and Altieri, 1992; Liebman and Gallandt, 1997) and allelopathy (Anaya et al., 1982) can shift the weed community toward less competitive species and that, although percentage of shade cover is significantly related to yield (Soto-Pinto et al., 2000), PAR and coffee yields under diversified and monospecific (*Inga* sp.) tree canopies might not differ statistically (Romero et al., submitted; also see Perfecto and Armbrecht, this volume, Chapter 6).

CONCLUSIONS

Almost any agroecosystem in the tropics contains a diversity of plants that the farmer wishes to control either to enhance their growth and optimize their benefit or to keep them under control and reduce the problems they generate. According to the features and purposes that the farmer wishes to favor, a number of ecological opportunities, limitations, and trade-offs derived from plant interactions have to be considered when managing the different plant components.

From an ecosystem perspective, the idea of creating a resource-tight system in which few nutrients escape has merit for sustainability purposes. However, complete closure contradicts the primary goal of both crop and livestock production systems: to harvest resources for sale or local consumption. A more realistic goal is to attain an efficient use of natural and human-made agricultural inputs and outputs, as well as a proper balance between them, at all scales. At the field level, all plant interactions directly or indirectly affect resource distribution among ecosystem components as well as resource capture, recycling, and conservation at the whole-system level. When manipulating plant–plant interactions, we should pursue optimum (not necessarily maximum) allocation toward components that are eventually extracted from the system. Optimality in this case means finding a proper balance between allocation toward components that are extracted and components that remain in the field and contribute to better resource capture, recycling, and conservation.

Identifying and obtaining the right amount and type of interactions within multispecies associations is a very difficult but necessary task for sustainable agroecosystem design and management. Diverse agroecosytems are intensive undertakings, difficult to design and manage. The complex web of interactions between planned components as well as with unplanned components must be reasonably understood if sustainable management is to be proactive instead of reactive.

REFERENCES

Abrams, P.A., Monotonic or unimodal diversity-productivity gradient: what does competition theory predict?, *Ecology,* 76:2019–2027, 1995.

Aiyer, A.K.N.Y., Mixed crops in India, *Indian J. Agric. Sci.,* 19:439–453, 1949.

Alemán, T., Los sistemas de producción forestal y agrícola de roza, in *El subdesarrollo agrícola en Los Altos de Chiapas,* Parra V.M.R., Ed., Universidad Autónoma de Chapingo/Centro de Investigaciones Ecológicas del Sureste, México, 1985.

Alsaadawi, I.S. et al., Effect of gamma irradiation on allelopathic potential of sorghum against weeds and nitrification, *J. Chem. Ecol.,* 12:1737–1745, 1985.

Altieri, M.A., *Biodiversidad, agroecología y manejo de plagas,* Cetal Ediciones, Chile, 1992.

Altieri, M.A., The impact, uses and ecological role of weeds in agroecosystems, in *Weed Management in Agroecosystems: Ecological Approaches,* Altieri, M.A. and Liebman, M., Eds., CRC Press, Boca Raton, FL, 1988, pp. 1–7.

Anaya, L.A. et al., Potencial alelopático de las principales plantas de un cafetal, in *Estudios Ecológicos en el Agroecosistema Cafetalero,* Jiménez, E. and Gómez-Pompa, A., Eds., Instituto Nacional de Investigaciones sobre Recursos Bióticos, Veracruz, México, 1982, pp. 85–94.

Anders, M.M., Sustainable crop production in the semi-arid tropics, in *Sustainable Agriculture: Issues, Perspectives and Prospects in Semi Arid Tropics,* Singh, R.P., Ed., paper presented at the proceedings of the First International Symposium on Natural Resources Management for a Sustainable Agriculture, New Delhi, Indian Society of Agronomy, 1990.

Andow, D.A., Management of weeds for insect manipulation in agroecosystems, in *Weed Management in Agroecosystems: Ecological Approaches,* Altieri, M.A. and Liebman, M., Eds., CRC Press, Boca Raton, FL, 1988, pp. 265–302.

Atkinson, M., The distribution and effectiveness of the roots of tree crops, *Hortic. Rev.,* 2:424–490, 1980.

Atsatt, P.R. and O'Dowd, D.J., Plant defense guilds, *Science,* 193:24–29, 1976.

Begon, M., Harper, J.L., and Townsend, C.R., *Ecology, Individuals, Populations and Communities,* Blackwell Scientific, Oxford, 1986.

Begon, M. and Mortimer, M., *Population Ecology. A Unified Study of Animals and Plants,* Sinauer Associates, Sunderland, MA, 1986.

Benjamin, L. and Sutherland, R., A comparison of models to simulate the competitive interactions between plants in even-aged monocultures, in *Individual-Based Models and Approaches in Ecology. Populations, Communities and Ecosystems,* DeAngelis, D. and Gross, L., Eds., Routledge, Chapman & Hall, 1992, pp. 455–471.

Bennema, J., Soils, in *Ecophysiology of Tropical Crops,* Alvim, P. de T. and Kosolowski, T.T., Eds., Academic Press, New York, 1977.

Berendse, F. and Aerts, R., Competition between *Erica tetralix* L. and *Molinia caerulea* (L.) Moench as affected by the availability of nutrients, *Acta Oecologica Oecol. Plant.*, 5:3–14, 1984.

Berkowitz, R.A., Competition for resources in weed-crop mixtures, in *Weed Management in Agroecosystems: Ecological Approaches,* Altieri, M.A. and Liebman, M., Eds., CRC Press, Boca Raton, FL, 1988, pp. 89–120.

Bertness, B. and Callaway, R., Positive interactions in communities, *Trends Ecol. Evol.,* 9(5):191–193, 1994.

Blair, B.C., The alley farming index: preliminary steps towards a more realistic model, *Agroforestry Syst.,* 40:19–27, 1998.

Boucher, D.H., *The Biology of Mutualism: Ecology and Evolution,* Croom Helm, London, 1985.

Briones, O., Montaña, C., and Ezcurra, E., Competition between three Chihuahuan desert species: evidence from plant size-distance relations and root distribution, *J. Veg. Sci.,* 7:453–460, 1996.

Buck, L.E., Lassoie, J. P., and Fernandes, E.C.M., *Agroforestry in Sustainable Agricultural Systems.* CRC Press, Boca Raton, FL, 1999.

Bunch, R., El potencial de coberturas muertas en el alivio de la pobreza y la degradación ambiental, in *TAPADO los Sistemas de Siembra con Cobertura,* Thurston, H.D. et al., Eds., CATIE y CIIFAD, 1994, pp. 5–10.

Cahill, J. F., Fertilization effects on interactions between above- and below-ground competition in an old field, *Ecology,* 80(2):466–480, 1999.

Callaway, R.M. and Walker, L.R., Competition and facilitation: a synthetic approach to interactions in plant communities, *Ecology,* 78(7):1958–1965, 1997.

Castiñeiras, A. et al., Efectividad técnico-económica del empleo de la hormiga leona *Phiedole megacephala* en el control del tetuán del boniato *Cylas formicarius elegantulus, Cien. Tecnol. Agric.* (Suppl.), Diciembre:103–109, 1982.

Chou, C.H. and Patrick, Z.A., Identification and phytotoxic activity of compounds produced during decomposition of corn and rye residues in soil, *J. Chem. Ecol.*, 2:369, 1976.

Christanty, L. et al., Traditional agroforestry in west Java: the pekarangan (homegarden) and kebun-talun (annual perennial rotation) cropping systems, in *Traditional Agriculture in Southeast Asia,* Marten, G.G., Ed., Westview Press, Boulder, CO, 1986, pp. 132–158.

Clements, F.E., Weaver, J.E., and Hanson, H.C., *Plant Competition: An Analysis of the Development of Vegetation,* Carnegie Institute, Washington, D.C., 1926.

Collins, W.W. and Qualset, C.O., *Biodiversity in Agroecosystems,* CRC Press, Boca Raton, FL, 1999.

Connell, J.H. and Slayter, R.O., Mechanisms of succession in natural communities and their role in community stability and organization, *Am. Nat.,* 111:1119–1144, 1977.

Cussans, G.W., Integrated weed management, in *Ecology and Integrated Farming Systems,* Glen, D.M., Greaves, M.P., and Anderson, H.M., Eds., John Wiley & Sons, New York, 1995, pp. 17–29.

Demo T.C. et al., El Banco Mundial y el desarrollo sustentable, Algunas reflexiones sobre sus perspectiva, Problemas del Desarrollo, *Rev. La. Econ.,* 30(118):9–34, 1999.

Dempster, J.P., Some affects of weed control on the numbers of the small cabbage white (*Pieris rapae* L.) on brussel sprouts, *J. Appl. Ecol.,* 6:339–345, 1969.

De Wit, C.T., *On Competition,* Instituut Voor Biologisch En Scheikundig Onderzoek Van Landbouwgewassen, Verslagen Landbouwkundige Onderzoekingen, Wageningen, 1960.

Donald, C.M., The interaction of competition for light and for nutrients, *Aust. J. Agric. Res.,* 9:421–435, 1958.

Edwards, C., The importance of integration in sustainable agricultural systems, in *Sustainable Agricultural Systems,* Edwards, C. et al., Eds., St. Lucie Press, Delray Beach, FL, 1990, pp. 249–264.

Edwards, C. et al., The role of agroecology and integrated farming systems in agricultural sustainability, *Agric. Ecosyst. Environ.,* 46:99–121, 1993.

Einhellig, F.A., Allelopathy: current status and future goals, in *Allelopathy, Organisms, Processes and Applications,* Inderjit, K.M., Dakshini, M., and Einhellig, F.A., Eds., American Chemical Society, Washington, D.C., 1995, pp. 1–23.

Estrada, L.E., Bello B.E., and Serralta, P.L., Dimensiones de la etnobotánica: el solar maya como espacio social, in *Lecturas en Etnobotánica,* Cuevas, S.J., Ed., Universidad Autónoma Chapingo, 1998, pp. 457–474.

Fath, B.D. and Patten, B.C., Network synergism: emergence of positive relations in ecological systems, *Ecol. Modelling,* 107:127–143, 1998.

Federer, W.T., *Statistical Design and Analysis for Intercropping Experiments, Three or More Crops,* Vol. 2., Springer-Verlag, Heidelberg, 1999.

Finlay, R.D. and Read, D.J., The structure and function of the vegetative mycelium of ectomycorrizal plants, *New Phytol.,* 103:143–165, 1986.

Firbank, L. and Watkinson, A., On the effects of competition: from monocultures to mixtures, in *Perspectives on Plant Competition,* Grace, J. and Tilman, D., Eds., Academic Press, San Diego, CA, 1990, pp. 165–192.

Fowler, N., Disorderliness in plant communities, in *Perspectives on Plant Competition,* Grace, J.B. and Tilman, D., Eds., Academic Press, San Diego, CA, 1990, pp. 291–308.

Francis, C., Sustainability issues with intercrops, in *Research Methods for Cereal/Legume Intercropping,* paper presented at the Proceedings of a Workshop on Research Methods for Cereal/Legume Intercropping in Eastern and Southern Africa, Waddington, S., Palmer A., and Edje, O., Eds., México, D.F. CIMMYT, 1990, pp. 194–199.

Francis, C.A. and Clegg, M.D., Crop rotations in sustainable production systems, in *Sustainable Agricultural Systems,* Edwards, C.A. et al., Eds., St. Lucie Press, Delray Beach, FL, 1990, pp. 107–122.

Fukai, S. and Trenbath, B., Processes determining intercrop productivity and yields of component crops, *Field Crops Res.,* 34:247–271, 1993.

García-Barrios, L., Desarrollo y Evaluación de un Modelo Dinámico Espacial del Crecimiento de Cultivos Asociados, Ph.D. thesis, Universidad Nacional Autónoma de México, 1998.

García-Barrios, L., Qualitative Change in Intercrop Land Equivalent Ratios in a Soil Moisture Tension Gradient, Research project for a wheat-fava bean intercrop experiment, El Colegio de la Frontera Sur, Chiapas, México, 2000.

García-Barrios, L. and García-Barrios, R., La modernización de la pobreza: dinámicas de cambio técnico entre los campesinos temporaleros de México, *Rev. Estud. Sociol. (El Colegio de México),* 10(29):263–288, 1992.

García-Barrios, L. and Kohashi, J., Efecto de la densidad de siembra sobre la asignación de materia seca aérea en un maíz criollo de Los Altos de Chiapas: consecuencias sobre el rendimiento máximo y la eficiencia del rendimiento, *Turrialba,* 44(4):205–219, 1994.

García-Barrios, L. et al., Development and validation of a spatially explicit individual based mixed, crop growth model, *Bull. Math. Biol.,* 63:507–526, 2001.

García-Espinosa, R., Quiroga, M.R., and Granados, A.N., Agroecosistemas de productividad sotenida de maíz en las regiones cálido húmedas de, in *TAPADO los Sistemas de Siembra con Cobertura,* Thurston, H.D. et al., Eds., CATIE/CIIFAD, 1994, pp. 5–10.

Gates, D.J., Competition and skewness in plantations, *J. Theor. Biol.,* 94:909–922, 1982.

Gause, G.F., *The Struggle for Existence,* Waverly Press, Baltimore, MD, 1934.

Gerig, T.M. and Blum, U., Effects of mixtures of four phenolic acids on leaf area expansion of cucumber seedlings grown in Portsmouth B soil materials, *J. Chem. Ecol.,* 17(1):29–40, 1991.

Gershenzon, J., Changes in the levels of plant secondary metabolites under water and nutrient stress, *Recent Adv. Phytochem.,* 18:273–320, 1984.

Giampetro, M., Socioeconomic constraints to farming with biodiversity, *Agric., Ecosyst. Environ.,* 62:145–167, 1997.

Gliessman, S.R., *Agroecology: Ecological Processes in Sustainable Agriculture,* Ann Arbor Science, Ann Arbor, MI, 1998.

Goldberg, D., Components of resource competition in plant communities, in *Perspectives on Plant Competition*, Grace, J.B. and Tilman, D., Eds., Academic Press, San Diego, CA, 1990, pp. 27–49.

Goldberg, D.E. and Barton, A.M., Patterns and consequences of interspecific competition in natural communities: a review of field experiments with plants, *Am. Nat.,* 139(4):771–801, 1992.

Goldberg, D. and Novoplansky, A., On the relative importance of competition in unproductive Environments, *J. Ecol.*, 85(4):409–418, 1997.

Goldberg, D.E. and Werner, P.A., Equivalence of competitors in plant communities: a null hypothesis and an experimental approach, *Am. J. Bot.,* 70:1098–1104, 1983.

Goldberg, D.A. et al., Empirical approaches to quantifying interaction intensity: competition and facilitation along productivity gradients, *Ecology,* 80(4):1118–1131, 1999.

Grace, J.B., On the relationship between plant traits and competitive ability, in *Perspectives on Plant Competition*, Grace, J.B. and Tilman, D., Eds., Academic Press, San Diego, CA, 1990, pp. 51–65.

Grace, J.B., The effects of habitat productivity on competition intensity, *Trends Ecol. Evol.,* 8(7):229–230, 1993.

Grace, J.B., In search of the Holy Grail: explanations for the coexistence of plant species, *Trends Ecol. Evol.,* 10(7):263–264, 1995.

Grime, J.P., Evidence for the existence of three primary strategies in plants and its relevance to ecological and evolutionary theory, *Am. Nat.,* 111:1169–1194, 1977.

Haines, S.G., Haines, L.W., and White, G., Leguminous plants increase sycamore growth in northern Alabama, *Soil Sci. Soc. Am. J.,* 42:130–132, 1978.

Hale, M.G. and Moore, L.D., Factors affecting root exudation, *Adv. Agron.*, 31:93–124, 1979.

Hall, R.L., Plant diversity in arable ecosystems, in *Ecology and Integrated Farming Systems,* Glen, D.M., Greaves, M.P., and Anderson, H.M., Eds., John Wiley & Sons, New York, 1995, pp. 9–15.

Hara, T., Dynamics of size structure in plant populations, *Trends Ecol. Evol.,* 3:129–133, 1988.

Harper, J.L., *Population Biology of Plants,* Academic Press, San Diego, CA, 1990.

Hartung, A.C. and Stephens, C.T., Effects of allelopathic substances produced by asparagus on incidence and severity of asparagus decline due to Fusarium crown rot, *J. Chem. Ecol.,* 9(8):1163–1174, 1983.

Hatfield, J. and Keeney, D., Challenges for the 21st century, in *Sustainable Agriculture Systems,* Hatfield, J. and Karlen, D., Eds., Lewis Publishers, Boca Raton, FL, 1994, pp. 194–199.

Hedin, P.A., Host Plant Resistance to Disease, paper presented at the ACS symposium series 62, American Chemical Society, Washington, D.C., 1977.

Holmgren, M., Scheffer, M., and Huston, M., The interplay of facilitation and competition in plant communities, *Ecology,* 78(7):1966–1975, 1997.

Hougthon, R., The worldwide extent of land-use change, *BioScience,* 44(5):305–313, 1994.

House, G.J. and Brust, G.E., Ecology of low-input no tillage agroecosystems, *Agric. Ecosyst. Environ.,* 27:331–345, 1989.

Hubbell, S.P. and Foster, R.B., Biology, chance and history and the structure of tropical rain forest tree communities, in *Community Ecology,* Diamond, J. and Case, T.J., Eds., Harper and Row, New York, 1986, pp. 314–329.

Hunter, A.F. and Aarsen, L.W., Plants helping plants, *Bioscience,* 38(1):34–40, 1988.

Huston, M. and DeAngelis, D., Competition and coexistence: the effects of resource transport and supply rates, *Am. Nat.*, 144(6):954–977, 1994.

Kass, D., Polyculture cropping systems: review and analysis, *Cornell Int. Agric. Bull.*, 32, 1978.

Kavamahanga, F., Bishnoi, U.R., and Aman, K., Influence of different N rates and intercropping methods on grain sorghum, common bean, and soya bean yields, *Trop. Agric.* (Trinidad), 72(4):257–260, 1995.

Keddy, P.A., *Competition,* Chapman and Hall, New York, 1989.

Kluson, R.A., Intercropping allelopathic crops with nitrogen fixing legume crops. A tripartite legume symbiosis perspective, in *Allelopathy, Organisms, Processes and Applications,* Inderjit, K.M., Dakshini M., and Einhellig, F.A., Eds., American Chemical Society, Washington, D.C., 1995, pp. 193–210.

Koyama, H. and Kira, T., Intraspecific competition among higher plants. VIII Frequency distribution of individual plant weight as affected by the interaction among plants, *J. Inst. Polytech. Osaka City Univ.,* 7:73–84, 1956.

Kremer, R.J., Management of weed seed banks with microorganisms, *Ecol. Appl.,* 3:42–52, 1993.

Kropff, M. and Lotz, L., Optimization of weed management systems: the role of ecological models of interplant competition, *Weed Technol.,* 6:462–470, 1992.

Kropff, M., Weaver, S., and Smits, A., Use of ecophysiological models for crop-weed interference: relations amongst weed density, relative time of weed emergence, relative leaf area, and yield loss, *Weed Sci.,* 40:296–301, 1992.

Kropotkin, P., *Mutual Aid: A Factor of Evolution,* Porter Sargent, Boston, 1902.

Lal, R. and Miller, F.P., Sustainable farming systems for the tropics, in *Sustainable Agriculture: Issues, Perspectives and Prospects in Semi Arid Tropics,* Singh, R.P., Ed., paper presented at the proceedings of the First International Symposium on Natural Resources Management for a Sustainable Agriculture, New Delhi, Indian Society of Agronomy, 1990.

Landauer, K. and Brazil, M., Eds., *Tropical Homegardens,* United Nations University Press, Tokyo, 1990.

Levine, J.M., Indirect facilitation: evidence and predictions from a riparian community, *Ecology,* 80:1762–1769, 1999.

Liebman, M., personal communication, 1999.

Liebman, M. and Gallandt, E.R., Many little hammers: ecology management of crop–weed interactions, in *Ecology in Agriculture,* Jackson, L.E., Ed., Academic Press, New York, 1997, pp. 291–343.

Liebman, M. and Stavers, C.P., Crop diversification for weed management, in *Ecological Management of Agriculture Weeds,* Liebman, M., Mohler, C.L., and Stavers, C.P., Eds., Cambridge University Press, New York, 2000.

Liebman, M. and Davis, A.S., Integration of soil, crop, and weed management in low-external-input farming systems, *Weed Res.*, 40:27–47, 2000.
MacArthur, J.D., Some general characteristics of farming in a tropical environment, in *Farming Systems in the Tropics,* Ruthenberg, H., Ed., Clarendon Press, Oxford, 1976a, pp. 19–28.
MacArthur, J.D., General tendencies in the development of tropical farm systems, in *Farming Systems in the Tropics,* Berg, H., Ed., Clarendon Press, Oxford, 1976b, pp. 326–329.
MacArthur, R.H. and Levins, R., The limiting similarity, convergence and divergence of coexisting species, *Am. Nat.,* 101:377–378, 1967.
Macías, F.A, Galindo, J.C., and Massanet, G.M., Potential allelopathic activity of several sesquiterpene lactone models, *Phytochemistry,* 31(6):1969–1977, 1992.
Mariaca, M.R. et al., Análisis estadístico de una milpa experimental de ocho añosde cultivo continuo bajo roza-tumba-quema en Yucatán, México, in *La Milpa en Yucatán. Un Sistema de Producción Agrícola Tradicional,* Hernandez, X.E., Bello, B.E., and Levy, T.S., Eds., Colegio de Postgraduados, México, 1995, pp. 339–365.
Melara, W. and Del Río, L., Uso de Labranza Mínima y Leguminosas de Cobertura en Honduras, in *TAPADO los sistemas de siembra con cobertura,* Thurston, H.D. et al., Eds., CATIE y CIIFAD, 1994, pp. 57–64.
Miller, T.E., Direct and indirect species interactions in an early old-field plant community, *Am. Nat.,* 143:1007–1025, 1994.
Miller, T.E., On quantifying the intensity of competition across gradients, *Ecology,* 77:978–981, 1996.
Mohler, C.L., Effect of tillage and mulch on weed biomass and sweet corn yield, *Weed Technol.,* 5:545–552, 1991.
Muller, C.H., Allelopathy as a factor in ecological processes, *Vegetatio,* 18:348, 1969.
Natarajan, M. and Willey, R.W., The effects of water stress on yield advantages of intercropping systems, *Field Crops Res.,* 13:117–131, 1986.
National Research Council, *Sustainable Agriculture and the Environment in the Humid Tropics,* National Academy Press, Washington, D.C., 1993.
Nestel, D. and Altieri, M.A., The weed community of Mexican coffee agroecosystems: effect of management on plant biomass and species composition, *Acta Oecol.,* 13:715–726, 1992.
Newman, E.I., Competition and diversity in herbaceous vegetation, *Nature,* 244:310, 1973.
Nissen, T.M., Midmore, D.J., and Keeler, A.G., Biophysical and economic tradeoffs of intercropping timber with food crops in the Philippine uplands, *Agric. Syst.,* 67(1):49–69, 2001.
Oelsligle, D.D. and Pinchinat, A.M., Effect of varying nitrogen levels on grain yields, energy and protein production and economic returns of corn and beans when grown alone and in different combinations, personal communication, 1975.
Okigbo, B.N., Sustainable agricultural systems in tropical Africa, in *Sustainable Agricultural Systems,* Edwards, C.A. et al., Eds., St. Lucie Press, Delray Beach, FL, 1990, pp. 323–352.
Oksanen, L., Predation, herbivory and plant strategies along gradients of primary productivity, in *Perspectives on Plant Competition,* Grace, J. and Tilman, D., Eds., Academic Press, New York, 1990, pp. 165–192.
Oksanen, L., Is the humped relationship between species richness and biomass an artifact due to plot size?, *J. Ecol.,* 84:293–295, 1996.
Ong, C.K., Alley cropping, ecological pie in the sky?, *Agrofor. Today,* 6:8–10, 1994.
Peltzer, D.A., Wilson, S.D., and Gerry, A.K., Competition intensity along a productivity gradient in a low diversity grassland, *Am. Nat.,* 151:465–476, 1998.

Pennings, S.C. and Callaway, R.M., The role of a parasitic plant in community structure and diversity in a western salt marsh, *Ecology,* 77:1410–1419, 1996.

Perfecto, I. et al., Shade coffee: a disappearing refuge for biodiversity, *BioScience,* 46(8):598–608, 1996.

Pugnaire, F., Haase, P., and Puigdefábregas, J., Facilitation between higher plant species in a semiarid environment, *Ecology,* 77(5):420–1426, 1996.

Putnam, A.R. and Weston, L.A., Adverse impacts of allelopathy in agricultural systems, in *The Science of Allelopathy,* Putnam, A.R. and Tang, C.S., Eds., John Wiley & Sons, New York, 1986, pp. 43–56.

Radosevich, S.R., Competition for resources in weed-crop mixtures, in *Weed Management in Agroecosystems: Ecological Approaches,* Altieri, M.A. and Liebman, M., Eds., CRC Press, Boca Raton, FL, 1988, pp. 89–120.

Radosevich, S.R. and Roush, M.L., The role of competition in agriculture, in *Perspectives on Plant Competition,* Grace, J.B. and Tilman, D., Eds., Academic Press, New York, 1990, pp. 341–366.

Ramakrishnan, P.S., Successional theory: implications for weed management in shifting agriculture, mixed cropping, and agroforestry systems, in *Weed Management in Agroecosystems: Ecological Approaches,* Altieri, M.A. and Liebman, M., Eds., CRC Press, Boca Raton, FL, 1988, pp. 183–196.

Ramírez, N., Gonzáles, M., and García-Moya, E., Establecimiento de *Pinus* spp. y *Quercus* spp. en matorrales y pastizales de Los Altos de Chiapas, México, *Agrociencia,* 30:249–257, 1996.

Rao, M.R., Nair, P.K.R., and Ong, C.K., Biophysical interactions in tropical agroforestry systems, *Agrofor. Syst.,* 38:3–50, 1998.

Rao, M.R. and Willey, R.W., Evaluation of yield stability in intercropping: studies on sorghum/pigeonpea, *Exp. Agric.,* 16:105–116, 1980.

Rejintjes, C., Haverkort, B., and Waters-Bayer, A., Experiences with improved fallow. Appendix A: Some promising LEISA techniques, in *Farming for the Future. An introduction to Low-External-Input and Sustainable Agriculture,* Rejintjes, C., Haverkort, B., and Waters-Bayer, A., Eds., Macmillan, New York, 1992, pp. 169–171.

Rice, E.L., *Allelopathy,* 2nd ed., Academic Press, New York, 1984.

Risch, J., Insect herbivore abundance in tropical monocultures and polycultures: an experimental test on two hypotheses, *Ecology,* 62:1325–1240, 1981.

Romero, A.Y. et al., Shade type effect on coffee yields, soil nutrients and environmental temperature in Chiapas, Mexico, submitted.

Root, R.B., Organization of a plant-arthropod association in simple and diverse habitats: the fauna of collards (*Brassica oleracea*), *Ecol. Monogr.,* 43:95–124, 1973.

Rose, S.L. et al., Allelopathic effects of litter on the growth and colonization of mycorrhizal fungi, *J. Chem. Ecol.,* 9(8):1153–1162, 1983.

Ruthenberg, H., *Farming Systems in the Tropics,* Clarendon Press, Oxford, 1976.

Santiago-Lastra, A. et al., Cambios en la competencia entre cultivos asociados en un gradiente de tensión hidríca del suelo, submitted.

Santiago-Vera, T., García-Barrios, L., and Santiago, L.A., Competencia y facilitación en un policultivo bajo diferentes proporciones de siembra y en un gradiente de tensión hídrica del suelo, submitted.

Schaller, N., The concept of agricultural sustainability, *Agric. Ecosys. Environ.,* 46:89–97, 1993.

Schoener, T.W., The controversy over interspecific competition, *Am. Sci.,* 70:586–595, 1982.

Schroth, G., A review of belowground interactions in agroforestry, focusing on mechanisms and management options, *Agrofor. Syst.,* 43:5–34, 1999.

Sembdner, G. and Parthier, B., The biochemistry and the physiological and molecular actions of jasmonates, *Annu. Rev. Plant Physiol. Plant Mol. Biol.,* 44:569–89, 1993.

Sharrow, S.H., Silvopastoralism: competition and facilitation between trees, livestock, and improved grass-clover pastures on temperate rainfed lands, in *Agroforestry in Sustainable Agricultural Systems,* Buck, L.E., Lassoie, J.P., and Fernandes, E.C.M., Eds., Lewis Publishers, Boca Raton, FL, 1999, pp. 111–130.

Silvertown, J. and Law, R., Do plants need niches? Some recent developments in plant community ecology, *Trends Ecol. Evol.,* 2(1):24–26, 1987.

Soemarwoto, O. and Soemarwoto, I., The Javanese rural ecosystem, in *An Introduction to Human Ecology on Agricultural Systems in Southeast Asia,* Rambo, A.T. and Sajise, P.E., Eds., Los Baños, University of the Philippines, 1984.

Soto-Pinto, L. et al., Shade effect on coffee production at the northern Tzeltal zone of the state of Chiapas, Mexico, *Agric. Ecosyst. Environ.,* 80:60–69, 2000.

Stern, W.R. and Donald, C.M., Light relationship in grass-clover sward, *Aust. J. Agric. Res.,* 13:599–614, 1962.

Stevens, M., Henry, H., and Carson, W.P., Plant density determines species richness along an experimental fertility gradient, *Ecology,* 80(2):455–465, 1999.

Stewart, B.A., Developing sustainable systems in semi-arid regions, in *Sustainable Agriculture: Issues, Perspectives and Prospects in Semi Arid Tropics,* Singh, R.P., Ed., paper presented at the proceedings of the First International Symposium on Natural Resources Management for a Sustainable Agriculture, New Delhi, Indian Society of Agronomy, 1990.

Tahvanien, J.O. and Root, R.B., The influence of vegetation diversity on the population ecology of a specialized herbivore, *Phyllotreta cruciferae* (Coleoptera:Chrysomelidae), *Oecologia,* 10:321–346, 1972.

Thomas, S. and Weiner, J., Including competitive asymmetry in measures of local interference in plant populations, *Oecologia,* 80:349–355, 1989.

Thrupp, L.A., *Cultivating Diversity: Agrobiodiverstiy and Food Security,* World Resources Institute, Washington, D.C., 1998.

Thurston, H.D., Historia de los Sistemas de Siembra con Cobertura Muerta o Sistemas de Tumba y Pudre en América Latina, in *TAPADO los Sistemas de Siembra con Cobertura,* Thurston, H.D. et al., Eds., CATIE y CIIFAD, 1994, pp. 1–4.

Tilman, D., *Resource Competition and Community Structure*, Princeton University Press, Princeton, NJ, 1982.

Tilman, D., Plant dominance along an experimental nutrient gradient, *Ecology,* 65:1445–1453, 1984.

Tilman, D., On the meaning of competition and the mechanisms of competitive superiority, *Functional Ecol.,* 1(4):304–315, 1987.

Tilman, D., *Plant Strategies and the Dynamics and Structure of Plant Communities,* Monographs on Population Biology, Princeton University Press, Princeton, NJ, 1988.

Tilman, D., Mechanisms of plant competition for nutrients: the elements of a predictive theory of competition, in *Perspectives on Plant Competition,* Grace, J. and Tilman, D., Eds., Academic Press, New York, 1990, pp. 117–141.

Tilman, D., Biodiversity: population versus ecosystem stability, *Ecology,* 77(2):350–363, 1996.

Tilman, D., Wedin, D., and Knops, J., Productivity and sustainability influenced by biodiversity in grassland ecosystems, *Nature,* 379:718–720, 1996.

Tongma, S., Kobayaski, K., and Usui, K., Allelopathic activity of Mexican sunflower *Titonia diversifolia* in soil, *Weed Sci.,* 46:432–437, 1998.

Trenbath, B., Biomass productivity of mixtures, *Adv. Agron.*, 26:177–210, 1974.

Trenbath, B., Plant interactions in mixed crop communities, in *Multiple Cropping,* Stelly, M., Ed., Special publication 127, American Society of Agronomy, Madison, WI, 1976, pp. 129–169.
Vandermeer, J., The interference production principle: an ecological theory for agriculture, *BioScience,* 31(5):361–364, 1981.
Vandermeer, J., The evolution of mutualism, in *Evolutionary Ecology,* Shorrocks, B., Ed., Blackwell Scientific, Oxford, 1984a.
Vandermeer, J., Plant competition and the yield-density relationship, *J. Theor. Biol.,* 109:393–399, 1984b.
Vandermeer, J., *The Ecology of Intercropping*, Cambridge University Press, Cambridge, 1989.
Vandermeer, J., The ecological basis of alternative agriculture, *Annu. Rev. Ecol. Syst.,* 26:201–24, 1995.
Vandermeer, J., Disturbance and neutral competition theory in rain forest dynamics, *Ecol. Modelling*, 85:99–111, 1996.
Vandermeer, J., Maximizing crop yields in alley crops, *Agrofor. Syst.,* 40:199–206, 1998.
Vandermeer, J., Efecto y respuesta, competencia y facilitación: conceptos claves para analizar policultivos, Documento inédito sometido a, *Rev. Agric. Org. Cuba*, in press.
Vandermeer, J., Agroecosystems and biodiversity, submitted.
Vandermeer, J., Hazelett, B., and Rathcke, B., Indirect facilitation and mutualism, in *The Biology of Mutualism: Ecology and Evolution,* Boucher, D.H., Ed., Croom Helm, London, 1985.
Vandermeer, J. et al., Global change and multispecies agroecosystems. Concepts and issues, *Agric. Ecosyst. Environ.,* 67:1–22, 1998.
Velásquez, H.J et al., Participación campesina en la gestión de tecnologías para la producción sustentable de maíz en Villaflores, Chiapas, *Agrociencia,* 33:217–225, 1999.
Wallace, A., Moisture levels, nitrogen levels — clue to the next limiting factor on crop production, *J. Plant Nutr.,* 13(3–4):451–457, 1990.
Weiner, J., Asymmetric competition in plant populations, *Trends Ecol. Evol.,* 5:360–364, 1990.
Weiner, J., Griepentrog, H., and Kristensen, L., Increasing the Suppression of Weeds by Cereal Crops, abstract presented at the Ecological Society of America 85th Annual Meeting, Snowbird, Utah, 2000, p. 228.
Weiner, J., Wright, D.B., and Castro, S., Symmetry of below-ground competition between *Kochia scoparia* individuals, *Oikos,* 79:85–91, 1997.
Westoby, M., The self-thinning rule, *Adv. Ecol. Res.,* 14:167–225, 1984.
Whittaker, R.H. and Feeny, P.P., Allelochemicals: chemical interactions between species, *Science*, 171:757, 1971.
Whittaker, R.H. and Levin, S.A., The role of mosaic phenomena in natural communities, *Theor. Pop. Biol.*, 12(2):117–139, 1977.
Willey, R., Intercropping — its importance and research needs. Part 2. Agronomy and research approaches, *Fields Crop Abstr.,* 32(2):72–85, 1979.
Willey, R. and Heath, S.B., The quantitative relation between plant population and crop yield, *Adv. Agron.,* 21:281–321, 1969.
Wilson, S.D. and Tilman, D., Components of plant competition along an experimental gradient of nitrogen availability, *Ecology,* 72:1050–1065, 1988.
Wyse, D.L., Future of weed science research, *Weed Technol.,* 6:162–165, 1992.
Zimdahl, R.L., The concept and application of the critical weed-free period, in *Weed Management in Agroecosystems: Ecological Approaches,* Altieri, M.A. and Liebman, M., Eds., CRC Press, Boca Raton, FL, 1988, pp. 145–155.
Zimdahl, R.L., *Fundamentals of Weed Science,* Academic Press, New York, 1993.

CHAPTER 3

Pest Management in Mesoamerican Agroecosystems

Luko Hilje, Carlos M. Araya, and Bernal E. Valverde

CONTENTS

0-8493-1581-6/03/$0.00+$1.50

INTRODUCTION

Any insect, pathogen, or undesirable plant that actually or potentially causes direct damage or competes with crops can be considered a pest. But this term should be restricted to those cases in which population abundance or severity of damage inflicted by such organisms is high enough to cause losses of economic importance. In other words, pest organisms are not intrinsically "bad." Instead, their status is an indication of disturbance of agroecosystem components that lead to undesirable increases in their population levels (Huffaker, Messenger, and DeBach, 1971; Spahillari et al., 1999).

In the past 60 years, chemical pesticides have been the preferred method to control pests worldwide. Nonetheless, the recognition and documentation of many unwanted agroecological, environmental, social, and economic problems resulting from pesticide overuse has led scientists to look for alternatives, among which integrated pest management (IPM) has been the most common (Stern et al., 1959; National Academy of Sciences, 1969; Bottrell, 1979; Kogan, 1998).

Since its origin (Stern et al., 1959), IPM has gained acceptance and support from research and educational institutions and scientists, extension agents, growers, the general public, and even agrichemical companies. In developed countries, there is an amazing wealth of conceptual and practical information, as well as of successful IPM programs (Kogan, 1998). Despite recent advances in the majority of tropical countries, IPM implementation is hindered by a limited understanding and documentation of agroecological and socioeconomic factors that constitute important constraints to the development of plant protection, as reflected by the few formal articles published on this topic (González, 1976; Vaughan, 1976, 1989; Brader, 1979; Bottrell, 1987; Hilje and Ramírez, 1992; Pareja, 1992a; Ramírez, 1994).

This chapter provides an overview of key biogeographical and agricultural features that determine pest distribution, abundance, and persistence in Mesoamerica, as well as the repercussions for implementing pest management programs. In addition, we discuss current critical agricultural issues as they relate to integrated pest management development in Mesoamerican countries.

THE BIOPHYSICAL AND AGRICULTURAL CONTEXT

Climate and Biogeography

Tropical areas of the world are those located between the Tropics of Cancer and Capricorn (between 23.5°N and 23.5°S). Mesoamerica (see Figure 3.1) extends from the Tehuantepec isthmus of Mexico (6°N) to the lowlands of the Atrato River (18°N) in Colombia (Dengo, 1973). This region exhibits varied climatic characteristics that strongly influence the biology and ecology of agricultural pests.

Temperature, rainfall, and air humidity are normally much higher than those of temperate areas, and photoperiod varies only slightly throughout the year (Portig, 1976). Temperature is fairly constant, owing to the narrow shape and small size of this land mass and the influence of oceans on its climate. Thus, it is the rainfall regime that determines the two seasons (dry and wet). But Mesoamerica, as well as

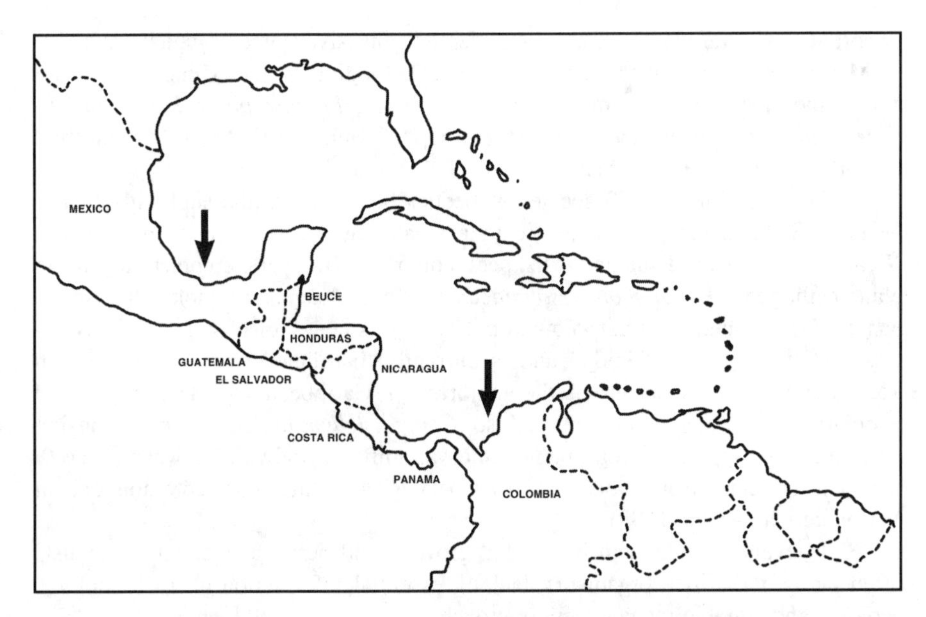

Figure 3.1 Map of Mesoamerica, a region that includes the entire territories of Guatemala, Belize, Honduras, El Salvador, Nicaragua, Costa Rica, and Panama, as well as part of Mexico and Colombia. Its extreme limits (Tehuantepec isthmus, in Mexico, and the lowlands of the Atrato River, in Colombia) are indicated by arrows.

the rest of neotropical areas, is climatically and ecologically diverse because of its geographical position and topographical and altitudinal features. Several life zones are recognized (Holdridge, 1978), including dry forests, thorn woodlands, natural pine and oak pure stands, broadleaf humid forests, deserts, and *paramos*.

The southern part of Mesoamerica has a common geological origin and appeared as a result of intense tectonic and volcanic activities that were completed some 3 million years ago (Dengo, 1973; Rich and Rich, 1983). In biogeographic terms, the Mesoamerican isthmus acted as a bridge between the two large North and South American land masses, allowing migration of organisms in both directions and thus favoring endemism (Rich and Rich, 1983).

Mesoamerica, as well as the rest of the neotropics, is exceptionally rich in number of species and endemism, that is, the presence of unique species for a given region. For example, four Mesoamerican countries (Mexico, Costa Rica, Panama, and Colombia) rank among the most species-rich in the world, especially considering estimated numbers of species of plants, mammals, and birds (Caldecott et al., 1994).

Considering only neotropical flowering plants, it has been estimated that about 22,000 species (ca. 25% of the total flora) will be new to science and await description (Thomas, 1999). For instance, on Barro Colorado Island (Panama), 180 out of 1316 existing plant species are exclusive to Central America (Croat, 1978). Globally, about 70% of all weed species belong to only 12 families; about 40% are in the Poaceae and Asteraceae families (Radosevich, Holt, and Ghersa, 1997). There are another ten important families, of which Amaranthaceae, Fabaceae, and Solanaceae are well represented in Mesoamerica. In addition to weed species of agricultural importance, several plants are also considered as invasive, some of which are native to Mesoamerica. Legumes, especially some belonging to the Mimosoideae, are among the most important invasive plants, including *Mimosa pigra, Leucaena leucocephala, Sesbania punicea,* and *Acacia* spp. (Cronk and Fuller, 1995); the first two are native to Mesoamerica.

Similarly, a third of 183 species of beetles belonging to the subfamily Scarabaeinae (Scarabaeidae) are endemic to a small region extending from southern Nicaragua to western Panama (Solís, pers. comm.; INBio, pers. comm.). Regarding plant pathogens, *Pseudomonas solanacearum*, causal agent of moko disease on bananas and plantains, is not known on *Musa* in its center of origin (Table 3.1); it was first described in Trinidad and is currently distributed from Brazil through Mexico but remains absent in the majority of Caribbean islands (French and Sequeira, 1970; Ploetz et al., 1994). Also, *Colletotrichum lindemuthianum*, causing bean anthracnose, has a broad pathogenic variability worldwide; however, race 9, which is endemic from Guatemala to Costa Rica, is the most common one in Mesoamerica (Araya, 1999).

High levels of endemism imply that growers and pest management specialists often face undescribed organisms, lacking essential information on their biology, ecology, and suitable management approaches.

Agronomic Characterization

Mesoamerican agriculture exhibits a wide variety of cropping systems, ranging from very small patches of polycultures to extensive monocultures. Small- and medium-size farms include an ample spectrum, from indigenous communities who plant small patches of crops surrounded by primary forest, for subsistence, to commercially oriented growers (either as individuals or in cooperatives) who plant several crops in different spatial and temporal schemes throughout the year. In some cases, as with vegetable production, these crop mosaics may act as functional monocultures on a regional scale, especially regarding pest management.

Table 3.1 Geographical Origin of Some Important Crops Planted in Mesoamerica

Common Name	Scientific Name	Center of Origin
Rice	*Oryza sativa*	India-Indochina
Bananas	*Musa paradisiaca*	Malaysian archipelago (?)
Cacao	*Theobroma cacao*	Andean equatorial slopes
Coffee	*Coffea arabica*	Ethiopia
Sugarcane	*Saccharum officinarum*	New Guinea
Beans	*Phaseolus vulgaris*	Mexico–Andean highlands
Corn	*Zea mays*	Mexico–Central America (?)
Potato	*Solanum tuberosum*	Andean altiplano
Cabbage	*Brassica oleracea*	SW Europe and England
Tomato	*Lycopersicon esculentum*	Peru–Ecuador

Source: From Purseglove, 1974, 1975; Singh et al., 1991.

Large monocultures are represented by single farms of hundreds to thousands of hectares. They involve both traditional (bananas, sugarcane, and coffee) and nontraditional export crops (melons, watermelons, and pineapple), as well as a few staple foods (rice and maize). The majority of these crops are intensively managed in agronomic terms (seeds, agrichemical inputs, and mechanization), similar to production systems in developed countries. An exception is coffee, which is generally planted in association with shade trees within diverse and complex agroforestry systems.

Agricultural lands are located between sea level and about 3000 m, although they are mainly concentrated below 1200 m, with bananas, rice, melons, cacao, and cotton usually planted below 300 m and sugarcane and pineapple between 0 and 900 m. Coffee extends its range between 500 and 1200 m, whereas maize and beans are planted from 0 to 2500 m. Some vegetables can be grown between 0 and 3000 m, depending on the crop; however, potato, onion, cabbage, broccoli, and snow peas are typically found above 1400 m. Within these altitudinal ranges, temperature is quite stable throughout the year and rainfall provides enough moisture for fostering pest development, reproduction, and dispersal, making pests a continuous threat to crops all year-round, the only exception being highly seasonal areas. This situation forces growers and specialists to invest large efforts and resources to deal with pests on a permanent basis.

Historically, exotic species, such as bananas, sugarcane, and coffee, have been the main crops planted in Mesoamerica, along with some native ones (Table 3.1). It is likely that along with the exotic crops, some of their associated pest organisms were also introduced (Table 3.2). But many of the main pests in the region, as well as the majority of occasional or secondary pests (those not normally reaching pest status, in economic terms), are native.

Native organisms stand out among the several thousand species affecting crops in Mesoamerica, a clear reflection of the high levels of biodiversity and endemism of plant-associated organisms and vegetation in this region. For example, some 1800 insect and 933 pathogen species are reported to affect crops in this region (Valerín, 1994; Coto et al., 1995). Moreover, about half of the most important weeds in the world are present as weeds in Mesoamerica (Holm et al., 1977), and some of the worst weeds are native to this area, including *Amaranthus spinosus, Ageratum conyzoides, Argemone mexicana, Axonopus compressus, Cenchrus echinatus, Chromolaena odorata, Lantana camara, Mimosa pudica,* and *Sida acuta.* Also, two native Rubiaceae have become extremely important as weeds in the two most important perennial crops in the region: *Spermacoce assurgens* (syn. *Borreria laevis*) in coffee and bananas, and *Spermacoce latifolia* (syn. *Borreria latifolia*) in coffee.

Native species can certainly become pests because of the establishment of large monocultures of their native host crops, which provide enough food or resources to support their population increase, but this conversion can also occur as a result of species recruitment in response to planted area, according to the concept of species-area (MacArthur and Wilson, 1967). For example, most insect species affecting cacao and sugarcane are native to each region where these crops have been introduced (Strong, 1974; Strong, McCoy, and Rey, 1977). This illustrates how native insects can adapt their feeding habits and development from native plants to exotic crops.

Table 3.2 The Top Ten Pest Species of Insects, Pathogens, and Weeds in Mesoamerica, Including Their Origin

Common Name	Scientific Name	Taxonomy	Origin
	Insects		
Whitefly	*Bemisia tabaci*	HOM: Aleyrodidae	Exotic
Coffee berry borer	*Hypothenemus hampei*	COL: Scolytidae	Exotic
Banana weevil	*Cosmopolites sordidus*	COL: Curculionidae	Exotic
Sugarcane borer	*Diatraea* spp.	LEP: Pyralidae	?
Army and cutworms	*Spodoptera* spp.	LEP: Noctuidae	Native
Fruit and bollworms	*Heliothis* spp.	LEP: Noctuidae	Native
Diamondback moth	*Plutella xylostella*	LEP: Plutellidae	Exotic
Rice delphacid	*Tagosodes orizicolus*	HOM: Delphacidae	?
Mediterranean fly	*Ceratitis capitata*	DIP: Tephritidae	Exotic
Whitegrubs	*Phyllopahaga* spp.	COL: Scarabaeidae	Native
	Pathogens		
Yellow sigatoka	*Mycosphaerella musicola*	Loculoascomycete	Exotic
Black sigatoka	*Mycosphaerella fijiensis*	Loculaoscomycete	Native
Rice blight	*Magnaporthe oryzae*	Pyrenomycete	Exotic
Bean anthracnose	*Colletotrichum lindemuthianum*	Coelomycete	Native
Coffee rust	*Hemileia vastatrix*	Hemibasidiomycete	Exotic
Potato late blight	*Phytophthora infestans*	Oomycete	Native
Cabbage black vein	*Xanthomonas campestris*	Pseudomonadeae	Exotic
Moko disease	*Pseudomonas solanacearum*	Pseudomonadae	Native
Root gall	*Meloidogyne incognita*	Heteroderidae	?
Burrowing root rot	*Radopholus similis*	Pratylenchidae	Exotic
	Weeds		
Purple nutsedge	*Cyperus rotundus*	Cyperaceae	Exotic
Itchgrass	*Rottboellia cochinchinensis*	Poaceae	Exotic
Junglerice	*Echinochloa colona*	Poaceae	Exotic
Hairy beggarticks	*Bidens pilosa*	Asteraceae	Native
Spreading dayflower	*Commelina diffusa*	Commelinaceae	Exotic
Bermuda grass	*Cynodon dactylon*	Poaceae	Exotic
Bushy buttonweed	*Spermacoce assurgens*	Rubiaceae	Native
Buttonweed	*Spermacoce latifolia*	Rubiaceae	Native
Goosegrass	*Eleusine indica*	Poaceae	Exotic
Crabgrass	*Digitaria* spp.	Poaceae	Exotic

Source: Selected according to authors' experience, as well as from informal assessments by colleagues.

Abbreviations: HOM (Homoptera), COL (Coleoptera), LEP (Lepidoptera), and DIP (Diptera).

PEST MANAGEMENT APPROACHES

Historical Pest Control Approaches

There are very few historical accounts about pest control approaches in Mesoamerica before the appearance of synthetic pesticides (Andrews and Quezada,

1989; Hilje, Cartin, and March, 1989; Ardón, 1993). But trends in pesticide use have very closely followed the general patterns observed in developed countries.

For example, in Costa Rica, synthetic pesticides became available just after their commercial introduction in Europe and the U.S. By 1950, six companies commercialized pesticides, and 19, 25, and 110 additional companies were established between 1950 and 1960, 1960 and 1970, and 1970 and 1985, respectively (Hilje et al., 1987). This boom in the pesticide market probably also occurred in the rest of the Mesoamerican countries and was a reflection of the promotion of development schemes geared to intensify agricultural production to increase productivity and per-capita income.

Two well-documented examples of the pesticide treadmill refer to cotton and bananas. In Nicaragua, by 1950 there were only two important cotton pests, the boll weevil (*Anthonomus grandis*, Curculionidae) and the leafworm (*Alabama argillacea*, Noctuidae), against which insecticides were sprayed on average up to five times during the growing season. But the number of insect pest species increased through time, to 5 in 1955, 9 to 10 in the 1960s, and 15 to 24 in 1979, when insecticide use averaged 30 sprays (ICAITI, 1977; Flint and van den Bosch, 1981). In Golfo Dulce, Costa Rica, before 1950 there were only two main banana pests, the banana weevil (*Cosmopolites sordidus*, Curculionidae) and the red rust thrip (*Chaetanophothrips orchidii*, Thripidae). Because of heavy dusting of dieldrin to control them, eight defoliating insect species became primary pests in less than a decade, two of them after 1954, and six more after 1958 (Stephens, 1984).

These cases seem to be extreme and unusual, as they refer to key export crops. But even in crops for domestic consumption, especially vegetables, insecticides and other pesticides are currently used in a unilateral, indiscriminate, and excessive way (Hilje, 1995). Their use is unilateral because growers seldom consider pest control tactics other than pesticides because of their perceived advantages (efficacy, profitability, and availability); indiscriminate because with a few exceptions, most pesticides are not specific, killing both pests and beneficial organisms (pest natural enemies, pollinators, and vertebrates); and excessive because they are generally applied at doses and frequencies higher than recommended, and on a calendar basis, regardless of pest density or crop damage levels.

In summary, it is rather common that Mesoamerican farmers overspray a given pesticide as long as it remains effective. As a result of intensive selection pressure and favored by short life cycles and suitable climatic conditions throughout the year, a number of important pest species have evolved resistance. Although pest resistance has been detected in insects, pathogens, and weeds, it has been underestimated in Mesoamerica, due to a paucity of systematic monitoring. Thus, it is not surprising that the few well-executed studies of resistance have confirmed previous presumptions (Table 3.3). In addition to rendering pesticides useless, as well as increasing production costs and risks of undesirable side effects, resistance represents a burden to agrichemical companies. For instance, only one out of some 20,000 substances tested for pesticidal activity reaches the market, after 7 to 10 years of research and development, and its production costs exceed $85 million (NACA, 1993; Marrone, 1999).

Table 3.3 Selected Cases of Pest Species Resistant to Pesticides in Mesoamerica

Common and Scientific Name	Countries[a]	Pesticides	Ref.
		Insects	
Whitefly (*Bemisia tabaci*)	G,N	13 insecticides[b]	Dittrich et al. (1990), Hruska et al. (1997)
Diamondback moth (*Plutella xylostella*)	N,CR	7 insecticides[c]	Blanco et al. (1990), Hruska et al. (1997), Carazo et al. (1999), Cartín et al. (1999)
Cotton weevil (*Anthonomus grandis*)	N	Methyl parathion	Swezey and Salamanca (1987)
Bollworm (*Heliothis zea*)	N	Methyl parathion	Wolfenbarger et al. (1973)
Armyworm (*Spodoptera exigua*)	N	4 insecticides[d]	Hruska et al. (1997)
		Pathogens	
Black sigatoka (*Mycosphaerella fijiensis*)	CR	Benomyl	Salas (1993)
Mango anthracnose (*Colletotrichum gloeosporioides*)	CR	Benomyl	Barquero and Arauz (1996)
Potato blight (*Phytophthora infestans*)	CR	Metalaxyl	Salas (1993)
		Weeds	
Junglerice (*Echinochloa colona*)	CR, Col, ES, G, H, M, N, P	Propanil	Fischer et al. (1993), Villa-Casáres (1998)
	CR, Col, N	Fenoxaprop	Valverde et al. (2000), Riches et al. (1996)
	Col	Quinclorac	Schmidt (2000, pers. comm.)
Saramollagrass (*Ischaemum rugosum*)	Col	Fenoxaprop	Almario (2000, pers. comm.)
Honduras grass (*Ixophorus unisetus*)	CR	Imazapyr	Valverde et al. (1993)
Goosegrass (*Eleusine indica*)	CR	Imazapyr	Valverde et al. (1993)

[a] G = Guatemala, N = Nicaragua, CR = Costa Rica, Col = Colombia, ES = El Salvador, H = Honduras, M = Mexico, P = Panama.
[b] Including organophosphates, pyrethroids, and organochlorines.
[c] Including pyrethroids, organophosphates, and *B. thuringiensis*.
[d] Cypermethrin, deltamethrin, chlorpyrifos, and methomyl.

IPM in Mesoamerica

Some 40 years ago, the undesirable side effects of pesticide overuse led scientists from developed countries to postulate IPM as an ecologically oriented alternative to increased agricultural production and productivity (Stern et al., 1959). Currently, there are at least 64 definitions of IPM (Kogan and Bajwa, 1999), but probably all of them are based upon three paramount concepts: prevention, coexistence with pests, and sustainability (Hilje, 1994). In short, IPM stands for both a philosophy and a strategy of preventative and long-standing nature that combines compatible tactics to reduce pest populations to levels of noneconomic importance, while avoiding or minimizing harm to people and to the environment. Tactics such as improved varieties (plant breeding) and cultural practices, as well as physical or mechanical, biological, and selective chemical control, are the means to achieve the IPM strategy.

Historically, the IPM philosophy and practices rapidly gained acceptance worldwide, especially through the support and endorsement by international organizations, such as the United Nations Food and Agriculture Organization (FAO). In fact, it was the FAO, along with local agricultural and financial entities, that promoted the first large-scale IPM program in Mesoamerica, in response to the economic and environmental crises caused by insecticide overuse in Nicaraguan cotton fields (Andrews and Quezada, 1989; Daxl, 1989). This program, established in 1971, was a cornerstone in the promotion of IPM in Mesoamerica.

In the 1970s and 1980s, there were important educational efforts regarding IPM in several of the local universities. These efforts involved sending abroad faculty for graduate training, as well as the inclusion of IPM topics in regular courses related to crop protection in both universities and regional centers, such as the Tropical Agricultural Research and Higher Education Center (CATIE) and the Panamerican School of Agriculture (EAP-Zamorano). The largest IPM project was launched in 1984 at CATIE, a regional organization based in Costa Rica, as an initiative promoted by the Consortium for International Crop Protection (CICP) and funded by the U.S. Agency for International Development (USAID) (Saunders, 1989; Pareja, 1992*b*).

This project developed a formal graduate *Magister Scientiae* program and provided short-term in-service training to several young scientists at CATIE, as well as demand-driven short courses in the region. Other IPM activities included pest diagnosis and identification; validation of IPM alternatives for vegetables in Mesoamerican countries; establishment of a Central American Plant Protection Network; several types of publications, including four detailed *IPM Guidelines* (tomato, bell pepper, cabbage, and corn), quarterly documents (*IPM Newsletter, IPM Current Contents,* and the *Pesticide Tolerances Bulletin for Export Crops*), the journal *Manejo Integrado de Plagas* (IPM Journal), and books. Additionally, it fostered the establishment of the International IPM Congress, in 1987. For the second phase of the project (1990–1995), EAP-Zamorano became CATIE's partner, playing a relevant role in promoting biological control and rational pesticide use in Mesoamerica.

Contributions of this project were truly remarkable, not only by endorsing and legitimizing IPM as a feasible alternative for crop protection in Mesoamerica, but also by giving rise to an endurable tradition on IPM in this region and accomplishing its institutionalization at both CATIE and EAP. About 100 IPM-major graduates from CATIE's M.Sc. program have made a significant contribution in this regard.

Currently CATIE, EAP, and a number of national universities and institutes provide graduate and in-service training, conduct formal research and IPM validation activities, promote networking, publish books and training materials, and organize the biannual IPM Congress. Also, the *IPM Journal* has become a recognized source of information and a forum on current plant protection topics. In addition to the substantial support provided by USAID, several international agencies, such as NORAD (Norway), SIDA (Sweden), NRI-DFID (England), DANIDA (Denmark),

GTZ (Germany), COSUDE (Switzerland), CARE (U.S.), and USDA (U.S.), have provided funding to develop and promote IPM.

CURRENT ISSUES RELATED TO IPM

IPM and the Paradigm of Sustainability

The old and misleading dichotomy between economic development and environmental conservation has been replaced by the paradigm of sustainability. Conceptual differences, some rather semantic, have given rise to many definitions (IICA, 1991). But all of them emphasize the fundamental principles of conservation and rational economic exploitation of basic natural resources to satisfy food and fiber needs of society, without jeopardizing those for future generations. Thus, sustainability involves key elements of environmental protection, economic viability, and social equity.

During the past decade, mostly as a result of the Rio 1992 Summit, sustainability has become a relevant issue in the governmental policies of Mesoamerican countries. The Central American Alliance for Sustainable Development (ALIDES), signed in 1994, is being implemented through a number of specific projects, such as the Mesoamerican Biological Corridor, to connect several protected areas, not only to preserve their biota, but also to benefit rural communities associated with them (Miller, Chang, and Johnson, 2001). But regional initiatives concerning pest management are still lacking, despite the recognition of the detrimental effects to public health and to the environment of excessive pesticide use.

Adverse effects of pesticides on essential resources (water, soils, and wildlife), as well as in the increase of production costs, rejection of export crops, evolution of pesticide resistance, acute poisonings in agricultural workers, and chronic illnesses among consumers, have been well documented for Mesoamerica (ICAITI, 1977; Hilje et al., 1987; Thrupp, 1990; Castillo de la Cruz, and Ruepert, 1997; Castillo, Ruepert, and Solís, 1998). These effects clearly demonstrate that conventional pest control approaches give rise to unsustainable production systems, both economically and environmentally (Pareja, 1992a). But of most concern is the limited attention paid to pesticide effects outside agriculture per se, especially in relation to fragile tropical ecosystems located inside national parks and reserves, as well as mangroves and coral reefs.

Conservation advocates, whose initiatives have greatly benefited from donor agencies, are often biased towards habitat and species preservation. Also, policy- and decision-makers have generally neglected the close relationships between pest management practices, environmental conservation, and poverty alleviation and sometimes, despite their rethorics in favor of IPM, they promote and enforce policies aimed at fostering pesticide use instead. For instance, the existing regulatory framework discourages IPM approaches in practice, as tax exemptions and even subsidies make pesticide use more attractive to farmers (Rosset, 1987; Agne, 1996; Ramírez and Mumford, 1996).

Therefore, IPM practitioners face the challenge of convincing donor agencies and conservation organizations, as well as government policy- and decision-makers, about the environmental and socioeconomic benefits of implementing sound IPM programs. Cost-effective IPM programs can contribute to developing sustainable production systems, as they can help in reducing poverty by diminishing production costs, as well as increasing income stability and access to services; also, they can reduce health risks for rural families and urban consumers, through minimizing pesticide exposure.

IPM as an Interdisciplinary Approach

In essence, IPM is based upon a holistic approach, which considers the agroecosystem as a whole, including pest interactions with the entire cropping system in a given farm, spontaneous vegetation associated with crops, and soils (Ruesink, 1976; Hart, 1985; Teng, 1985; Altieri, 1987).

Therefore, interaction and complementarity between plant protection disciplines (entomology, plant pathology, and weed science) are critical for developing successful IPM programs. Also essential is the involvement of specialists in plant breeding, soil science, and plant nutrition, as well as of experts on specific crops. Crop management aimed solely to increase yields can make crops more prone to pest damage, especially in the presence of susceptible cultivars, inappropriate soil types, or nutrition imbalances.

Thus, a comprehensive approach to deal with pests calls for an expansion of the IPM concept to make it less pest focused and more system oriented. Therefore, IPM programs should evolve into integrated production systems that consider all production factors to deal with pests and abiotic constraints. This conceptual framework is essential for the development of sustainable production systems that merge and optimize productivity, stability (managerial, economic and cultural), elasticity, and diversity (Altieri, 1987; Hamblin, 1995; Arauz, 1996; Meerman et al., 1996; Rabbinge and van Oijen, 1997).

This kind of approach implies that other types of experts, such as agricultural economists and social scientists, need to join efforts with plant protection and agronomy specialists. Because IPM programs are always oriented to benefit farmers in both economic (income level and stability, purchasing power, etc.) and social terms (family labor, access to services, gender equity, etc.), analytical procedures from economics, as well as organization methodologies from rural sociology and social anthropology, can be very helpful in documenting the economic advantages of IPM programs, as well as in organizing farmers.

In the authors' experience, when economists and social scientists team up with plant protection and agronomy experts, accomplishments in on-farm IPM implementation are, by far, superior to those achieved through IPM-disciplinary efforts. In fact, IPM programs promoted by CATIE and EAP have involved economists and social scientists, and these disciplines have been incorporated into the plant protection curricula of these centers and in those of several universities throughout Mesoamerica. Nonetheless, scientific research is still mostly disciplinary in all academic institutions, leading to isolated and fragmentary contributions.

Although disciplinary contributions are important and necessary for generating IPM-oriented tactics for specific pests, in practice what makes possible the integration of tactics is the crop or cropping system. Thus, interdisciplinary work can be fostered by the establishment of on-farm validation plots in which available IPM alternatives are appraised after discussion and selection by growers, extension agents, and researchers (Calvo et al., 1994, 1996). Appraisal includes plant protection and agronomic aspects, economics, and grassroots organization. Therefore, continuous and coordinated interactions between experts from all disciplines becomes feasible and enriched by growers' perceptions, practices, and needs.

Decision Criteria in IPM

Decision criteria, such as economic thresholds, initially used for insect management, have been one of the foundations of the IPM philosophy since its origin (Stern et al., 1959). Despite being a sound approach for coexistence with pests by allowing pesticides only when pest damage or density justifies their use, practical experience with thresholds reveals complexities and shortcomings (Andow and Kiritani, 1983; Pedigo, Hutchins, and Higley, 1986; Rosset, 1991; Arauz, 1998; Ramírez and Saunders, 1998).

Aside from biological data (yield response to pest intensity), thresholds depend on economic variables, such as current production costs, produce supply and demand, and inflation rates. Therefore, thresholds vary between cropping seasons, which makes their use unpractical for both extension agents and farmers. Moreover, thresholds are usually developed experimentally for specific pests, but under field conditions several pest species may affect a particular plant structure, so that damage could be synergistic rather than additive. A further level of complexity arises when a particular pest affects several plant structures, forcing consideration of multiple thresholds according to each affected structure. Conceptually, it seems that balancing benefits and costs is the main goal of economic thresholds, instead of maximizing economic benefits from control measures. Finally, thresholds are rather static, as they refer to decisions made at specific times, underestimating the agroecosystem dynamics.

Action thresholds can help circumvent some of these shortcomings. These are pest levels based on yield response to pest intensity, resulting from a combination of scientific data taken from the literature and practical experience of field researchers, extension agents, and growers. Action thresholds are conceptually simpler for growers, as they consist of a single figure that is rather constant. They are also flexible enough to be easily modified by experience, according to crop season and location.

Action thresholds have been used in Costa Rica in validation plots to manage key tomato and potato insect pests (Table 3.4), and have proved successful in largely reducing pesticide applications and production costs (Calvo et al., 1994, 1996). Resource-poor farmers in Nicaragua have successfully used action thresholds at commercial scale as a decision tool to manage fall armyworm (*Spodoptera frugiperda*, Noctuidae) in corn (Hruska, pers. comm.; CARE, pers. comm.), and

Table 3.4 Action Thresholds Used for Key Tomato and Potato Insect Pests, in IPM Validation Plots in Costa Rica

Pest Species	Action Threshold
	Tomato
Leafminers (*Liriomyza* spp.)	20 fresh mines/30 plants
Tomato pinworm (*Keiferia lycopersicella*)	20 larvae/30 plants, or 4 fruits with larvae/30 plants
Fruitworms (*Spodoptera* spp.)	1 egg mass/30 plants, or 2 fruits with recent damage/30 plants[a]
Fruitworms (*Heliothis* spp.)	4 eggs or young larvae/30 plants, or 2 fruits with recent damage/30 plants[a]
	Potato
Leafminer (*Liriomyza huidobrensis*)	300 adults/yellow sticky trap/week
Tubermoths[b]	100 adults of both species/pheromone trap/week

[a] The second threshold for fruitworms applies from fruit set on.
[b] Tubermoths include *Scrobipalpopsis solanivora* and *Phthorimaea operculella*.
Source: From Calvo et al., 1994, 1996.

fruitworms (*Heliothis* spp. and *Spodoptera* spp., Noctuidae) in tomato (Guharay, pers. comm. CATIE, pers. comm.). These growers even scout their fields and record pest levels to make decisions by themselves.

In the case of pathogens, because of their intrinsic behavior, dissemination means, environmental influence on their biology, and the difficulty to quantify individuals, decision criteria for management of plant diseases are based on incidence, severity, and distribution of pathogen populations. Monitoring disease development through growing seasons, fields, cultivars, and phenological stages, along with climatological data, can provide enough information to build useful disease progress curves for decision-making purposes. In Costa Rica, these criteria have been successfully used for managing black sigatoka (*Mycosphaerella fijiensis* var. defformis) in bananas (Romero, 1984) and for gummy stem (*Didymella melonis*) in muskmelon (Araya, unpubl. data).

Another important decision criterion is the critical periods of susceptibility to specific pests, as crop plants can vary in their response to pests according to their physiological state or phenological stage (Trumble, Kolodyn-Hirsch, and Ting, 1993; Arauz, 1998). Although critical periods are conceptually and operationally simpler than thresholds, there has been only limited research toward their development and implementation, except for weed control. Many crops have periods, especially at the beginning of the life cycle, when weeds can be tolerated without a yield penalty (Zimdahl, 1980; Radosevich et al., 1997). Critical periods could be especially suited for illiterate growers, who are common in Mesoamerica — illiteracy rates in 1996 were as high as 44, 34, 29, and 27%, in Guatemala, Nicaragua, El Salvador, and Honduras, respectively (Proyecto Estado de la Nación, 1999). In fact, critical periods and action thresholds can complement one another, allowing control measures to be applied only when pests reach damaging levels within those critical periods.

IPM Implementation

The few regional and successful IPM programs implemented in Mesoamerica have been focused on cotton, sugarcane, and banana insect pests. These are primarily export crops, commonly planted over extensive areas, and owned by either a few local entrepreneurs and cooperatives or transnational companies. High profitability of these large agricultural operations makes it possible to support IPM programs (trained personnel, scouting, local research, etc.), some of which, especially in the case of bananas, are induced by consumers' strong pressures in international markets.

In Nicaragua, for example, a large IPM program against the cotton weevil, based on a sound combination of cultural practices, pheromones, and insecticides, comprised 34,000 ha in 1983 (Daxl, 1989). Since 1978 in Panama (Narváez 1986) and 1985 in Costa Rica (Badilla, Solís, and Alfaro, 1991), the sugarcane borer (*Diatraea tabernella*, Pyralidae) has been managed with periodic releases of the parasitoid *Cotesia flavipes* (Braconidae). In Costa Rica this IPM program has produced, up to date, over 330 million wasps for 26,600 ha, with an outstanding benefit:cost ratio of 9:1 (Sáenz, pers. comm.; DIECA, pers. comm.).

In banana plantations of Costa Rica and Honduras, a number of practices were established to manage insect pests. Higher damage levels were tolerated to favor natural biological control and fruit-protecting plastic bags impregnated with an insecticide were introduced (Stephens, 1984; Ostmark, 1989), an approach that is now widely disseminated throughout Mesoamerica. Management of key banana diseases relies on monitoring, damage quantification, eradication, and exclusion methods for moko (*P. solanacearum*) since 1977 (González, 1987) and black sigatoka since 1982 (Romero, 1984).

Another successful example is the IPM program developed in Costa Rica and Honduras for the red ring disease in oil palm. This disease is caused by the nematode *Bursaphelenchus cocophilus*, which is transmitted by the weevil *Rhynchophorus palmarum* (Curculionidae). The program is based on scouting, removal of wilted palms, reduction of favorable sites for insect reproduction, and monitoring in prospective areas to be planted to reduce both the nematode and the insect vector (Chinchilla, 1996). Oil palm production in both countries also represents a system in which weed control mostly relies on the use of cover crops, with herbicide applications being directed only to the palm circles. Weeds associated with oil palm plantations are also valued as important reservoirs for beneficial insects (Mexzón and Chinchilla, 1999).

Substantial research is also being conducted toward the development and implementation of integrated management of itchgrass (*Rottboellia cochinchinensis*) in maize, which affects more than 3.5 million ha in Central America and the Caribbean (FAO, 1992). In seasonally dry areas of Central America, itchgrass infests both annual and perennial crops, causing significant yield losses in maize, sugarcane, upland and rainfed rice, beans, and sorghum (Herrera, 1989). Integrated management of this weed is based on intersowing the legume cover crop *Mucuna deeringiana* to suppress itchgrass growth and reproduction. Recommendations have been developed and validated at farmers' fields on planting arrangements, densities, and time of planting of mucuna for weed suppression (Valverde et al., 1999). Additional control of the weed is obtained by combining preemergence control with herbicides, weed elimination during the fallow period, and zero tillage. Integrated itchgrass management also proved economically feasible for small holders. A promising alternative is biological control with the itchgrass head smut, *Sporisorium ophiuri*, which prevents seed set and is host specific (Ellison, 1987, 1993; Reeder, Ellison, and Thomas, 1996). A request to introduce the smut as a classical biocontrol agent has been filed with the Costa Rican plant health authorities, and the dynamics and possible impact of the smut have been explored with a modeling approach (Smith, Reeder, and Thomas, 1997; Smith et al., 2001).

A mechanism that has fostered IPM adoption is that, for a number of crops, farmers with small- and medium-size farms commit to sell their harvest to large agroindustrial companies that specify in advance the cropping practices to be used, including those related to pest management. These companies are forced by consumer pressure in international markets to implement IPM-oriented practices, a process that makes growers aware of IPM tactics, favors their rapid adoption, and allows spontaneous technology transfer to other neighboring farmers. The lesson learned is that strong support to IPM from agroindustrial companies effectively contributes to a wider IPM implementation, as has occurred with snow peas and broccoli in Guatemala (Pareja, 1992b).

In Mesoamerica, adoption and implementation of IPM programs are still perceived as scanty and the aforementioned successful cases as exceptional. But it should be emphasized that farmers are often reluctant to accept technology "packages" as a whole and prefer to adopt specific, isolated tactics to improve the plant protection component in their production systems (Pareja, 1992a). Therefore, actual IPM implementation in the region may be underestimated, making it almost impossible to quantify the derived benefits from adoption of this strategy. Preferred IPM tactics more readily adopted by farmers are improved cultivars adapted to local conditions and resistant to pests; effective and affordable microbial (e.g., *Bacillus thuringiensis*) and botanical insecticides (e.g., derivatives of neem tree, *Azadirachta indica*, Meliaceae); and cultural practices, such as crop rotation, high-quality seeds, and improved drainage for disease control.

Realistically, no single control tactic outweighs pesticides, which possess several intrinsic comparative advantages. Field experimentation with pesticides is also rather straightforward, for there are well-established protocols to conduct field trials (experimental designs, dose ranges, methods for interpretation, etc.), and short-term experiments allow rapid gathering of information. Additionally, agrichemical companies provide both strong technical assistance and commercial promotion services that allow them to easily reach and maintain a permanent contact with farmers. In contrast, the complexity of IPM as a strategy makes it difficult to compete with pesticide-based control programs. Several constraints also limit IPM adoption, when compared to conventional pesticides (González, 1976; Brader, 1979; Bottrell, 1987; Vaughan, 1976, 1989; Andrews, 1989; Pareja, 1992a; Hilje, 1994).

Finally, complexity of IPM as a strategy makes it difficult to compete with pesticide-based control programs. Because of its multitactic nature, IPM requires higher involvement of farmers in the processes of generation, validation, and technology transfer. In this regard, several innovative conceptual models of *participatory research* (as well as training and learning) have been developed and applied in a number of tropical countries, worldwide, as an alternative to conventional extension models. These include participatory platforms, such as *Farmers' Field Schools* (FFS) and *Local Agricultural Research Committees* (LARC), which are aimed at enhancing decision-making capacity and local innovation (Braun et al., 1999). These approaches emphasize farmer *empowerment*, meaning that farmers' capabilities and abilities for both ecological reasoning and economic decision-making are continuously increased, regardless of the crops or pests they have to deal with. In the last decade, a *farmer first* methodology has been widely applied in Nicaragua (Nelson, 1996; SIMAS, 1996). This concept has evolved into an approach geared to strengthen farmers' planning and decision-making capabilities, which will be promoted throughout Central America during this decade (CATIE, 1999).

Farmers are trained through a participatory process which is based upon particular crop stages, or critical phenological events, during the cropping season. Before the planting season begins, a group of 10 to 25 farmers gather to discuss their former experience with the crop to be planted. They identify and prioritize their main pest problems and their current or possible control practices and design a plan to manage the crop and its pests. They also decide on a simple method to collect data and on dates for future meetings. Farmers volunteer to evaluate alternative practices in test plots. At each meeting, farmers describe the situation in their fields using their own data, analyze why pests have increased or decreased, and justify the decisions they have made. They also visit fields, including the test plots, learn new methods for scouting pests and natural enemies, and evaluate alternative practices. Once the crop is harvested, the group meets to review production costs, yields, pest dynamics during the season, and what they have learned. In such diverse crops as coffee, tomatoes, and cooking bananas, field studies have shown that, through participatory training and experimentation by crop stage, farmers have reduced pest losses and pesticide applications and improved crop yields (CATIE, 1999).

In addition, in Mesoamerica, public agricultural institutions are quite fragile, unable to offer job stability to qualified specialists and continuity to research and development programs, so as to undertake long-term, permanent IPM initiatives. Also, although there are IPM alternatives for most pests, some tactics are inapplicable because of either their high cost, labor intensiveness, incompatibility with some components of production systems, or consumer preferences (unacceptable characteristics in produce, cosmetic damage, etc.). Moreover, a number of IPM tactics or tools, such as economic thresholds and sampling methods, are cumbersome for many farmers. Additionally, it is not possible to establish a single or universal IPM program for wide regions because of the typical agroecological and socioeconomic heterogeneity of the neotropics. Finally, in many cases, quantitative information on the potential value of nonchemical alternatives is scarce.

IPM and the Agrichemical Industry

As a coherent philosophy, IPM arose in response to excessive pesticide use. But the IPM concept and practices are not aimed to completely eliminate pesticides, but rather to rationalize their use (Bottrell, 1979). Until recently, agrichemical companies were often belligerent toward IPM, but their attitude has changed and IPM is now perceived as a means to ensure long-term effectiveness and profitability for their agrichemicals. Past experience with the pesticide treadmill demonstrates that pesticide overuse, although quite profitable initially, sacrifices long-term economic benefits. Agrichemical companies fully understand that the economic sustainability of their products depends upon their rational use, as proposed by IPM. Likewise, several companies are currently involved in manufacturing nonconventional pesticides (biopesticides or biorationals), selective to nontarget organisms, generally perceived as environmentally benign, and best suited for IPM (Hall and Menn, 1999).

There are several novel insecticides available, made from viruses, protozoa, bacteria, fungi, marine worms, and plants, as well as a variety of oils, detergents, and waxes that are also used for insect control (Rodgers, 1993). The case of azadirachtin is worth emphasizing, a limonoid obtained from seeds of the neem tree, already marketed under several brand names and formulations. Regarding pathogens, Kilol, which is a fungicide and bactericide extracted from citrus seeds, is widely used in Costa Rica (Picado and Ramírez, 1998). Also, several mycoparasites have shown potential as biological control agents of plant pathogens, but extensive research on both the epidemiology of the disease and the target pathogen is still needed before widespread commercialization (Adams, 1990).

In the case of herbicides, very few of those commercially available have been derived from naturally occurring compounds, although in some cases there is a striking resemblance between commercial herbicides and natural phytotoxins (Duke et al., 2000). Only three mycoherbicides have been registered and used commercially: DeVine (*Phytophthora palmivora*), Collego (*Colletotrichum gloesporioides* f. sp. *aeschynomene*), and BioMal (*Colletotrichum gloesporioides* f. sp. *malvae*), but their market share was always limited (Auld and Morin, 1995) and none have been used in Mesoamerica.

The agrichemical industry also offers other products suitable for IPM programs, including growth regulators and soil conditioners such as algae extracts and humic acids. Agrichemical companies are also involved in the production of insect pheromones and attractants, which can be used either as monitoring tools or as direct control methods. More recently, the agrichemical industry has introduced several genetically modified crops, most of them with traits that facilitate pest control; this topic is considered in detail separately in this chapter.

In Mesoamerica, both microbial and botanical pesticides have been widely accepted. In addition, farmers' interest in them has paved the way to use other nonconventional materials. This is the reason that there is an increasing number of small- and medium-size companies involved in biopesticide production, such as neem products in Nicaragua (i.e., Investigaciones Orgánicas S.A.), entomopathogens in Guatemala (i.e., Agrícola El Sol), as well as botanical products (i.e., PROQUIVA) and pheromones (i.e., ChemTica International) in Costa Rica. Because of their high adoption potential by farmers, CATIE recently began a long-term project strengthening and promoting the involvement of small and medium-size companies in manufacturing nonconventional products in Central America (Röttger, pers. comm.; CATIE-GTZ, pers. comm.).

IPM and Biodiversity

The concept of biodiversity includes species richness in a given ecological community, as well as two other levels of complexity (genes, and ecosystems or habitats). Biodiversity has received considerable attention in recent years because of the worldwide concern about the high rates of destruction of some of the last major forest masses on Earth, which are mainly located in the tropics (Wilson, 1988). These forests contain many organisms not yet described, some of them potentially useful for humankind.

In Mesoamerica, a major accomplishment regarding biodiversity was the establishment of the National Institute of Biodiversity (INBio) in Costa Rica in 1989 (Gámez et al., 1993). As a national institution, INBio has focused its activities on inventorying the Costa Rican biota and on the search for compounds of pharmaceutical value. But few resources have been allocated to search for biodiversity applications to agriculture, including IPM programs.

Biodiversity and IPM are closely interrelated (Hilje and Hanson, 1998). For example, new products (genes, natural pesticides, and beneficial organisms) for IPM programs can be obtained from tropical species and, conversely, implementation of IPM programs can have beneficial effects on both terrestrial and aquatic biodiversity. Some representative examples follow.

In the U.S., 98% of crops planted (rice, corn, potatoes, tomatoes, tobacco, oranges, and peanuts) have tropical ancestors (Plotkin, 1988). Thus, historically, tropical regions have been sources of genes from wild crop relatives for plant breeding programs, including those for pest resistance. Perhaps the best examples of wild ancestors contributing to genetic improvement of crops of tropical origin in temperate zones are perennial teosinte (*Zea diploperennis*), a wild relative of maize found in Mexico, and wild tomato (*Lycopersicon chmielewskii*), found in Peru (Iltis, 1988).

As for the exploration of natural pesticides, it has focused mainly on plants and microorganisms showing either pathogenic or antagonistic activity toward pest species. A large number of tropical plants offer an untapped potential as pesticide sources (Grainge and Ahmed, 1988; Stoll, 1989). The interest in discovering, characterizing, and exploiting natural products has prompted scientists to revisit ethnobotanical knowledge and to pursue screening of crude extracts for biological activity. Most research in this field is very preliminary, some anecdotal, but a body of information is building up, as mentioned in local publications (Sabillón and Bustamante, 1996; Arauz, 1998) and presented in national and regional scientific meetings.

An outstanding example is bitterwood (*Quassia amara*, Simaroubaceae), a shrub growing within several indigenous reservations in Mesoamerica (Ocampo, 1995), whose extracts can either deter or kill very important pests, such as the mahogany shootborer (*Hypsipyla grandella*, Pyralidae) (Mancebo et al., 2000) and whiteflies (*Bemisia tabaci*) (Cubillo, Sanabria, and Hilje, 1997). Also, the *chaperno* (*Lonchocarpus felipei*, Fabaceae) contains DMDP (2R,5R-dihydroxymethyl-3R,4R-dihydroxypyrrolidine), a nematicidal compound, and is being developed for nematode control in Costa Rica (INBio, 1998). Perennial peanuts (*Arachis pintoi*, Fabaceae) also have nematicidal activity (Marabán, Dicklow, and Zuckerman, 1992).

Concerning pesticides derived from microorganisms and wild animals, insecticides made of *B. thuringiensis* occur worldwide, with at least 12 brand names. Nonetheless, there is a large potential for biopesticide development in other tropical organisms. For example, in Costa Rica, the insecticidal activity of extracts from plants, bryophytes, and even insects, is currently being investigated (INBio, 1998). Similarly, chitinolitic microorganisms for the control of black sigatoka disease on bananas are being selected (González et al., 1996; Ruiz-Silvera et al., 1997), as well as antagonistic bacteria against the potato early blight in tomato (Sánchez, Bustamante, and Shattock, 1998, 1999).

To be better able to advance biocontrol as part of IPM in Mesoamerica, more effort should be devoted to the inventory, conservation, evaluation, and use of pests' natural enemies (parasitoids, predators, pathogens, and antagonists). For instance, parasitoid populations are generally quite low in wild areas, as a result of their efficiency in regulating herbivore host populations, which suggests that the most useful parasitoids for biological control programs are the most vulnerable and prone to extinction (LaSalle and Gauld, 1991). Thus, deforestation could cause an irreversible loss of valuable resources for IPM, most of which are still unknown to science. A preliminary inventory of the order Hymenoptera, which includes the majority of parasitoid species, estimated some 20,000 to 40,000 species in Costa Rica, 70 to 80% of which are not yet described (Hanson and Gauld, 1995).

In addition to protected areas, vegetation present within and around crop fields can also be an important source of biocontrol agents and pollinators (Vandermeer and Perfecto, 1997; Hilje and Hanson, 1998).Thus, patches of wild or planted vegetation such as those commonly found in Mesoamerican agricultural areas could contribute to improving pest management and, consequently, agricultural production.

Coffee, commonly planted in association with shade trees as an agroforestry system (Beer et al., 1998), represents an interesting example, as discussed in a related chapter in this volume (Perfecto and Armbrecht, Chapter 6 this volume). Coffee plantations provide refuge for insects, including several groups of Hymenoptera (Perfecto and Vandermeer, 1994; Perfecto and Snelling, 1995; Perfecto et al., 1996, 1997), Coleoptera (Nestel, Dickschen, and Altieri, 1993), and Homoptera (Rojas et al., 2001). When comparing insect biodiversity associated with three types of coffee plantations in Costa Rica (a traditional polyculture, a coffee plantation where shade was intensively managed, and a coffee monoculture), Perfecto et al. (1996) found the highest species diversity in the traditional polyculture, which was quantitatively similar to that found in some pristine tropical forests.

From an IPM perspective, perhaps the main challenge is to study and understand biodiversity from a functional viewpoint (Vandermeer and Perfecto, 1997), that is, the mechanisms by which certain organisms reduce pest problems. Such understanding could allow manipulation of associated vegetation and even its enrichment, to favor populations of beneficial organisms in plantation crops, agroforestry systems, and crop mosaic areas in the region.

Finally, implementing sound IPM programs could assist in conservation of biodiversity by reducing pesticide use and through other indirect benefits. Improved weed control without herbicides, for example, could reduce soil erosion and preclude damage to tropical aquatic ecosystems (Hilje and Hanson, 1998). The few documented cases of adverse effects of pesticides and sediments on aquatic biota depict worrisome situations for both rivers (Castillo, de la Cruz, and Ruepert, 1997; Castillo, Ruepert, and Solís, 1998) and coral reefs (Cortés and Risk, 1984; Glynn et al., 1984).

IPM and Genetically Modified Crops

Biotechnology, particularly genetically modified (GM) crops, is perceived as a key element for attaining higher yields and improved quality of food to fulfill the requirements of the world's increasing population (Dunwell, 1999; Mann, 1999). GM crops that exhibit resistance to insects or pathogens, as well as herbicide tolerance, are a new option in IPM, whose implications are just beginning to be considered in Mesoamerica.

Worldwide, over 52 million ha of GM crops were planted in 2001, which represents a more than a 30-fold increase in relation to 1996 (James, 2001). Currently, the most important traits being commercially exploited are herbicide tolerance and insect resistance based on the production of *B. thuringiensis* (Bt) insecticidal proteins by Bt-transgenic plants. By 2001, herbicide-tolerant and insect-resistant crops represented about 77% and 15% of the global transgenic crop area, respectively. By area planted, the most important transgenic crops were soybeans (63%), maize (19%), and cotton (13%) and canola/oilseed rape (5% each). Other transgenic crops, each representing less than 1% of the planted area, were potato, squash, and papaya (James, 2001).

Concerning Latin America, about 11.8 million hectares of herbicide-tolerant soybean were planted in Argentina in 2001, representing almost the entire soybean planted area (James, 2001). But in Mesoamerica, release or commercial planting of transgenic crops is very limited. An estimated less than 100,000 ha of GM crops was planted in Mexico in 1999. In Costa Rica, field releases of glyphosate-resistant soybean and glyphosate- and bromoxynil-resistant cotton have been made mostly for seed production for export (May, 1997). GM crops have also been tested in Belize and Guatemala, but some of these are unconfirmed releases (de Kathen, 1996). Also, research is being conducted to transform locally adapted crop varieties (Valdez et al., 1998).

GM crops have given rise to acrimonious debate of sensitive issues, which are beyond the scope of this chapter. These include potential health risks for people, as a result of ingestion of substances mediated by transgene expression, as well intellectual property rights. Among the important benefits attributed to GM crops are increased flexibility of crop management, high productivity, and delivery of a safer environment by substitution of conventional pesticides. But there are also concerns about their field use. Relevant to Mesoamerica in ecological terms are those related to gene flow and escape of the transgenes (biosafety), the impact of GM crops on pest population dynamics, the likelihood of new developments of pesticide resistance, and problems with volunteers of herbicide-tolerant crops (Madsen et al., 2002; Riches and Valverde, 2002).

Regarding biosafety, for a transgene of a GM crop to escape into wild relatives, there should be hybridization followed by the subsequent establishment and persistence of the hybrid. There is no doubt that genetically modified crops could cross with crop ancestors or wild relatives in Mesoamerica, as well as with wild or weedy compatible plants. But the relevant aspect about gene flow is whether the hybrids will have a fitness advantage in the wild as to become better suited for the natural conditions or to become more invasive. Thus, ecological risk is higher for transgenes that may confer ecological advantages to the plant, particularly genes for insect and pathogen resistance, which could increase the abundance of the transgene in natural habitats (Hails, 2000). Although there may be some physical and biological barriers to hybridization, the typical mosaic of agricultural fields found in Mesoamerica could be conducive to gene flow. Crops are often sown at different times during the cropping cycle, providing plants of different ages across short distances. Physical barriers such as differential flowering times could be overcome by this circumstance.

But risk of transgene escape should be evaluated on a case-by-case basis. Soybean, for example, the main transgenic crop grown in Latin America, is an exotic cultigen without wild relatives in Mesoamerica. Rice (*Oryza sativa*) and sorghum (*Sorghum bicolor*) are also exotic crops, but both have weedy relatives in Mesoamerica. Other GM crops may require more careful consideration regarding risks of transgene escape. For example, there are wild species capable of hybridizing with the tetraploid cultivated cotton in Mesoamerica. In some southern states of the U.S., proximity of wild relatives to cotton-producing areas has precluded the use of Bt cultivars (Halford, 1999), but they have been allowed in Mexico (Sage, 1999). In addition, Central Mexico is one of the centers of diversity of potatoes, the other comprising Peru, Bolivia, and northwest Argentina. Several wild and cultivated tuber-bearing *Solanum* species are found in Mesoamerica. Hanneman (1994) concluded that there is minimal risk for the potential introduction of transgenes from genetically modified potatoes to wild species in Costa Rica, the southern edge of the Mexican center of diversity, but it is more likely to occur in Mexico, especially into hexaploid wild species and cultivated tetraploid species.

Of most concern for gene flow and possible crop gene erosion is maize (*Zea mays mays*), a crop that originated from Mesoamerica with wild relatives and many landraces found in Mexico and Guatemala. In fact, no Bt maize has been grown legally in Mexico because of concerns about gene introgression (Halford, 1999; Macilwain, 1999). But there has been a wide divergence of opinions among scientists about the possible impact of GM maize as to its center of origin and diversity that has been fueled with the recent report of transgene flow and possible introgression into maize landraces in southern Mexico (Quist and Chapela, 2001). Maize was probably domesticated from teosinte (*Zea mays parviglumis*) of southern Mexico about 7000 years ago (White and Doebley, 1998; Wang et al., 1999) and contains 75% of the variation found in the latter (Eyre-Walker et al., 1998). Other important wild relatives of maize are three additional annual teosintes (*Z. mays mexicana, Z. mays luxurians*, and *Z. mays huehuetenangensis*), as well as perennial teosintes (*Z. diploperennis* and *Z. perennis*).

Despite maize and "teosintes" being completely interfertile (Wilkes, 1977; Lubberstedt, Dussle, and Melchinger, 1998; White and Doebley, 1998), some researchers state that the risk of transgene flow from GM maize to "teosinte" is negligible, because in most cases "teosinte" populations are spatially isolated from maize. The largest "teosinte" populations are located in sparsely settled areas, and there are differential flowering times between maize and "teosinte" (Wilkes, 1977). It has also been argued that transgenes cannot become established in natural populations of "teosinte" and that GM maize poses no risk in the center of origin of the crop (Martínez-Soriano and Leal-Klevezas, 2000).

These statements, however, have been strongly criticized because reciprocal gene flow between maize and "teosintes" is possible, transgenic traits could easily move from maize to "teosinte" given that pollen flow most likely would occur from the crop to the wild relative and, especially, if the trait is dominant (Ningh et al., 2000; Abbo and Rubin, 2000). In many areas "teosinte" grows along maize field margins and in the Valley of Mexico it grows exclusively as a weed in maize fields, a condition under which hybrids between the two species are always present (Wilkes, 1977); hybrids, however, do not have superior fitness (Doebley, 1990; De Kathen, 1996). But the greatest controversy has come from the report of transgenes into landraces in the Mexico states of Oaxaca and Puebla (Quist and Chapela, 2001; Hodgson, 2002b) and criticism about possible methodological pitfalls in their detection and opposing arguments about its implications (Christou, 2002; Hodgson, 2002a; Kaplinsky et al., 2002; Mann, 2002; Metz and Fütterer, 2002). As a result of the discussion, *Nature*, the journal that published the initial report by Quist and Chapela (2001), concluded that the publication of the paper was not justified based on the available evidence.*

On the other hand, the impact of GM crops on pest population dynamics and exacerbation of pesticide resistance problems could also be important in IPM programs. For instance, concerns have been raised about the possibility of insect pests becoming resistant to Bt by exposure to transgenic plants, especially considering the high selection pressure imposed by season-long expression of the transgene over wide areas planted with the Bt crop. Widespread resistance to Bt under field conditions woud jeopardize the use of the most successful biological insecticide ever used (Wallimann, 2000). So far, several Bt-resistant populations of insects have been developed under controlled conditions (Frutos, Rang, and Royer, 1999) but only *Plutella xylostella* has evolved resistance under field selection pressure (Tabashnik et al., 1998), including in Mesoamerica (Cartín et al., 1999). In some developed countries, regulatory action has been taken to decrease this risk, by requiring refuges (areas planted with a non-transgenic, normal variety) of about 20% of the planted area (Halford, 1999). There is also concern about the possible indirect negative effect of Bt crops on non-target organisms, especially parasitoids, but research on the subject is scarce and, in some cases, has been criticized for being conducted under "worst case scenarios" that could be unrealistic (Poppy, 2000).

In the case of weed control, overreliance on new herbicides to which crops have been transformed for tolerance may impose enough selection pressure for resistance evolution in weeds. Two of the most important herbicides, glyphosate and glufosinate, to which GM crops have been produced and commercialized, are considered low risk for resistance. Glufosinate resistance has not evolved in weeds under cropping conditions, and there are only two cases of glyphosate resistance in weeds worldwide: *Lolium rigidum* (Powles et al., 1998) and *Eleusine indica* (Tran et al., 1999).

* Editorial note. *Nature*, 416:600, 2002.

Finally, in the case of herbicide-tolerant crops, volunteers may become a problem in rotation crops or when varieties are changed. This will be of particular concern with crops that establish readily from seed lost during harvest, such as rice and soybean. For example, in Mesoamerica, red rice (*Oryza sativa*) and volunteer rice are frequently controlled by glyphosate before planting the rice crop. If glyphosate-tolerant rice is introduced, it would be more difficult to control volunteer rice by chemical means, as very few additional, nonpersistent herbicides are available (Olofsdotter, Valverde, and Madsen, 2000). The situation would be aggravated if movement of transgenes occurs among cultivars, resulting in "gene stacking," as has been illustrated in Canada where volunteer oilseed rape (canola) resistant to imidazolinones, glyphosate, and glufosinate as a result of gene flow from adjacent fields was confirmed within a few growing seasons (Hall, Huffmann, and Topinka, 2000).

In our opinion, the implications of releasing GM crops into an important center of origin such as Mesoamerica have not been thoroughly addressed, as to decide if the alleged benefits of transgenic crops outweight their risks. Although IPM can benefit from the positive attributes of GM crops for food and fiber production, most pests relevant to crops of economic importance under the prevailing cropping systems in Mesoamerica can be managed without the introduction of GM crops. In technologically-intensive, high-input crop production systems, transgenic crops with resistance to important pests could become quite useful in helping to alleviate pest problems. A good example would be transgenic bananas resistant to black sigatoka and nematodes, for which traditional breeding programs have been only partially successful because of the limited crop variability and "clonal" nature of production. The usefulness of GM crops also has to be evaluated considering the socioeconomic conditions of the region, especially addressing the impact of possible inceased dependency on imported technologies and inputs relative to local, adapted agricultural practices perhaps more suitable to resource-poor farmers.

IPM and Novel Models for Crop Production

In the past 12 years, the so-called nontraditional export crops (NTAEs), such as pineapple, melons, sesame, snow peas, and broccoli, as well as flowers and ornamental plants, have been added to the list of crops historically predominant in Mesoamerica (Kaimowitz, 1991). NTAEs that were promoted by USAID agricultural programs (Kaimowitz, 1991), are normally planted in small- and medium-size farms. Commonly, farmers sign contracts with agroindustrial export companies that promote the use of imported technology packages of agronomic practices, including those related to pest management (Rosset, 1992; Conroy, Murray, and Rossett, 1996). Because NTAEs are short-span crops and their planting and harvesting dates are closely tied to opportunities in the international markets (export windows), farmers are forced to rely heavily on pesticides to decrease pest damage risks. These exported goods are subject to stringent cosmetic standards by consumers and to strict quarantine inspection at ports of entry. Thus, Mesoamerican farmers are faced with the dilemma of exporting perfect products while complying with strict pesticide tolerances.

In response to this dilemma, farmers often reduce economic uncertainty by applying pesticides in excess. High concentrations of pesticide residues or presence of unauthorized residues in exported produce has resulted in frequent detentions and rejections of shipments at U.S. ports of entry. For example, between 1984 and 1994, there were 3168 detentions of shipments of snow peas and broccoli from Guatemala, with losses amounting to almost U.S.$18 million (Thrupp, 1995). Chlorotalonil and metamidophos residues were the main pesticides responsible for these rejections. Because a tolerance for chlorothalonil was not established in the U.S., shipments of snow peas containing traces of the fungicide were considered adultered and thus destroyed at the port of entry.

To deal with this situation, IPM programs for both snow peas and broccoli were promoted (Pareja, 1992b; Thrupp, 1995). It was soon demonstrated that it was possible to maintain acceptable broccoli yields if conventional insecticides were substituted by *B. thuringiensis* formulations. But these breakthroughs were not good enough to comply with quality standards in U.S. markets (Salguero, pers. comm.). Thus, stringent cosmetic standards imposed by gourmet consumers precluded the implementation of a biologically sound and environmentally safe solution to a pest problem. We suggest that the success of IPM programs for export vegetables requires consumer education in the importing country, toward which both public and private institutions should play a relevant role.

The feasibility of this type of approach is demonstrated by how consumers' attitudes in developed countries have sparked a notorious change in Mesoamerican agriculture toward organic production. The number of organic products in the U.S. increased from 98 to 510 between 1988 and 1992, while sales increased from U.S. $812 million in 1988 to $1400 million in 1992 (Conroy, 1996). Similarly, the area planted with organic crops in Europe increased from 115,080 to 1,169,192 ha between 1985 and 1995 (García, 1997).

Wide acceptance of organic products in international markets opens important niches for tropical products. Organic products usually carry a premium value compared to the price paid for conventional produce that has stimulated a remarkable expansion of organic agriculture in Mesoamerica, especially in coffee, bananas, and vegetables (Gabriela Soto, 2001, CATIE, pers. comm.). Several European and Japanese groups, especially nongovernmental organizations (NGOs), have promoted organic agriculture in the past two decades in this region (García and Monge-Nájera, 1995; Soto-Muñoz, 1999). In Costa Rica some 30 crops, including fruits, vegetables, legumes, spices, and plants used to produce cosmetics, are produced under organic principles, for both domestic and international markets (García, 1998). But, in general, organic production is very limited in Mesoamerica, lacking governmental support — for some time, Nicaragua was the only country with a national certification agency (Soto-Muñoz, 1999).

The conceptual approaches and practices of organic agriculture and IPM are quite convergent, despite a local misleading belief that they are incompatible. The core of organic agriculture is the crop plant's health, resulting from its interaction with the soil. Organic farming regards soil organic matter as a source of minerals for increased soil fertility and improved soil structure, and for its role in conserving soil microbiota and macrobiota (García and Monge-Nájera, 1995; García, 1997).

Thus, IPM is consistent with organic agriculture because it also prioritizes cultural practices, biological control, and resistant plant cultivars. The only obvious discrepancy with organic farming is that IPM allows the rational use of synthetic inputs (fertilizers and pesticides).

Clear evidence of this convergence in approaches and interests is that an increasing number of IPM practitioners in universities and research institutions in Mesoamerica now support low-input agriculture in order to make a smooth transition from conventional to organic agriculture. This is so because it is generally agreed upon that information and technologies for producing all crops in accordance to an organic approach are not yet available.

CONCLUDING REMARKS

Since the IPM philosophy was formally proposed about 40 years ago, there have been rather dramatic changes in terms of environmental matters and consciousness worldwide. Today, new challenges have emerged, especially with regard to the pressing need to reconcile both agricultural production and economic development with environmental conservation. Then, sustainability does not seem to be a buzzword, but rather a useful paradigm to guide efforts in that direction.

Within this context, the outlook for IPM in Mesoamerica is quite promising. Nonetheless, in order for IPM practitioners to make meaningful contributions for the development of sustainable production systems, they have to strongly persuade conservation-minded advocates and political decision makers about the benefits to be gained from IPM programs. This could be achieved not only by conveying the relevance of this crop protection strategy for the conservation of water, soil, and wildlife, but also by participating with them in projects and programs geared at optimizing local resources for crop production. Tropical biodiversity use and organic agriculture for both export and domestic crops are critical issues in exploiting local resources.

Certainly, national economies of Mesoamerican countries remain highly dependant on developed countries, and local agriculture is heavily influenced by international market preferences and by technologies developed abroad. In fact, pesticide overuse, expansion of nontraditional export crops, and introduction of genetically modified crops, have been promoted by government and financial agencies and private enterprises from developed countries. But foreign organizations also have made remarkable contributions by promoting IPM, biodiversity conservation and utilization, and organic agriculture in developing countries.

Therefore, in today's world, when globalization trends are so powerful and determinant, IPM practitioners in the neotropics should continue their efforts to strengthen both regional and international partnerships and to further develop sustainable production systems based upon farmers' empowerment. Thus, novel systems to be developed should embody the concepts of environmental protection, economic viability, and social equity.

ACKNOWLEDGMENTS

Thanks to Joseph L. Saunders (CATIE, Turrialba), Charles Staver (CATIE, Nicaragua), Ulrich Röttger (CATIE-GTZ), Ricardo Labrada (FAO), Luis Felipe Arauz (Universidad de Costa Rica), Luis Carlos González (CINDE), Mario Pareja (CARE), and Angel Chiri (EPA) for their valuable criticisms of a first version of this chapter. Thanks to Charlotte M. Taylor (Missouri Botanical Garden), Carlos Sáenz (DIECA), Gabriela Soto (CATIE), and Angel Solís (INBio) for providing helpful recent information. Also, thanks to Jaime García (UNED), Laura Rodríguez (CATIE), Guiselle Brenes (CATIE), and Hildara Araya for their logistic support.

REFERENCES

Adams, P.B., The potential of mycoparasites for biological control of plant diseases, *Annu. Rev. Phytopathol.,* 28:59–72, 1990.

Agne, S., *Economic Analysis of Crop Protection Policy in Cost Rica*, Pesticide Policy Project, Publ. Series No. 4, Hannover, Germany, 1996, p. 70.

Altieri, M.A., *Agroecology: The Scientific Basis of Alternative Agriculture*, Westview Press, Boulder, CO, 1987.

Andow, D.A. and Kiritani, K., The economic injury level and the control threshold, *Jpn. Pest. Inf.,* 43:3–9, 1983.

Andrews, K.L., Modelos de investigación y transferencia de tecnología en manejo integrado de plagas, *Manejo Integrado de Plagas (Costa Rica),* 13:65–82, 1989.

Andrews, K.L. and Quezada, J.R., Antecedentes entomológicos del manejo integrado de plagas en Centroamérica, in *Manejo Integrado de Plagas Insectiles en la Agricultura. Estado Actual y Futuro,* Andrews, K.L. and Quezada, J.R., Eds., Departamento de Protección Vegetal, Escuela Agrícola Panamericana, El Zamorano, Honduras, 1989, p. 21–28.

Arauz, L.F., La protección de cultivos en la agricultura sostenible: perspectivas para Costa Rica, *Manejo Integrado de Plagas (Costa Rica),* 41:29–36, 1996.

Arauz, L.F., *Fitopatología. Un Enfoque Agroecológico.* Edit. Universidad de Costa Rica, San José, Costa Rica, 1998.

Araya, C.M., unpublished data.

Araya, C.M., Diversidad Patogénica en Poblaciones de *Colletotrichum Lindemuthianum* en Frijol, paper presented at Mem. X Congr. Nac. Agron. Rec. Nat., San José, Costa Rica, 1999.

Ardón, M., Aproximación a la fitoprotección en Mesoamérica durante el siglo 16, *Manejo Integrado de Plagas (Costa Rica),* 30:35–44, 1993.

Auld, B.A. and Morin, L., Constraints in the development of bioherbicides, *Weed Technol.,* 9:638–652, 1995.

Badilla, F., Solís, A.I., and Alfaro, D., Control biológico del taladrador de la caña de azúcar *Diatraea* spp. (Lepidoptera: Pyralidae) en Costa Rica, *Manejo Integrado de Plagas (Costa Rica),* 20–21:39–44, 1991.

Barquero, C. and Arauz, L.F., Relación de la sensibilidad *in vitro* a benomyl de *Colletotrichum gloeosporiodes* y el combate de antracnosis en frutos de mango, in Memoria, X Congeso Nacional de Agronomia y Recursos Naturales, San José, Costa Rica, 1996, p. 132.

Beer, J. et al., Shade management in coffee and cacao plantations, *Agrofor. Syst.,* 38:139–164, 1998.

Blanco, H., Shannon, P.J., and Saunders, J.L., Resistencia de *Plutella xylostella* (*Lep.: Plutellidae*) a tres piretroides sintéticos en Costa Rica, *Turrialba (Costa Rica),* 40(2):159–164, 1990.

Bottrell, D.G., *Integrated Pest Management,* Council on Environmental Quality, U.S. Government Printing Office, Washington, D.C., 1979.

Bottrell, D.G., Applications and problems of integrated pest management in the tropics, *J. Plant Prot. Tropics,* 4(1):1–8, 1987.

Brader, L., Integrated pest control in the developing world, *Annu. Rev. Entomol.,* 24:225–254, 1979.

Braun, A.R., Thiele, G., and Fernández, M., La escuela de c ampo para MIP y el comité de investigación agrícola local: plataformas complementarias para fomentar decisiones integrales en la agricultura sostenible, *Manejo Integrado de Plagas (Costa Rica),* 53:1–23, 1999.

Caldecott, J.O. et al., *Priorities for Conserving Global Species Richness and Endemism,* World Conservation Monitoring Centre, Biodiversity Series No. 3, World Conservation Press, Cambridge, U.K., 1994.

Calvo, G. et al., Un esquema comprensivo y funcional para el manejo integrado de plagas del tomate en Costa Rica, in *Lecturas sobre Manejo Integrado de Plagas,* Serie Técnica, Hilje, L., Comp., Informe Técnico No. 237, CATIE, Turrialba, Costa Rica, 1994, p. 58–73.

Calvo, G. et al., Informe de Avance sobre la Validación de Tecnologías de Manejo Integrado de Plagas en Papa en las Estribaciones del Volcán Irazú, 1994, MAG-CATIE-UNA, 1996.

Carazo, E. et al., Resistencia de *Plutella xylostella* a *Bacillus thuringiensis* en Costa Rica, *Manejo Integrado de Plagas (Costa Rica),* 54:31–36, 1999.

Cartín, V.M. et al., Resistencia de *Plutella xylostella* a deltametrina, metamidofós y cartap en Costa Rica, *Manejo Integrado de Plagas (Costa Rica),* 53:52–57, 1999.

Castillo, L.E., de la Cruz, E., and Ruepert, C., Ecotoxicology and pesticides in tropical aquatic ecosystems of Central America, *Environ. Toxicol. Chem.,* 16:41–51, 1997.

Castillo, L.E., Ruepert, C., and Solís, E., Pesticides in the water bodies influenced by banana production, in *Book of Abstracts, International Conference on Pesticide Use in Developing Countries: Impact on Health and Environment,* San José, Costa Rica, 1998, p. 55

CATIE, Informe final Proyecto CATIE/INTA-MIP, Centro Agronómico Tropical de Investigación y Enseñanza (CATIE) e Instituto Nicaragüense de Tecnología Agropecuaria (INTA), Managua, Nicaragua, 1999.

Chinchilla, C.M., Epidemiologia y manejo del anillo rojo en palma aceitera. Memorias. X Cong. Nac. Agron. y Rec. Nat. San José, Costa Rica. p. 37–41, in *Memoria. X Congeso Nacional de Agronomia y Recursos Naturales,* San José, Costa Rica, 1996, p. 37–41.

Christou, P., No credible evidence is presented to support claims that transgenic DNA was introgressed into traditional maize landraces in Oaxaca, Mexico, *Transgenic Research,* 11:iii-v, 2002.

Conroy, M.E., Murray, D.L., and Rosset, P.M., *A Cautionary Tale: Failed U.S. Development Policy in Central America,* Lynne Rienner, Boulder, CO, 1996.

Cortés, J. and Risk, M.J., El arrecife coralino del Parque Nacional Cahuita, Costa Rica, *Rev. Biol. Trop. (Costa Rica),* 32(1):109–121, 1984.

Coto, D. et al., *Plagas Invertebradas de Cultivos Tropicales, con Énfasis en América Central; un inventario,* CATIE, Serie Técnica, Manual Técnico No. 12, Turrialba, Costa Rica, 1995.

Croat, T.B., *Flora of Barro Colorado Island,* Stanford University Press, CA, 1978.

Cronk, Q.C.B. and Fuller, J.L., *Plant Invaders: The Threat to Natural Ecosystems*, Chapman & Hall, London, 1995, p. 241.

Cubillo, D., Sanabria, G., and Hilje, L. Mortalidad de adultos de *Bemisia tabaci* con extractos de hombre grande (*Quassia amara*), *Manejo Integrado de Plagas (Costa Rica),* 45:25–29, 1997.

Daxl, R., Manejo integrado de plagas del algodonero, in *Manejo Integrado de Plagas Insectiles en la Agricultura. Estado Actual y Futuro,* Andrews, K.L and Quezada, J.R., Eds., Departamento de Protección Vegetal, Escuela Agrícola Panamericana, El Zamorano, Honduras, 1989, pp. 397–421.

De Kathen, A., The impact of transgenic crop releases on biodiversity in developing countries, *Biotechnol. Dev. Monitor,* 28:10–14, 1996.

Dengo, G., *Estructura Geológica, Historia Tectónica y Morfología de América Central,* 2 ed., Centro Regional de Ayuda Técnica, AID, México-Buenos Aires, 1973.

Dittrich, V., Uk, S., and Ernst, G.H., Chemical control and insecticide resistance of whiteflies, in *Whiteflies: Their Bionomics, Pest Status and Management,* Gerling, D., Ed., Athenaeum Press, New Castle, U.K., 1990, p. 263–285.

Doebley, J., Molecular evidence for gene flow among *Zea* species, *Bioscience,* 40:443–448, 1990.

Duke, S.O. et al., Natural products as sources of herbicides: current satus and future trends, *Weed Research 2000*, 40:99–111, 2000.

Dunwell, J.M., Transgenic crops: the next generation, or an example of 2020 vision, *Ann. Bot.,* 84:269–277, 1999.

Ellison, C.A., Preliminary studies to assess the potential of fungal pathogens for biological control agents of the graminaceous weed *Rottboellia cochinchinensis* (Lour) W.D. Clayton, M.S. thesis, Imperial College, University of London, 1987.

Ellison, C.A., An evaluation of fungal pathogens for biological control of the tropical graminaceous weed *Rottboellia cochinchinensis*, Ph.D. thesis, University of London, 1993.

Eyre-Walker, A. et al., Investigation of the bottleneck leading to the domestication of maize, *Proc. Natl. Acad. Sci., USA,* 95:4441–4446l, 1998.

FAO, Memoria. Taller Regional Manejo de la Maleza Caminadora *Rottboellia cochinchinensis* (Lour.) Clayton, Managua, Nicaragua, FAO Dept. of Agriculture, Rome, Italy, 1992.

Fischer, A.J., Grandos, E., and Trujillo, D., Propanil resistance in populations of junglerice (*Echinochloa colona*) in Colombia rice fields, *Weed Sci.,* 41:201–206, 1993.

Flint, M.L. and van den Bosch, R., *Introduction to Integrated Pest Management,* Plenum Press, New York, 1981.

French, E. and Sequeira, L., Strains of *Pseudomonas solanacearum* from Central and South America: a comparative study, *Phytopathology,* 60:506–512, 1970.

Frutos, R., Rang, C., and Royer, M., Managing insect resistance to plants producing *Bacillus thuringiensis* toxins, *Crit. Rev. Biotechnol.,* 19:227–276, 1999.

Gámez, R. et al., El programa de conservación de Costa Rica y el Instituto Nacional de Biodiversidad (INBio), in *La Prospección de la Biodiversidad: el Uso de los Recursos Genéticos para el Desarrollo Sostenible,* Reid, W.V. et al., Eds., WRI-INBio-Rainforest Alliance-ACTS, 1993, p. 61–113.

García, J.E., Agricultura orgánica en Costa Rica, *Agron. Costarricense (Costa Rica),* 21(1):9–17, 1997.

García, J.E., *La Agricultura Orgánica en Costa Rica,* EUNED, San José, Costa Rica, 1998.

García, J.E. and Monge-Nájera, J., Eds., *Agricultura Orgánica: Memoria sobre el Simposio Centroamericano,* EUNED, San José, Costa Rica, 1995.

Glynn, P.W. et al., The occurrence and toxicity of herbicides in reef building corals, *Mar. Pollut. Bull.,* 15:370–374, 1984.

González, M., *Enfermedades del Cultivo del Banano,* Universidad de Costa Rica-Asociación Bananera Nacional, 1987.

González, R., Crop protection in Latin America, with special reference to integrated pest control, *FAO Plant Prot. Bull.,* 24(3):65–75, 1976.

González, R. et al., Selección de mircroorganismos quitinolíticos en el control de sigatoka negra (*Mycosphaerella fijiensis*) en banano, *Manejo Integrado de Plagas (Costa Rica),* 40:6–11, 1996.

Grainge, M. and Ahmed, S., *Handbook of Plants with Pest-Control Properties,* John Wiley & Sons, New York, 1988.

Hails, R.S., Genetically modified plants — the debate continues, *Trends Ecol. Evol.,* 15:14–18, 2000.

Halford, N.G., GM crops — is there a future?, *Pest. Outlook,* 10:246–251, 1999.

Hall, F.R. and Menn, J.J., *Biopesticides: Use and Delivery,* Humana Press, Totowa, NJ, 1999.

Hall, L.M., Huffmann, J., and Topinka, K., Pollen flow between herbicide tolerant canola (*Brassica napus*) is the cause of multiple resistant canola volunteers, *Weed Sci. Soc. Am. Abstr.,* 40:48, 2000.

Hamblin, A., The concept of agricultural sustainability, *Adv. Plant Pathol.,* 11:1–19, 1995.

Hanneman, R.E., Jr., The testing and release of transgenic potatoes in the North American center of diversity, in *Biosafety for Sustainable Agriculture: Sharing Biotechnology Regulatory Experiences of the Western Hemisphere,* Krattiger, A.F. and Rosemarin, A., Eds., ISAAA (Ithaca, New York) and SEI (Stockholm, Sweden), 1994, p. 47–67.

Hanson, P.E. and Gauld, I.D., Eds., *The Hymenoptera of Costa Rica,* Oxford University Press, Oxford, 1995.

Hart, R.D., Ed., *Conceptos Básicos sobre Agroecosistemas,* CATIE, Turrialba, Costa Rica, 1985, p. 112–120.

Herrera, F., Situación de *Rottboellia cochinchinensis* en Costa Rica, in *Seminario-Taller sobre Rottboellia Cochinchinensis y Cyperus Rotundus. Distribución, Problemas e Impacto Económico en Centroamérica y Panamá,* Proyecto MIP-CATIE, Honduras, 1989, p. 14.

Hilje, L., El manejo integrado de plagas como noción y estrategia para enfrentar los problemas de plagas, in *Lecturas sobre manejo integrado de plagas,* Hilje L., Ed., Colección Temas de Fitoprotección para Extensionistas, Serie Técnica, Informe Técnico No. 237, CATIE, 1994, p. 1–23.

Hilje, L., Siete preguntas de actualidad sobre el manejo integrado de plagas en América Central, *Agron. Mesoamericana (Costa Rica),* 6:169–178, 1995.

Hilje, L., Cartín, V., and March, E., El combate de plagas agrícolas dentro del contexto histórico costarricense, *Manejo Integrado de Plagas (Costa Rica),* 14:68–86, 1989.

Hilje, L. and Hanson, P., La biodiversidad tropical y el manejo integrado de plagas, *Manejo Integrado de Plagas (Costa Rica),* 48:1–10, 1998.

Hilje, L. and Ramirez, O., Una propuesta comprensiva para el desarrollo de programas de manejo integrado de plagas (MIP) en América Central, *Manejo Integrado de Plagas (Costa Rica),* 24–25:63–71, 1992.

Hilje, L. et al., *El Uso de los Plaguicidas en Costa Rica*, EUNED-Ed., Heliconia, San José, Costa Rica, 1987.

Hodgson, J., Doubts linger over Mexican corn analysis, *Nature Biotechnology,* 20:3–4, 2002a.

Hodgson, J., Maize uncertainties create political fallout, *Nature Biotechnology,* 20:106–107, 2002b.

Holdridge, L.R., *Ecología Basada en Zonas de Vida,* IICA, Costa Rica, 1978.

Holm, L. et al., *The World's Worst Weeds: Distribution and Biology,* University of Hawaii Press, Honolulu, 1977.

Hruska, A.J., Vanegas, H.N., and Pérez, C.J., *La Resistencia de Plagas Agrícolas a Insecticidas en Nicaragua: Causas, Situación Actual y Manejo,* Escuela Agrícola Panamericana, Zamorano, Honduras. 1997.

Huffaker, C.B., Messenger, P.S., and DeBach, P., The natural enemy component in natural control and the theory of biological control, in *Biological Control,* Huffaker, C.B., Ed., Plenum Press, New York, 1971, p. 16–67.

ICAITI, *An Environmental and Economic Study of the Consequences of Pesticide Use in Central American Cotton,* Guatemala, UNEP-ICAITI, 1977.

IICA (Instituto Interamericano de Cooperación para la Agricultura), *Bases para una Agenda de Trabajo para el Desarrollo Agropecuario Sostenible,* Serie Documentos de Programas No. 25, IICA, San José, Costa Rica, 1991.

Iltis, H.H., Serendipity in the exploration of biodiversity. What good are weedy tomatoes?, in *Biodiversity,* Wilson, E.O., Ed., National Academy Press, Washington, D.C., 1988, p. 98–105.

INBio (Instituto Nacional de Biodiversidad), *Mem. Annu.,* INBio, Costa Rica, 1998.

James, C., *Global Review of Commercialized Transgenic Crops, ISAAA Briefs*, 24:Preview ISAAA, Ithaca, NY, 2001.

Kaimowitz, D., Cambio tecnológico y la promoción de exportaciones agrícolas no tradicionales en América Central, in *Taller Regional Centroamericano y Consulta sobre Planificación de Investigación Hortícola,* AVRDC-IICA, 5–8 Noviembre, 1991, Coronado, Costa Rica, s.p., 1991.

Kaplinsky, N. et al., Maize transgene results in Mexico are artifacts, *Nature*, 416:601, 2002.

Kogan, M., Integrated pest management: historical perspectives and contemporary developments, *Annu. Rev. Entomol.,* 43:243–270, 1998.

Kogan, M. and Bajwa, W.I., Integrated pest management: a global reality?, *Ann. Soc. Entomol. Brasil,* 28(1):1–25, 1999.

LaSalle, J. and Gauld, I.D., Parasitic Hymenoptera and the biodiversity crisis, *Redia (Appendice),* 74(3):315–334, 1991.

Lubberstedt, T., Dussle, C., and Melchinger, A.E., Application of microsatellites from maize to teosinte and other relatives of maize, *Plant Breeding,* 117:447–450, 1998.

MacArthur, R.H. and Wilson, E.O., The theory of island biogeography, *Monogr. Pop. Biol.,* 1, Princeton University Press, New Jersey, 1967.

Macilwain, C., Access issues may determine whether agri-biotech will help the world's poor, *Nature,* 402:341–345, 1999.

Madsen, K.H., Valverde, B.E., and Jensen, J.E., Risks assessment of herbicide resistant crops: a Latin American perspective using Oryza sativa as a model, *Weed Technology,* 16:215–223, 2002.

Mancebo, F. et al., Antifeedant activity of *Quassia amara* (Simaroubaceae) extracts on Hypsipyla grandella (Lepidoptera: Pyralidae) larvae, *Crop Prot.*, 19:301–305, 2000.

Mann, C.C., Crop scientists seek a new revolution, *Science,* 283:310–314, 1999.

Mann, C.C., Has GM corn 'invaded' Mexico? *Science*, 295:1617–1619, 2002.

Marabán, M., Dicklow, M.B., and Zuckerman, M.B., Control of *Meloidogyne incognita* on tomato by leguminous plants, *Fundam. Appl. Nematol.,* 12:409–412, 1992.

Marrone, P.G., Microbial pesticides and natural products as alternatives, *Outlook Agric.,* 28:149–154, 1999.

Martínez-Soriano, J.P.R. and Leal-Klevezas, D.S., Transgenic maize in Mexico: no need for concern, *Science*, 287:139, 2000.

May, A., Transgenic plants in Costa Rica: legislation, regulation and enforcement applicable to transgenic plants, in *Transgenic Plants in Mesoamerican Agriculture,* Hruska, A.J. and Lara, M. Eds., Zamorano Academic Press, Honduras, 1997, p. 58–61.

McCarthy, R. et al., Eds., *Buscando Respuestas: Nuevos Arreglos para la Gestión de Áreas Protegidas y del Corredor Biológico en Centroamérica,* UICN, San José, Costa Rica. 1997.

Meerman, F. et al., Integrated crop management: an approach to sustainable agricultural development, *Int. J. Pest Manage.,* 42:13–24, 1996.

Metz, M. and Fütterer, J., Suspect evidence of transgenic contamination, *Nature,* 416:600–601, 2002.

Mexzón, R.G. and Chinchilla, C.M., Plant species attractive to beneficial entomofauna in oil palm (*Elaeis guineensis* Jacq.) plantations in Costa Rica, *ASD Oil Palm Papers,* 19:1–39, 1999.

Miller, K., Chang, E., and Johnson, N., *Defining Common Ground for the Mesoamerican Biological Corridor,* World Resources Institute, Washington, D.C., 2002, p. 45.

Narváez, L., Resultados agro-industriales y económicos de siete años del programa de control biológico de *Diatraea* spp. en caña de azúcar, in *Memorias Seminario-Taller de Entomología,* CATIE, Serie Técnica, Informe Técnico 72, 1986, p. 72–79.

National Academy of Sciences, *Insect-Pest Management and Control,* NAS, Washington, D.C., 1969.

Nelson, K., Estudio comparativo de la generación te tecnología MIP en el cultivo de tomate, Nicaragua, *Manejo Integrado de Plagas (Costa Rica),* 41:16–28, 1996.

Nestel, D., Dickschen, F., and Altieri, M., Diversity patterns of soil macro-Coleoptera in Mexican shaded and unshaded coffee agroecosystems: an indication of habitat perturbation, *Biodiversity Conserv.,* 2:70–78, 1992.

Ocampo, R.A., Ed., *Potencial de Quassia amara como Insecticida Natural,* Serie Técnica, Informe Técnico No. 267, CATIE, Turrialba, Costa Rica, 1995.

Olofsdotter, M., Valverde, B.E., and Madsen, K.H., Herbicide resistant rice (*Oryza sativa* L.): global implications for weedy rice and weed management, *Ann. Appl. Biol.,* 137:279–295, 2000.

Ostmark, H.E., Banano, in *Manejo Integrado de Plagas Insectiles en la Agricultura. Estado Actual y Futuro,* Andrews, K.L. and Quezada, J.R., Eds., Departamento de Protección Vegetal, Escuela Agrícola Panamericana, El Zamorano, Honduras, 1989, p. 445–456.

Pareja, M.R., El manejo integrado de plagas: componente esencial de los sistemas agrícolas sostenibles, *Manejo Integrado de Plagas (Costa Rica),* 24–25:44–50, 1992a.

Pareja, M.R., Generación, adaptación y validación de programas de manejo integrado de plagas para hortalizas en Centroamérica: la experiencia del CATIE, *Manejo Integrado de Plagas (Costa Rica),* 24–25:51–57, 1992b.

Pedigo, L.P., Hutchins, S.H., and Higley, L.G., Economic injury levels in theory and practice, *Annu. Rev. Entomol.,* 31:341–368, 1986.

Perfecto, I. and Snelling, R., Biodiversity and the transformation of a tropical agroecosystem: ants in coffee plantations, *Ecol. Appl.,* 5(4):1084–1097, 1995.

Perfecto, I. and Vandermeer, J., Understanding biodiversity loss in agroecosystems: reduction of ant diversity resulting from transformation of the coffee ecosystem in Costa Rica, *Entomol. (Trends Agric. Sci.),* 2:7–13, 1994.

Perfecto, I. et al., Shade coffee: a disappearing refuge for biodiversity, *BioScience,* 46(8):598–608, 1996.

Perfecto, I. et al., Arthropod biodiversity loss and the transformation of a tropical agroecosystem, *Biodiversity Conserv.,* 6:935–945, 1997.

Picado, J.L. and Ramírez, F., *Guía de Agroquímicos,* Desarrollo de Agroquímicos Agrocontinental S.A. San José, Costa Rica, 1998.

Ploetz, R.C. et al., *Compendium of Tropical Fruit Diseases,* APS Press, St. Paul, MN, 1994, p. 2–22.

Plotkin, M.K., The outlook for new agricultural and industrial products from the tropics, in *Biodiversity,* Wilson, E.O., Ed., National Academy Press, Washington, D.C., 1988, p. 106–116.

Poppy, G., GM crops: environmental risks and non-target effects, *Trends Plant Sci.,* 5:4–6, 2000.

Portig, W.H., The climate of Central America, in *Climates of Central and South America,* Schwerdtfeger W., Ed., Elsevier, Amsterdam, 1976, p. 405–478.

Powles, S.B. et al., Evolved resistance to glyphosate in rigid ryegrass (*Lolium rigidum*) in Australia, *Weed Sci.,* 46:604–607, 1998.

Proyecto Estado de la Nación E., *Informe Estado de la Región en Desarrollo Humano Sostenible,* Proyecto Estado de la Nación, Editorama S.A., San José, Costa Rica, 1999.

Purseglove, J.W., *Tropical Crops: Dicotyledons,* Longman, London, 1974.

Purseglove, J.W., *Tropical Crops: Monocotyledons,* John Wiley & Sons, New York, 1975.

Quist, D. and Chapela, I.H., Transgenic DNA introgressed into traditional maize landraces in Oaxaca, Mexico, *Nature*, 414:541–543, 2001.

Rabbinge, R. and Van Oijen, M., Scenario studies for future agriculture and crop protection, *Eur. J. Plant Pathol.,* 103:197–201, 1997.

Radosevich, S.R., Holt, J., and Ghersa, C., *Weed Ecology. Implications for Management,* John Wiley & Sons, New York, 1997.

Ramírez, O.A., Generación de tecnologías de manejo integrado de plagas (MIP) para su implementación en América Central, *Manejo Integrado de Plagas (Costa Rica),* 34:31–35, 1994.

Ramírez, O.A. and Mumford, J.D., Formulación de políticas fítosanitarias en América Central, *Manejo Integrado de Plagas (Costa Rica)*, 40:24–34, 1996.

Ramírez, O.A. and Saunders, J.L., Una metodología para determinar criterios económicos de decisión en el control de plagas (MIP), *Manejo Integrado de Plagas (Costa Rica),* 50:1–18, 1998.

Reeder, R.H., Ellison, C.A., and Thomas, M.B., Population dynamic aspects of the interaction between the weed *Rottboellia cochinchinensis* (itch grass) and the potential biological control agent *Sporisorium ophiuri* (head smut), *Proc. IX International Symposium on Biological Control of Weeds*, University of Cape Town, South Africa, Moran, V.C. and Hoffman, J.H., Eds., 1996, p. 205–211.

Rich, P.V., and Rich, T.H., The Central American dispersal route: biotic history and paleogeography, in *Costa Rican Natural History,* Janzen, D.H., Ed., The University of Chicago Press, Chicago, 1983, p. 12–34.

Riches, C.R. et al., Resistance of *Echinochloa colona* to ACCase inhibiting herbicides, paper presented in *Proc. International Symposium on Weed and Crop Resistance to Herbicides*, De Prado, R. et al., Eds., University of Cordoba, Spain, 1996, p. 14–16.

Riches, C.R. and Valverde, B.E., Agricultural and biological diversity in Latin America: implications for development, testing and commercialization of herbicide resistant crops, *Weed Technology,* 16:200–214, 2002.

Rodgers, P.B., Potential of biopesticides in agriculture, *Pest. Sci.,* 39:117–129, 1993.

Rojas, L. et al., Hopper (Homoptera: Auchenorrhyncha) diversity in shaded coffee systems of Turrialba, Costa Rica, *Agroforestry Systems*, 53:171–177, 2001.

Romero, R., Observaciones sobre la incidencia de sigatoka negra (*Mycosphaerella fijiensis* var. *defformis*) en el cultivo del banano en la zona Atlántica de Costa Rica, *ASBANA*, 25:22–25, 1984.

Rosset, P.M., Precios, subvenciones y los niveles de naño económico, *Manejo Integrado de Plagas (Costa Rica)*, 6:27–35, 1987.

Rosset, P., Umbrales económicos: problemas y perspectivas, *Manejo Integrado de Plagas (Costa Rica),* 19:26–29, 1991.

Rosset, P., ¿Es factible el manejo integrado de plagas en el contexto de la producción campesina de los cultivos no tradicionales de exportación?, *Ceiba (Honduras),* 33(1):75–90, 1992.

Ruesink, W.G., Status of the systems approach to pest management, *Annu. Rev. Entomol.,* 21:27–44, 1976.

Ruiz-Silvera, C. et al., Sustratos y bacterias antagonistas para el manejo *Mycosphaerella fijiensis* en banano, *Manejo Integrado de Plagas (Costa Rica),* 45:9–17, 1997.

Sabillón, A. and Bustamante, M., *Guía Fotográfica para la Identificación de Plantas con Propiedades Plaguicidas. Parte I,* Zamorano Academic Press, Honduras, 1996.

Sage, G.C.M., The role of DNA technologies in crop breeding, in *Gene Flow and Agriculture: Relevance for Transgenic Crops*, Lutman, P.J.W., Ed., Farnham, U.K., British Crop Protection Council, 1999, p. 23–31.

Salas, J.A., Resistencia a fungicidas, paper presented in Mem. IX Congr. Nac. Agron. y Rec. Nat., San José, Costa Rica, p.i., 1993.

Sánchez, V., Bustamante, E., and Shattock, R., Selección de antagonistas para el control biológico de *Phytophthora infestans* en tomate, *Manejo Integrado de Plagas (Costa Rica),* 48:25–34, 1998.

Sánchez, V., Bustamante, E., and Shattock, R., Control microbiológico de *Phytophthora infestans* en tomate, *Manejo Integrado de Plagas (Costa Rica),* 51:47–58, 1999.

Saunders, J.L., *Manejo Integrado de Plagas en Centroamérica*, el Proyecto Regional MIP del CATIE, 1984–1989, 1989.

Schmutterer, H., Ascher, K.R.S., and Rembold, H., Eds., *Natural Pesticides from the Neem Tree (*Azadirachta indica *A. Juss),* GTZ Eschorn, Germany, 1982.

SIMAS (Mesoamerican Information Service for Sustainable Agriculture), Nicaragua: Experiences with IPM in the CATIE-INTA/IPM Project, in *New Partnerships for Sustainable Agriculture,* Thrupp, L.A., Ed., World Resources Institute, 1996, p. 75–84.

Singh, S.P., Gepts, P., and DeBouck, D., Races of common bean (*Phaseolus vulgaris* Fabaceae), *Econ. Bot.,* 45:379–396, 1991.

Smith, M., Reeder, R.H., and Thomas, M., A model of the biological control of *Rottboellia cochinchinensis* with the head smut *Sporisorium ophiuri, J. Appl. Ecol.,* 34:388–398, 1997.

Smith, M.C. et al., Integrated management of itchgrass in a corn cropping system: modeling the effect of control tactics, *Weed Science*, 49:123–143, 2001.

Soto-Muñoz, G., Organic farming in Central America, in *Managed Ecosystems. The Mesoamerican Experience*, Hatch, L.P. and Swisher, M.E., Eds., Oxford University Press, Oxford, 1999, p. 155–160.

Spahillari, M. et al., Weeds as part of agrobiodiversity, *Outlook Agric.,* 28:227–232, 1999.

Stephens, C.S., Ecological upset and recuperation of natural control of insect pests in some Costa Rican banana plantations, *Turrialba (Costa Rica),* 34(1):101–105, 1984.

Stern, V.M et al., The integrated control concept, *Hilgardia,* 29(2):81–101, 1959.

Stoll, G., *Protección Natural de Cultivos en las Zonas Tropicales,* Alemania Federal, Ed., Científica Josef Margraf., 1989.

Strong, D.R., Rapid asymptotic species accumulation in phytophagous insect communities, *Science,* 185:1064–1066, 1974.

Strong, D.R., McCoy, E.D., and Rey, J.R., Time and the number of herbivore species: the pests of sugarcane, *Ecology,* 58(1):167–175, 1977.

Swezey, S. and Salamanca, M., Susceptibility of the boll weevil (*Coleoptera: Curculionidae*) to methyl parathion in Nicaragua, *J. Econ. Entomol.,* 80(2): 358–361, 1987.

Tabashnik, B.E. et al., Insect resistance to *Bacillus thuringiensis:* uniform or diverse?, *Philos. Trans. R. Soc. London, B,* 353:1751–1756, 1998.

Teng, P.S., Integrating crop and pest management: the need for comprehensive management of yield constraints in cropping systems, *J. Plant Prot. Trop.,* 2(1):15–26, 1985.

Thomas, W.W., Conservation and monographic research on the flora of tropical America, *Biodiversity and Conservation*, 8:1007–1015, 1999.

Thrupp, L.A., Entrapment and escape from fruitless insecticide use: lessons from the banana sector of Costa Rica, *Int. J. Environ. Stud.,* 36(1):173–189, 1990.

Thrupp, L.A., *Bittersweet Harvests for Global Supermarkets: Challenges in Latin America's Agricultural Export Boom,* World Resources Institute, Washington, D.C., 1995.

Tran, M. et al., Characterization of glyphosate resistant *Eleusine indica* biotypes from Malaysia, paper presented in the *Proceedings of the 17th Asian-Pacific Weed Science Society Conference — Weeds and Environmental Impact*, Bangkok, Thailand, Vol. 1B:527–536, 1999.

Trumble, J.T., Kolodny-Hirsch, D.M., and Ting, I.P., Plant compensation for arthropod herbivory, *Annu. Rev. Entomol.,* 38:93–113, 1993.

Universidad de Costa Rica, personal communication.

Valdez, M. et al., Transgenic Central American, West African and Asian elite rice varieties resulting from particle bombardment of foreign DNA into mature seed-derived explants utilizing three different bombardment devices, *Ann. Bot.,* 82:795–801, 1998.

Valerín, M., Lista de Enfermedades de los Cultivos Agrícolas de Costa Rica, Ministerio de Agricultura y Ganadería, Dirección General de Sanidad Vegetal, Convenio MAG-GTZ, San José, Costa Rica. 1994.

Valverde, B.E., Riches, C.R., and Caseley, J.C., Prevention and management of herbicide resistant weeds in rice: experiences from Central America with *Echinochloa colona,* Cámara de Insumos Agropecuarios de Costa Rica, San José, Costa Rica, 2000.

Valverde, B.E. et al., Field-evolved imazapyr resistance in *Ixophorus unisetus* and *Eleusine indica* in Costa Rica, paper presented in the *Proceedings of the Brighton Crop Protection Conference — Weeds*, Brighton, U.K., 3:1189–1194, 1993.

Valverde, B.E. et al., Integrated management of itchgrass (*Rottboellia cochinchinensis*) in maize in seasonally-dry Central America: facts and perspectives, paper presented in the *Proceedings of the Brighton Crop Protection Conference — Weeds*, Brighton, U.K., 1:131–140, 1999.

Vandermeer, J. and Perfecto, I., The agroecosystem: a need for the conservation biologist's lens, *Conserv. Biol.,* 11(3):591–592, 1997.

Vargas, L., Eficacia de productos bactericidas contra *Erwinia* sp. en mango cv. Tommy Atkins, paper presented in *Mem. X Congr. Nac. Agron. y Rec. Nat.*, San José, Costa Rica, 1996, p. 80.

Vaughan, M., Problems and limitations of the development of integrated pest control in Latin America, *FAO Plant Prot. Bull.,* 24:165–171, 1976.

Vaughan, M., Transferencia de programas de manejo integrado de plagas, in *Manejo Integrado de Plagas Insectiles en la Agricultura. Estado Actual y Futuro,* Andrews, K.L. and Quezada, J.R., Eds., Departamento de Protección Vegetal, Escuela Agrícola Panamericana, El Zamorano, Honduras, 1989, p. 371–393.

Villa-Casáres, J.T., Respuesta de *Echinochloa colona* (L.) Link a Propanil en el Cultivo de Arroz (*Oryza sativa* L.) en Áreas Selectas de México, tesis M.Sc., Universidad Autónoma de Chapingo, Mexico, 1998.

Wallimann, T., Bt toxin: assessing GM strategies, *Science,* 287:41, 2000.

Wang, R.L. et al., The limits of selection during maize domestication, *Nature,* 398:236–238, 1999.

White, S. and Doebley, J., Of genes and genomes and the origin of maize, *Trends Genet.,* 14:327–332, 1998.

Wilkes, H.G., Hybridization of maize and teosinte, in Mexico and Guatemala and the improvement of maize, *Econ. Bot.,* 31:254–293, 1977.

Wilson, E.O. Ed., *Biodiversity,* National Academy Press, Washington, D.C., 1988.

Wolfenbarger, D.A., Lukefahr, M.J., and Graham, H.M., LD_{50} values of methyl parathion and endrin to tobacco budworm and bollworms collected in the Americas and a hypothesis on the spread of resistance in these lepidopterans to these insecticides, *ESA Bull.,* 66:211–216, 1973.

Zimdahl, R.L., *Weed Crop Competition: A Review,* International Plant Protection Center, Corvallis, OR, 1980.

CHAPTER 4

Managing Mycorrhizae for Sustainable Agriculture in the Tropics

Chris Picone

CONTENTS

0-8493-1581-6/03/$0.00+$1.50

INTRODUCTION

One of the paradoxes of tropical agriculture is that native systems — even those on poor soils — can maintain tremendous plant productivity, while agricultural systems on those same soils are often degraded after only a few years. A truly sustainable agriculture must learn to mimic and incorporate the biological mechanisms found in natural systems that can maintain high plant productivity despite having nutrient-poor soils.

One of those mechanisms is the association between mycorrhizal fungi and plant roots. Mycorrhizae are critical, ubiquitous symbioses between roots and soil fungi. The fungi are best known for their role in improving nutrient uptake. Given the growing global impacts of chemical fertilizers (e.g., Tilman et al., 2001), such "biofertilizing" soil microbes must be an important component of any sustainable agriculture. They have especially great potential in tropical agriculture, where phosphorous deficiencies and severe nutrient leaching frequently inhibit crop production, and where economic barriers prevent many small farmers from accessing synthetic fertilizers. But mycorrhizae are more than just biofertilizers. In certain conditions, the fungi also can help resist root pathogens, suppress nonhost weeds, reduce damage from toxic metals, and improve soil structure. This chapter will review strategies to incorporate and manage communities of mycorrhizal fungi as part of a sustainable agriculture in the tropics.

BACKGROUND ON MYCORRHIZAE

The term *mycorrhiza,* or *fungus-root*, encompasses several distinct types of associations (Smith and Read, 1997). The rarest are orchid mycorrhizae and ericoid mycorrhizae, formed exclusively in the Orchidaceae and Ericaceae, respectively. More common are ectomycorrhizae, which are formed predominantly by trees in the Pinaceae, Fagaceae, Myrtaceae (e.g., *Eucalyptus*), Dipterocarpaceae, and Caesalpiniaceae.* The fungi involved in these mycorrhiza types come from many lineages of Ascomycetes, Basisiomycetes, and a few Zygomycetes (Molina, Masicotte, and Trappe, 1992; Smith and Read, 1997).

This chapter will deal exclusively with the mycorrhizae most important to sustainable agriculture — arbuscular mycorrhizae (AM). In contrast to the above associations, AM are distinct in terms of their diverse range of host plants, their fungi, and their anatomy.

AM Host Plants

AM are ubiquitous: they are found in virtually all terrestrial ecosystems, in the roots of 70 to 80% of plant species, in all plant subclasses, and in most crops (Trappe, 1987; Sieverding, 1991).

Plants can be divided into three categories according to their dependence on mycorrhizal infection: obligate, facultative, and nonmycorrhizal (Figure 4.1). A plant's dependence is determined by the threshold level of soil fertility at which it no longer benefits from mycorrhizae (Janos, 1988). Obligate plants cannot grow beyond seed reserves if their roots are not colonized by AM fungi, even in a very fertile soil. Such plants include many tropical trees and some crops such as cassava (Janos, 1980; Sieverding, 1991). Facultative plants receive some benefit from colonization when grown in soil with low fertility, but not with high fertility. Most crops are facultative, but their dependence varies along a continuum from almost obligate to weakly facultative. Finally, nonmycorrhizal plants receive no benefit from the fungi, and they even may be suppressed if colonized. Most nonmycorrhizal plants belong to the families Brassicaceae, Amaranthaceae, Chenopodiaceae, Cyperaceae, Caryophyllaceae, or Polygonaceae, and they include crops in those families such as crucifers, spinach, and beets (Giovannetti and Sbrana, 1998). Not coincidentally, these same families comprise some of the most pernicious agricultural weeds. The causes and implications of this pattern are explored below.

AM Fungi

Compared to the tremendous diversity of AM plant species, the taxonomy of their fungus symbionts is quite simple (Figure 4.2). All AM fungi belong to an

* Janos (1988, 1996) reviews the role of mycorrhizae in tropical forestry. This chapter will focus on agriculture.

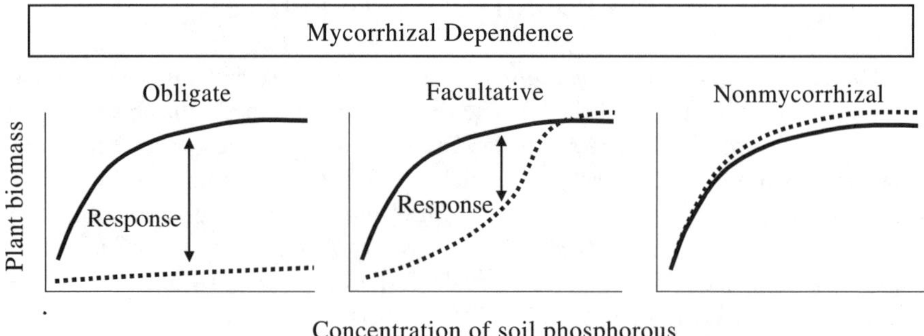

Figure 4.1 Three levels of mycorrhizal dependence in plants. Plant species are not restricted to these three discrete categories; plant dependence forms a continuum from obligately dependent to nonmycorrhizal plants. The magnitude of a plant's response to mycorrhizae will depend on the plant's dependence, soil fertility (especially phosphorous), and the effectiveness of the fungi (Figure 4.3). Solid lines represent inoculated plants, and dashed lines are noninoculated.

ancient order of Zygomycetes called the Glomales (Morton and Benny, 1990). Only about 150 morphotypes, or species, are recognized, and they currently comprise only five genera and three families.* Because these are asexual fungi, they produce none of the sexual structures (e.g., mushrooms) typically used in fungus taxonomy. Identification is based mostly on their large spores (40–800 μm). Genera are distinguished by the morphological and developmental traits listed in Figure 4.2. Within genera, species are distinguished by their spore size, color, ornamentations, staining patterns, and inner walls (Schenck and Perez, 1990; also http://invam.caf.wvu.edu/).

All Glomalean fungi are called arbuscular because they form arbuscles, finely divided hyphal tips. Arbuscles are the sites where most nutrients and carbohydrates are exchanged between fungus and host. They are produced inside root cells by penetrating the cell walls while remaining outside the host plasma membrane.

Most AM fungus species also form vesicles, spore-like storage organs inside plant roots, but the two genera in the Gigasporaceae lack them (Figure 4.2). This distinction is thought to be responsible for the different responses among the genera to various forms of agricultural disturbance, as discussed below.

FUNCTIONS OF MYCORRHIZAE IN THE AGROECOSYSTEM

Improved Uptake of Nutrients and Water

The best-known role of arbuscular mycorrhizae is to increase their host's ability to take up nutrients, especially phosphorous (Marschner and Dell, 1994). Many crops that are stunted in sterile (noninoculated) soils will exhibit robust growth if either P or mycorrhizal inoculum is applied (Figure 4.1). The benefits of improved nutrient uptake make mycorrhiza management especially critical in tropical soils where P

* Five primitive species may soon be relocated to two new genera in two new families, offshoots of ancestral lines. See Redecker, Morton, and Bruns (2000*b*) and http://invam.caf.wvu.edu/.

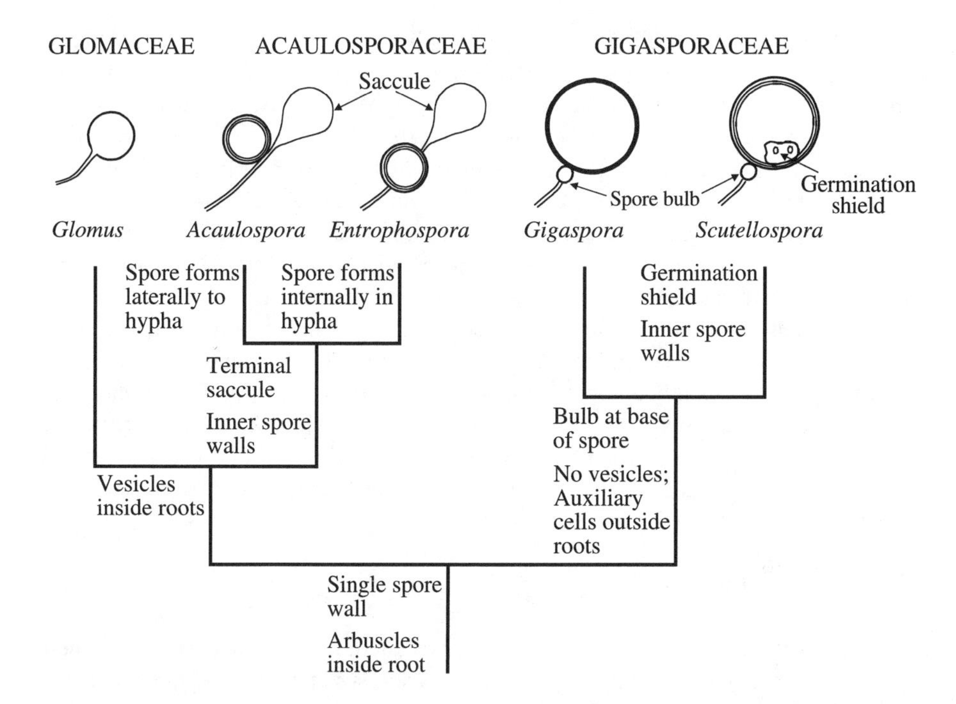

Figure 4.2 Taxonomy of arbuscular mycorrhizal fungi — the Glomales — based on their spherical spores. The figure is adapted from Bentivenga (1998), and Morton and Benny (1990). Note that the former genus *Sclerocystis* that forms small sporocarps is now included within *Glomus* (Redecker, Morton, and Bruns, 2000c). Recent developmental and genetic evidence indicates that five primitive species currently in the Glomaceae and Acaulosporaceae probably belong in two new, basal families (http://invam.caf.wvu.edu/; Redecker, Morton, and Bruns, 2000b). However, this chapter will use only three families in order to be consistent with the studies reviewed here.

deficiency is so common (Sanchez, 1976; Janos, 1987). Many weathered tropical soils have high P adsorption and they bind P to Al and Fe oxides. As a consequence, nearly 80% of the P applied as fertilizers is not immediately available, and much of it eventually converts to unavailable forms (Diederichs and Moawad, 1993). Mycorrhizae can improve the efficiency of P inputs and thereby reduce the amount of fertilizer required for optimal plant growth (Sieverding, 1991; Domini, Lara, and Gomez, 1997).

In addition to phosphorus, nutrition of other elemental nutrients is also improved by mycorrhizal colonization. Nitrogen (as ammonium) is taken up better by colonized than by noncolonized plants (Marschner and Dell, 1994). Nitrogen is also made more available in mycorrhizal soils indirectly by legumes: increased P uptake significantly improves nodulation and N fixation (Diederichs, 1990). Relative to noncolonized hosts, mycorrhizal plants can also take up more K, Ca, Fe, Mg, S, Cu, and Zn when these nutrients are deficient (Marschner and Dell, 1994; Saif, 1987).

AM fungi absorb nutrients from the same inorganic pool that roots access, but the fungi are apparently more efficient than plant roots (Diederichs and Moawad,

1993; Marschner and Dell, 1994; Bagyaraj and Varma, 1995). AM fungus hyphae are much finer than plant roots — between 2 and 7 microns (Abbott, 1982) — so their surface area to volume ratio is much higher than that of roots. The hyphae also extend several centimeters out beyond the root depletion zone. Furthermore, AM hyphae have a high affinity for soil P, although it is comparable to that in some higher plant roots (Marschner and Dell, 1994; Smith and Read, 1997).

Mycorrhizal colonization generally improves water uptake and drought resistance in crops, although results have been mixed (Augé, 2001). Improved water relations are attributed to both indirect and direct mechanisms. Indirectly, mycorrhizae increase water uptake through better nutrition: healthier roots grow larger and explore more soil volume, and thus absorb more water. But independent of improved nutrition and root length, colonized roots are often better at exploiting soil moisture than noncolonized roots (Augé, 2001). Roots with AM hyphae can explore a greater percent of soil volume than noncolonized roots, and AM roots also can access moisture at lower water potentials. These mechanisms could be critical in low-input tropical systems that experience drought stress.

Soil Aggregation

One of the most important, yet underappreciated, roles of mycorrhizae is their ability to stabilize soil aggregates and improve soil structure. Stable aggregates are critical for soil aeration, and they help resist erosion from wind and water (Miller and Jastrow, 1992). When soil structure is improved, roots and earthworms can penetrate more easily, rainfall infiltrates more rapidly and deeply, soil holds more moisture for a given volume, and runoff is reduced.

The biological mechanisms of soil aggregation are best modeled as a "sticky string bag" (Oades and Waters, 1991). Networks of fine roots and fungus hyphae physically entangle soil particles and cement them together into macroaggregates (>1 mm) by secreting polysaccharides and glycoproteins. In some grassland soils (Mollisols), arbuscular mycorrhizal fungi are the most important agent binding soil particles, even more important than fine roots and organic matter (Miller and Jastrow, 1990; Jastrow, Miller, and Lussenhop, 1998). In such soils, AM hyphae can reach over 100 m per gram of soil (Miller, Reinhardt, and Jastrow, 1995). The glycoprotein glomalin, secreted only by AM fungi, is a primary cementing agent associated with high aggregate stability (Wright and Upadhyaya, 1998). AM fungi thus comprise a keystone functional group that determines the structure of some soils and thereby influences many ecosystem properties.

Unfortunately, it is not yet clear how important AM fungi are to the structure of tropical soils because little research has been done there. Most research on their role in aggregation is from temperate mollisols (e.g., Jastrow, Miller, and Lussenhop, 1998) and Australian red-brown earths (Tisdall and Oades, 1980). In some types of oxisols, ultisols, alfisols, and inceptisols, iron and aluminum oxides may be the dominant stabilizing agents, not organic matter, roots, or hyphae (Sanchez, 1976; Oades and Waters, 1991; Picone, 1999). On the other hand, weathered, acidic soils lose their structural stability following tillage (Sanchez, 1976; Beare and Bruce, 1993), so AM fungi could be important agents for restoring their structure. This is one neglected area of mycorrhizal research in the tropics that needs to be addressed.

Alleviation of Effects from Heavy Metals

One of the limitations to plant productivity in acid, tropical soils is the high concentration of heavy metal ions and their oxides (Sanchez, 1976). In some studies, mycorrhizal colonization has been shown to reduce the detrimental effects of Al, Fe, Zn, Ni, and Cu, by reducing their concentrations in plant tissues (Koslowsky and Boerner, 1989; Heggo, Angle, and Chaney, 1990; Kaldorf et al., 1999). The most likely mechanism for this effect is improved P nutrition, because simply increasing soil P can alleviate stress from heavy metals (Sylvia and Williams, 1992). In addition, the fungi can reduce damage by immobilizing some metals (Fe, Zn, Ni) within crop roots (Kaldorf et al., 1999). It has also been suggested that the fungi generate insoluble metal-phosphate complexes, or chelate the metals to organic acids, and then deposit them in the soil or hyphal walls (Koslowsky and Boerner, 1989; Heggo, Angle, and Chaney, 1990).

Defense against Root Pathogens

AM colonization has been demonstrated to defend tropical crops against root pathogens, including nematodes (Jaizme-Vega and Pinochet, 1997; Rivas-Platero and Andrade, 1998) and root fungi (Azcón-Aguilar and Barea, 1996). AM inoculation of transformed carrot roots in axenic culture reduced populations of burrowing nematodes by almost 50% (Elsen, Declerck, and Waele, 2001). In addition, AM fungi are effective against the soil fungi that cause peanut pod rot (*Fusarium solani* and *Rhizoctonia solani*, in Abdalla and Abdel-Fattah, 2000). But mycorrhizal colonization is not a defense against pathogens on aboveground plant structures (Feldmann et al., 1995).

Although the effects are well demonstrated, we know little about the mechanisms behind AM defense of roots (Azcón-Aguilar and Barea, 1996). Improved nutrition is likely a factor, but even under high nutrient conditions, AM colonization can still defend roots effectively. More likely, mycorrhizal fungi compete against pathogens for photosynthate and colonization sites on roots. In addition, AM colonization can induce local defenses, such as chitinases. These induced responses are very weak, but they may prepare a plant for pathogen attack and thus make its defense response faster and stronger. Finally, AM-colonized roots can produce exudates that affect the microbial community in the rhizosphere. Extracts from AM roots reduce the production of sporangia and zoospores of the common root pathogen *Phytophthora cinamomi*. Likewise, the rhizosphere of AM plants has lower populations of pathogenic *Fusarium* but higher populations of pathogen-antagonisitc actinomycetes (Azcón-Aguilar and Barea, 1996).

Suppression of Nonmycorrhizal Weeds

Many of the plants that receive no benefit from mycorrhizae are from annual, weedy plant genera.* The presence of mycorrhizal fungi can suppress these weeds, both indirectly and directly. Indirect suppression derives from a higher order interaction: mycorrhizae can mediate the competition between nonhost species (or weakly dependent species) and plants that are more dependent (Grime et al., 1987; Hartnett et al., 1993; Jordan, Zhang, and Huerd, 2000). That is, mycorrhizal plants — including most crops — compete better against nonhost weeds when the soil has abundant AM fungi. In addition to mediation of competition, mycorrhizae can directly antagonize and inhibit nonhost plants. In pot studies with nonhost weeds (*Amaranthus, Chenopodium, Polygonum, Rumex, Portulacca,* and *Brassica*), soil inoculation with AM fungi reduced weed biomass by an average of 60% (Jordan, Zhang, and Huerd, 2000). Presence of AM fungi can also reduce weed germination (Francis and Read, 1995).

The direct negative effects of AM fungi on nonhosts are caused by their carbon cost and chemical exudates (Francis and Read, 1995). AM fungi can infect roots of some nonhosts and even form vesicles, thus conferring a carbon cost on the plant while returning no nutrient benefit (Allen, Allen, and Friese, 1989; Giovannetti and Sbrana, 1998). In addition, extracts from mycorrhizal soil will inhibit root development of nonhost plants (Francis and Read, 1994). Therefore, mycorrhizal fungi are a potential agent to employ in the battle against certain weeds, although in this role they are poorly understood and — as seen below — often suppressed.

QUANTITY VS. QUALITY OF MYCORRHIZAE: THE IMPORTANCE OF COMMUNITY COMPOSITION

Because of the importance of mycorrhizae in agroecosystems, research has often sought ways to increase their abundance as spores, root colonies, or soil hyphae, but the quality of the community has received little emphasis (Johnson, Tilman, and Wedin, 1992*b*). For example, many studies have assessed the mycorrhiza response of particular crops, but they rarely acknowledge that they are only assessing one, or at best a few, AM fungus species. In fact, AM fungus species are not equally effective at improving plant growth (Figure 4.3). While some species are tremendously beneficial, others have only minor effects, and some can even be functional parasites in certain situations (Johnson, Graham, and Smith, 1997). In order to incorporate AM fungi into a sustainable tropical agriculture, we must better understand how the community composition of the fungi affects their importance, effectiveness, and functions.

Unfortunately, it is difficult to determine from the literature what is a high-quality, or effective, community of fungi for any particular agroecosystem. A study in one region may find that a particular species is an effective fungus, but that result

* Although many nonhost plants are weeds, that does not imply most weeds are nonhosts. Many weeds benefit from AM colonization (e.g., Sanders and Koide, 1994; Marler, Zabinski, and Callaway, 1999), so the points in this section would not apply to them.

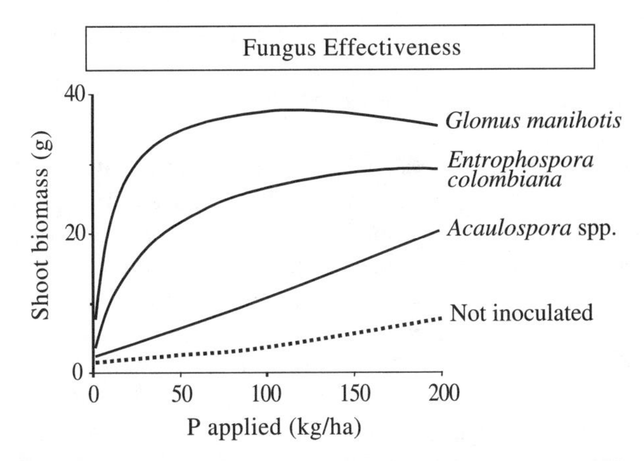

Figure 4.3 Response of a plant to mycorrhizae varies with both the effectiveness of the fungi and soil fertility. In this case, the plant is cassava (*Manihot esculenta*), an obligately mycorrhizal crop. *Glomus manihotis* is a consistently effective fungus, even at low soil P. *Entrophospora colombiana* is effective at intermediate levels of soil P and above. Less effective species cause little growth response unless P application is excessively high. (Based on Howeler, Sieverding, and Saif, 1987.)

does not mean the same species will be optimal when it is isolated from a different region. On the contrary, morphologically identical isolates of the same fungus species can have very different effects on plant growth (Howeler, Sieverding, and Saif, 1987; Siqueira et al., 1998). The identification of these asexual fungi is based on spore morphology — not their physiology — so effectiveness can vary considerably among isolates within a species. This limitation has been a barrier to precise application of mycorrhizae in agriculture.

Another barrier is the conditional behavior of any particular fungus isolate. The performance of an effective isolate is dependent on environmental factors, including soil pH, temperature, moisture, nutrient status, and salinity (Howeler, Sieverding, and Saif, 1987; Sieverding, 1989, 1991; Diederichs and Moawad, 1993). Effectiveness may also vary among different host plants: a fungus that promotes the growth of one plant species may be ineffective with another plant species (Sieverding, 1989; van der Heijden et al., 1998a). This conditional nature means that effective species should be determined for particular, local agricultural systems, and not simply extrapolated from other systems or regions.

One way to improve our ability to predict which species should be promoted is to focus research on the traits that make species (or isolates) most effective. That will require a better understanding of the fungus traits that are associated with particular AM functions, such as nutrient uptake and soil aggregation. If we can emphasize these traits — rather than simply relying on species identities — we should improve our ability to manage AM fungi effectively. The following section reviews current understanding of the relationships between fungus traits and agronomic functions.

Traits of Effective Fungi and Their Associated Functions

Rapid and Extensive Colonization of Roots

Rapid root colonization has been considered a prerequisite for an AM fungus species to be effective at nutrient uptake (Abbott and Robson, 1982, 1991). However, surveys of tropical species have found no general relationship between root colonization and nutrient uptake (Simpson and Daft, 1990a; Sieverding, 1991; Boddington and Dodd, 1998, 2000b). For example, species of *Gigaspora* can colonize roots extensively and be the least effective, while *Acaulospora* species can exhibit the poorest root colonization and be the most effective. It should be noted, however, that within the genus *Glomus* in Colombia, species that colonize roots best include those that provide the most phosphorous, while poor root colonizers in this genus are typically ineffective (Sieverding, 1989, 1991). Perhaps relatively high root colonization could be associated with increased nutrient uptake within a genus, but not across genera.

The benefits of rapid root colonization may be most evident in the presence of root pathogens. If pathogens arrive at roots before colonization by AM fungi, they can reduce AM colonization and its benefits (Linderman, 1994). Moreover, the ability to defend roots against pathogens may be associated with percent root colonization (Linderman, 1994). Therefore, rapidly colonizing fungus species may seem especially effective when soil pathogens are a problem. Species of *Glomus* and *Gigaspora* tend to colonize roots more extensively than species in other genera (Simpson and Daft, 1990a; Sieverding, 1991; Boddington and Dodd, 1998, 1999, 2000b; Brundrett, Abbott, and Jasper, 1999a; Dodd et al., 2000). These two genera should provide good candidates for isolates that are effective at defending roots against soil pathogens. But very little research has been done to compare the defense capabilities among AM fungus species that are in different genera, or that vary in growth strategies (Linderman, 1994; Azcón-Aguilar and Barea, 1996). Nor have studies addressed interactions between different fungi and different types of soil pathogens (Linderman, 1994).

Rapid and Extensive Production of Extraradical Mycelium

No general relationship has yet emerged between production of extraradical hyphae and nutrient uptake. Intuitively, fungus species that produce extensive networks of extraradical mycelium should be most proficient at taking up nutrients and water. They should be most adept at finding the soil patches where roots and other hyphae have not already depleted soil resources. Indeed, Jakobsen, Abbott, and Robson (1992) found that among *Acaulospora laevis, Glomus* sp., and *Scutellospora calospora*, the *Acaulospora* species produced the most mycelium and likewise took up the most soil P. On the other hand, with the tropical species *A. tuberculata, G. manihotis,* and *Gigaspora rosea*, the *Gigaspora* species produced the most extraradical hyphae but was the least effective fungus (Boddington and Dodd, 1998, 2000b). In the same study, *A. tuberculata* was highly effective despite having a poorly developed extraradical mycelium. Perhaps in short-term pot studies, fungi that grow extensive mycelia generate a carbohydrate cost on the host that exceeds the benefits from increased nutrient uptake.

A relationship is likely to exist between extraradical hyphae and soil aggregation. AM fungi that produce the most soil hyphae are the best candidates for improving and maintaining healthy soil structure. Both *Gigaspora* and *Scutellospora* tend to produce thick hyphae that span long distances between roots, while *Glomus* species tend to produce fewer extraradical hyphae (Boddington and Dodd, 1998; Dodd et al., 2000). This variation in growth strategies may have significant implications for soil structure. In one pot study (Schreiner and Bethlenfalvay, 1995), a species of *Gigaspora* was more effective at stabilizing soil aggregates than a *Glomus* species. Likewise, field soils with high spore populations of *Gigaspora* have higher densities of soil hyphae and better aggregate structure (Miller and Jastrow, 1992). Finally, some species of *Gigaspora* produce more glomalin than species of *Glomus* (Wright, Upadhyaya, and Buyer, 1998). Glomalin is the glycoprotein produced by AM fungi that is an important cementing agent of soil particles (Wright and Upadhyaya, 1998).

Rapid Nutrient Absorption and Transfer to the Host

Species in the Gigasporaceae family are typically less effective at improving plant growth than species in other families (Howeler, Sieverding, and Saif, 1987; Sieverding, 1989, 1991; Hetrick and Wilson, 1991; Jaizme-Vega and Azcon, 1995; Boddington and Dodd, 1998, 2000b; Clark, Zeto, and Zobel, 1999; but see Diederichs, 1990 for the contrary). Poor host response to the Gigasporaceae may result from reduced nutrient uptake, especially of phosphorous. Even when they have colonized roots well, some species of *Gigaspora* and *Scutellospora* transfer phosphorous to their hosts at reduced or delayed rates relative to *Glomus* and *Acaulospora* species (Jakobsen, Abbott, and Robson, 1992; Pearson and Jakobsen, 1993; Boddington and Dodd, 1999).

Apparently, members of the Gigasporaceae family regulate phosphorous transfer to their hosts by storing P as polyphosphate in extraradical mycelium (Boddington and Dodd, 1999). Storage of P may help maximize the carbohydrate transfer to these fungi. Once facultatively dependent plants have received sufficient P, they can reduce colonization (Figure 4.4) and carbohydrate availability (e.g., Pearson, Abbott, and Jasper, 1994). By delaying P transfer to their host plants, these fungi may ensure that carbon flow is maximized. The Gigasporaceae may require this strategy because of the high carbon cost of producing spores in this AM fungus family — its spores are larger than those of the other families, and they take longer to produce (Boddington and Dodd, 1999; Struble and Skipper, 1988). If delayed phosphorous transfer is common in the Gigasporaceae, then isolates from other families are likely to provide better candidates for alleviating P deficiency in tropical crops.

Ability to Suppress Nonhost Weeds

Although AM fungi have repeatedly been shown to inhibit many agricultural weeds, the traits that make fungi effective in this role have not been studied. Species that aggressively colonize roots should provide the best weed control if they can infect nonhost roots and inflict a carbon cost (Giovannetti and Sbrana, 1998). Species of *Glomus* and *Gigaspora* may be good candidates because they typically colonize roots rapidly and extensively relative to other genera (Simpson and Daft, 1990a; Sieverding, 1991; Boddington and Dodd, 1998, 1999, 2000b; Brundrett, Abbott, and Jasper, 1999a; Dodd et al., 2000). It is not known whether the species that best suppress weeds via chemical exudates belong to a particular taxonomic group or whether any other traits are associated with these exudates.

Competitive Ability and Persistence

In order to benefit sustainable agriculture, effective fungus species must be able to compete well and persist amid other AM fungus species. There is little point in augmenting the populations of a fungus species — either by management or inoculation (see below) — if it cannot survive and spread among the native AM fungus community.

The rate of root colonization is a poor indicator of a fungus' competitive ability. On one hand, a rapidly colonizing species of *Glomus* can outcompete its slower colonizing competitors, *Gigaspora margarita* and *Glomus tenue* (Lopez-Aguillon and Mosse, 1987). On the other hand, roots in the field can be colonized rapidly by *Glomus* species, which are then displaced by slower growing — but more competitive — *Scutellospora* species (Brundrett, Jasper, and Ashwath, 1999b). Likewise, *Scutellospora calospora* has a slightly slower colonization rate than a *Glomus* species, but *S. calospora* can inhibit colonization by its competitor by depleting carbohydrate availability throughout the entire root system (Pearson, Abbott, and Jasper, 1993, 1994).

Rate of sporulation may be associated with competitive success in disturbed conditions. Disturbance probably gives an advantage to species that sporulate rapidly and abundantly. For example, soil tillage should select species that produce spores quickly, while it reduces or eliminates species that take longer to sporulate (see below). Rapid sporulation also may be an advantage in crop systems with bare-soil fallows or a long dry season. Effective fungus communities for crop systems with these kinds of stresses may need to include species that produce abundant spores.

AGRICULTURAL MANAGEMENT OF THE AM FUNGUS COMMUNITY

Agricultural management of fungus communities should have two goals: to encourage diverse mixtures of fungi and to promote the most effective species in those mixtures (Sieverding, 1991).

The Role of AM Fungus Diversity

Sustainable agricultural systems should be designed to maintain a high diversity of AM fungi. Diversity is important to a healthy soil because of the many functions of AM fungi and their associated traits discussed in the previous section. No single fungus species could exhibit all of these traits together. Indeed, the fungi that best improve plant growth can be the least effective at improving soil structure, and vice versa (Schreiner and Bethlenfalvay, 1995). Members of the Gigasporaceae are often deemed ineffective for improving plant growth via nutrient uptake (e.g., Sieverding, 1991), but they are likely to be most effective at other functions that affect plant growth and soil quality (Miller and Jastrow, 1992; Schreiner and Bethlenfalvay, 1995). Those other functions require investment into structures that may come at a cost to nutrient uptake, such as exudates and aggregate-binding mycelia.

Furthermore, the performance of any particular species is dependent on abiotic soil and climate factors, such as pH, temperature, moisture, and nutrient availability (Howeler, Sieverding, and Saif, 1987; Sieverding, 1989, 1991; Diederichs and Moawad, 1993). As these factors vary seasonally and from year to year, a diverse fungus community is most likely to contain effective species under those different conditions. Thus, diversity reduces the risk of crop failure from unexpected stresses.

Fungus diversity is especially important to agrosystems with multiple crop species. Sustainable forms of agriculture require diverse crop systems instead of monocultures (e.g., Vandermeer, 1989; Soule and Piper, 1992; Zhu et al., 2000). In such plant communities, a diverse mycorrhizal community is most likely to contain effective species for all crops present. The relative performance of different fungus species depends on the species of crops. Indeed, the most beneficial fungus with one crop can be ineffective with another crop (Howeler, Sieverding, and Saif, 1987; Sieverding, 1991; Johnson et al., 1992a; Bever, 1994; van der Heijden et al., 1998a). In a Brazilian oxisol, both *Gigaspora margarita* and *Scutellospora verrucosa* improved the growth of the legume *Cajanus cajans*, but the same fungi had no effect on maize (Diederichs, 1991). Besides reducing risk for individual crops, AM fungus diversity can increase total crop productivity. In experiments where fungus diversity has no effect on the productivity of any single plant species growing alone, it increases total productivity of plant mixtures (van der Heijden et al., 1998b). The high diversity of AM fungi found in natural systems may be one key to those ecosystems' productivity and resilience to stress.

In contrast, high AM fungus diversity could be a disadvantage in some simplified crop systems. In pot studies, plant growth is often greater with the single most effective fungus species than with fungus mixtures (Sieverding, 1991; Boddington and Dodd, 2000b). In some field studies, the benefits of the most effective species can be diluted if they are mixed with less useful fungi (Howeler, Sieverding, and Saif, 1987; Sieverding, 1989, 1991).

Several explanations account for the apparent cost of high AM fungus diversity in simplified experimental systems. Using controlled conditions, most mycorrhizal studies simply base fungus effectiveness on the ability to take up nutrients. Such an emphasis can overlook the influence of diversity on other functions, such as improving soil structure and inhibiting pathogens and weeds. Indeed, yield of peas in pots is no greater with three species than with one, but aggregate stability is greatest in the fungus mixture (Schreiner and Bethlenfalvay, 1995). In addition, fungus diversity may not be a benefit in a homogeneous environment, such as a high-input, annual crop monoculture (Janos, 1988). In such environments, a single effective species that is adapted to the particular conditions may be optimal. In contrast, low-input systems experience a myriad of environmental stresses and heterogeneous soil conditions over time, especially farms that include perennial crops. Because of their heterogeneity, such systems would benefit most from a diverse fungus community (Janos, 1988). Finally, because of host-dependent effectiveness noted above, simplified systems that use only a single host species can mask the benefits of fungus diversity to plant mixtures. Therefore, increased diversity of AM fungi is most likely to benefit complex agricultural systems that increase crop diversity.

Promoting the Most Effective Species

Mycorrhiza management should augment populations of the most effective fungus species and decrease the least effective. A few species have been found to consistently enhance plant growth under a variety of conditions. For example, the CIAT program in Colombia has promoted a few isolates of *Glomus manihotis* (= *G. clarum*) because they remain very effective under broad ranges of soil types, fertility, acidity, and host plants (Howeler, Sieverding, and Saif, 1987; Sieverding, 1991). But the performance of most species will vary with different abiotic factors and host plants, so optimal species will need to be found for the particular soil, climate, and crops of interest.

In the long term, it would be advantageous to target the AM fungus community for particular agricultural situations. Ideally, farmers or extension agents should know which specific problems are to be addressed by optimizing the fungus community. Species could then be promoted that have the traits that best address those problems. For example, if phosphorous deficiency is limiting crop growth, fungi that best increase P uptake should be emphasized. In degraded soils with poor structure, some species of *Gigaspora* may be known to produce copious soil hyphae that most rapidly bind soil aggregates. If plants are stressed from soil pathogens, management may promote certain *Glomus* species that are known to be rapid root colonizers.

Optimizing the fungus community is a far more difficult goal than simply increasing diversity. Very few AM fungus communities have been characterized for their relative effectiveness, even at a single function like nutrient uptake. The process of evaluating and selecting effective strains is tremendously labor intensive: different species must be isolated and cultured, then screened for their effectiveness under a range of abiotic stresses and host crops, and in regard to distinct AM functions (see below). This process will occupy many years for any single region or crop system, and the results might not apply to other regions or crop systems. Knowing the traits that make AM fungi effective at particular functions might expedite this selection process, but our understanding of these traits is also in its infancy. Therefore, it will be some time before AM fungus communities can be optimized to address specific agricultural problems. Such long-term research goals are appropriate, however, for programs that are designing truly sustainable agricultural systems (e.g., Cox, Picone, and Jackson, in press).

Agricultural Effects on the AM Fungus Community

To effectively manage diversity and community composition of mycorrhizal fungi, we must first understand how different agricultural practices affect them.

Native and Agricultural Systems

Because mycorrhizae are important contributors to plant productivity in native ecosystems, it is worthwhile to compare AM fungi in native versus agricultural systems. Clearly, the impacts of agriculture depend on the type of agriculture that is practiced.

When native tropical forest is converted to low-input systems that have little soil disturbance, such as pastures and tree crops, there is typically no long-term decline in fungus diversity or abundance (Wilson et al., 1992; Cuenca and Meneses, 1996; Johnson and Wedin, 1997; Picone, 1999, 2000; but see possible reduction in pasture species richness in Allen et al., 1998). Likewise, in native savanna, areas with introduced forage grasses and legumes can have greater soil infectivity than the native system (Howeler, Sieverding, and Saif, 1987).

On the other hand, fungus abundance and diversity will decline dramatically when native systems experience severe soil disturbance (Cuenca, De Andrade, and Escalante, 1998). In Colombia, Brazil, and Zaire, declines in fungus richness have been recorded when native tropical systems are converted to low-input crops, and the decline is even more severe with the conversion to intensive agriculture (Sieverding, 1991). Highly degraded pastures that are filled with sedges (Cyperaceace) have insufficient AM inoculum for tree growth (Janos, 1988), and eroded pastures have low fungus diversity (Carpenter, Mayorga, and Quintero, 2001). Finally, conversion of native tropical systems to agricultural ones can shift the relative abundances among species, favoring the Glomaceae and Acaulosporaceae while reducing the Gigasporaceace. This community shift consistently occurs whether the soil is highly disturbed (Rose and Paranka, 1987; Siqueira et al., 1989 cited in Cuenca and Meneses, 1996; Cuenca, De Andrade, and Escalante, 1998) or not (Picone, 1999, 2000).

Soil Tillage

Soil preparation through tillage is the probably the most significant direct impact humans have on AM fungi. These fungi have been associated with plants for over 460 million years (Redecker, Kodner, and Graham, 2000a), so they are adapted to many environmental stresses and changes to the plant community; but tillage is a very recent phenomenon. Tillage has been shown repeatedly to reduce AM fungus spore populations (Kabir et al., 1998; Boddington and Dodd, 2000a), percent root colonization (Abbott and Robson, 1991; McGonigle and Miller, 1993; Galvez et al., 1995; Kabir et al., 1997), and density of soil hyphae (Kabir et al., 1997, 1998; Boddington and Dodd, 2000a).

The most likely mechanism for the effect of tillage is the physical breakup of the fungus mycelium. Unlike some fungi that can use small hyphal fragments as sources of inoculum, AM fungi are generally negatively affected by the disruption of their mycelium. When soil is chopped into small fragments (6–40 mm), root colonization is dramatically reduced compared to soil with larger fragments (>70 mm) (Bellgard, 1993; Picone, 1999).

Effects of tillage can manifest themselves in ways that increase dependence on fertilizers and herbicides. Reduced root colonization can result in lower P uptake by mycorrhizal crops such as maize (Evans and Miller, 1988; Entry et al., 1996). Applying synthetic fertilizers typically compensates for this problem, so it is likely to go unnoticed. By reducing abundance of AM fungi, tillage also benefits nonhost weeds. Nonmycorrhizal plants have a competitive advantage over more AM-dependent plants in soils with low inoculum (Allen and Allen, 1984; Hartnett et al., 1993). Tillage thus becomes part of a cycle of dependence found in industrial agriculture (Figure 4.4). One input (soil preparation) induces the need for other inputs: increased fertilizer application to overcome the deficiency of nutrient uptake, and herbicide application to combat outbreaks of nonhost weeds.

In addition to impaired nutrient uptake, tillage negatively affects the biotic mechanisms for improving soil structure. First, tillage disrupts and crushes soil aggregates, and it reduces the abundance of hyphae, roots, and organic matter that bind soil particles together. The compacted, poor soil structure of tilled crop systems is a testament to these negative effects. Second, tillage may select a fungus community that is ineffective at soil aggregation. Soil disturbance will favor fungus species that rapidly sporulate, produce vesicles, and/or invest little in soil hyphae; it will reduce the relative abundance of species that take a long time to sporulate, lack vesicles, and/or invest heavily in soil hyphae. With tillage, rapid production of spores and vesicles is an advantage because these are the propagules that will survive soil disruption, unlike hyphal networks. By shifting community composition away from species that invest in extensive soil hyphae, tillage may undermine the biological mechanisms required to restore a healthy soil structure. As a consequence, the agrosystem is further dependent on tillage to loosen and aerate its poorly structured soil (Figure 4.4).

Evidence suggests that tillage shifts the community away from species in the Gigasporaceae and toward species in the Glomacae and Acaulosporaceae. Species in the Gigasporaceae are ideal candidates to be damaged by tillage: they take the longest to sporulate, they produce few, large spores, they lack vesicles, and they often produce extensive hyphal networks. In a pot study, soil disruption reduced subsequent colonization by *Gigaspora rosea* while it increased colonization of *Glomus manihotis* (Boddington and Dodd, 2000b). Tropical field studies repeatedly indicate that spore populations of *Gigaspora* and *Scutellospora* are reduced by soil cultivation far more than *Glomus*, *Acaulospora*, and *Entrophospora* (Rose and Paranka, 1987; Johnson and Pfleger, 1992; Cuenca, De Andrade, and Escalante, 1998; Boddington and Dodd, 2000a). As a consequence, tillage reduces both species richness and evenness, because a few species in the latter genera eventually dominate the posttillage AM community (Sieverding, 1991). Similar community shifts away from the Gigasporaceae are reported in tilled temperate soils (Wacker, Safir, and Stephenson, 1990; Miller and Jastrow, 1992; Hamel et al., 1994). Given the proficiency of the Gigasporaceae at stabilizing soil aggregates, tillage does seem to undermine the ability to restore soil structure.

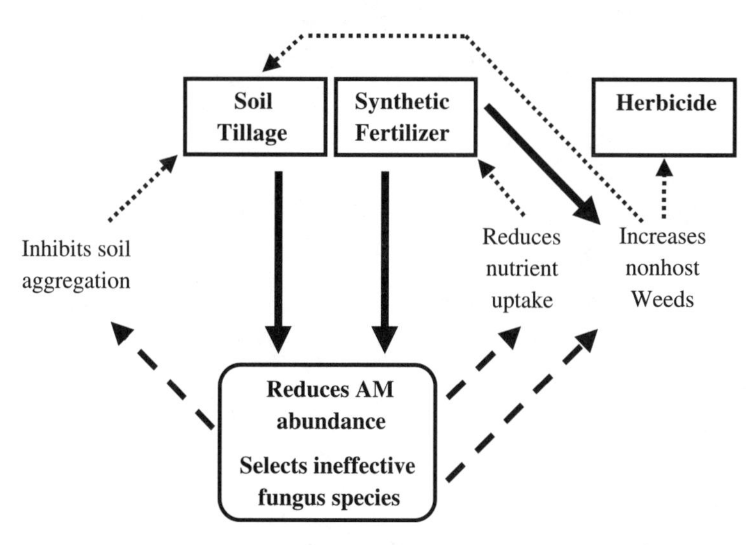

Figure 4.4 How industrial inputs induce the need for more inputs, mediated by impacts on AM fungi. Solid arrows lead to direct effects of industrial inputs; e.g., both tillage and synthetic fertilizers reduce mycorrhizal abundance and select inferior species. Dashed arrows point to negative consequences of impairing the AM fungus community (center row). Those consequences call for further inputs (dotted arrows). Note that synthetic fertilizers also directly benefit nonhost weeds (solid arrow). See text for details.

Fertilizers

Because P uptake is one of the fundamental functions of arbuscular mycorrhizae, effects of P fertilization have been the most studied. In general, AM root colonization, soil hyphae, and spores are most abundant at low to intermediate levels of soil fertility (Figure 4.5; also Abbott, Robson, and De Boer, 1984; Sieverding and Leihner, 1984; Salinas, Sanz, and Sieverding, 1985; Johnson and Pfleger, 1992; Brundrett, Jasper, and Ashwath, 1999b). If soil P concentration is extremely low, then some fertilization can increase the amount of mycorrhizal fungi. In such low-nutrient soils, poor plant productivity may inhibit AM colonization. But those positive effects of added P can quickly reach a maximum at around 2 to 17 mg P/kg (Abbott, Robson, and De Boer, 1984; Brundrett, Jasper, and Ashwath, 1999b). When fertilizers boost P concentrations well above these levels — as is the case for most synthetic fertilizers — they can dramatically reduce the abundance of AM fungi. Plants that are bathed in excess soil nutrients have little need to pay the carbon cost required to form mycorrhizae, so many hosts reduce fungal colonization.

Although the relationship in Figure 4.5 is fairly common, the exact shape and location of the peak are both contingent on the species of host plants and fungi. More specifically, the P concentration at which colonization rates decline depends on where the plant species falls in the facultative–obligate continuum. For example, when some legumes and grasses are treated with a range of soil P, the grasses display

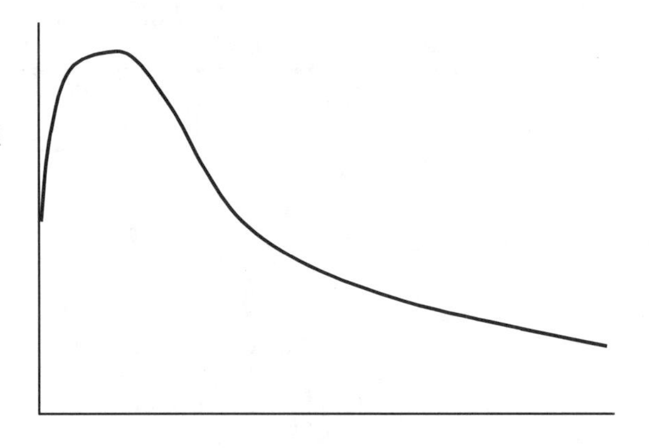

Figure 4.5 Generalized response of mycorrhizal fungi to increasing concentrations of soil phosphorous. The response is fairly consistent whether fungus abundance is measured as the amount of root colonization, density of soil hyphae, or numbers of soil spores. The exact location and shape of the curve's apex will depend on both the fungus species and the mycorrhizal dependence of the plant host (see text). Examples of peak values are 2–6 mg P/kg soil (Brundrett, Jasper, and Ashwath, 1999*b*) and 17 mg P/kg soil (Abbott, Robson, and De Boer, 1984). When expressed as kilograms of phosphate applied per hectare (soil P concentrations were not given), peaks of mycorrhizal measurements have occurred at 9–18 kg/ha (Salinas, Sanz, and Sieverding, 1985) and 0–50 kg/ha (Sieverding, 1991) in nutrient-poor tropical soils.

reduced AM colonization at much lower P concentrations than the legumes (Salinas, Sanz, and Sieverding, 1985; Arias et al., 1991). The legumes remain well infected even with P applications over 140 kg/ha. The legume species in these studies are apparently much more dependent on mycorrhizae than the grasses.

Likewise, different species of AM fungi will respond uniquely to fertilization. Colonization by *Glomus manihotis* (= *G. clarum*) remains high under a wide range of soil fertility, even at excessive P levels when other fungus species have declined (Salinas, Sanz, and Sieverding, 1985; Howeler, Sieverding, and Saif, 1987; Vaast, Zasoski, and Bledsoe, 1996). Isolates of this species can produce extraradical mycelium at high P concentrations, when the mycelium of other species is greatly reduced (Boddington and Dodd, 1998). Other studies have found that P fertilization can reduce colonization by other *Glomus* species more than *Scutellopora* species (Pearson, Abbott, and Jasper, 1994; Brundrett, Jasper, and Ashwath, 1999b). In sum, the effects of P fertilization will depend on the initial soil fertility, crop species' dependence, and fungus species.

Effects of nitrogen fertilization are not well understood. Nitrogen application generally decreases AM colonization, but it can also increase it (Abbott and Robson, 1991; Sieverding, 1991). As with P application, these results may vary, depending on the initial soil fertility. In addition, application of N as part of a balanced NPK fertilizer may stimulate root colonization, whereas application of excess N alone can have the opposite effect (Johnson and Pfleger, 1992).

The solubility of a fertilizer can determine whether it promotes or inhibits mycorrhizae. In contrast to chemical fertilizers, adding slow-releasing fertilizers, such as compost and organic matter, consistently increases root colonization, soil hyphae, and spore counts (Sieverding, 1991; Noyd, Pfleger, and Norland, 1996; Kabir et al., 1997). This effect is also consistent across different genera, including *Glomus*, *Acaulospora*, and *Gigaspora* (Boddington and Dodd, 2000b). Because AM fungi are susceptible to rapid nutrient release, fresh mulches and manure may decrease AM colonization if they have not yet composted sufficiently (Sieverding, 1991). Slow-releasing sources of P, such as rock phosphate and bonemeal, have minimal negative effects on AM fungi relative to more soluble fertilizers like triple-superphosphate (Howeler, Sieverding, and Saif, 1987; Munyanziza, Kehri, and Bagyarai, 1997).

In addition to the short-term consequences above, synthetic fertilizers initiate two long-term changes to the community composition of the fungi. First, P or N fertilization can selectively reduce populations of the Gigasporaceae relative to other families (Johnson, 1993; Egerton-Warburton and Allen, 2000). Species in the Gigasporaceae require more time to sporulate because they must first build up sufficient carbohydrates to produce their large spores. Fewer carbohydrates are made available to AM fungi as soil nutrient concentrations increase (e.g., Pearson, Abbott, and Jasper, 1994). Fertilization thus gives an advantage to smaller spored species because they can quickly sporulate despite low carbohydrate availability. This effect is compounded by the lack of vesicles in the Gigasporaceae — spores are their only means of surviving between growing seasons. With each year of fertilization, species in the Gigasporaceae may become less abundant than other species.

Second, fertilization increases the relative abundance of ineffective fungus species (Johnson, 1993). In Colombia, balanced (NPK) fertilizer increased abundance of an ineffective species of *Acaulospora* while reducing the most effective species of *Glomus* and *Acaulospora* (Sieverding, 1991). Even within a species, tropical isolates of *G. clarum* from low-phosphorus soils were more effective at P uptake than isolates from high-phosphorus soils (Louis and Lim, 1988). As above, this trend may result from reduced carbon flow to the roots as a consequence of fertilization, in which case only the most aggressive AM fungi are able to colonize roots (Abbott and Robson, 1991; Johnson, 1993). Because the plants have sufficient P, the fungi are taking carbohydrates without providing any apparent benefit.

As in the case with tillage, the use of one input — chemical fertilizer — increases the need for other inputs (Figure 4.4). First, as fertilization reduces total mycorrhizal colonization and soil hyphae and as it shifts the community toward ineffective species, plants require more fertilizer inputs. Like a drug pusher creating addicts, industrial agriculture creates conditions where increasing fertilizers are needed to simply maintain plant growth, and crops become fertilizer addicts. In fact, modern crop varieties of wheat (Manske, 1990; Hetrick, Wilson, and Cox, 1993) and perhaps soy (Khalil, Loynachan, and Tabatabai, 1994) are more dependent on fertilizers — and less responsive to mycorrhizae — than their landrace ancestors. Second, the low abundance of mycorrhizae in fertilized fields will promote nonmycorrhizal weeds. Nonhost weeds can take advantage of the excess nutrients without paying the carbon cost that mycorrhizal hosts must pay. Meanwhile, the reduction in AM fungus populations may limit their ability to inhibit such weeds. As a result, weed control through herbicides and tillage is required to compensate for a problem exacerbated by the use of excess chemical fertilizer. Once again, inputs induce the need for more inputs.

Cropping Strategies

AM fungi are obligate symbionts, so they require the presence of host-plant roots to survive for more than a few months in the field. Growth of nonmycorrhizal crops, such as crucifers, beets, amaranths, or spinach, will reduce fungus populations relative to host crops (Bagyaraj, 1992; Bagyaraj and Varma, 1995). Nonhost plants in mixed cropping systems can also reduce mycorrhizal colonization in adjacent hosts (Abbott and Robson, 1991; Fontenla, Garcia-Romera, and Ocampo, 1999). The root exudates of *Brassica* species can inhibit spore germination in some AM fungi (Schreiner and Koide, 1993).

Bare-soil fallows have even more severe effects than nonhost crops (Harinikumar and Bagyaraj, 1988; Singh and Tilak, 1992). With nonhost crops, the fungi can persist for a while by colonizing dead roots and/or by establishing vesicles in live roots without forming arbuscles (Giovannetti and Sbrana, 1998). Bare-soil fallows, by contrast, provide no carbohydrate sources for the fungi to persist. As a result, many crops suffer from low productivity after long fallows. A deficiency of mycorrhizal inoculum is a likely cause of this long fallow disorder because crop productivity can be restored by fertilizers (Johnson and Pfleger, 1992). Therefore, high populations of AM fungus propagules are best maintained by a constant presence of host plants, either as cover crops or continuous cropping systems (Harinikumar and Bagyaraj, 1988; Abbott and Robson, 1991; Galvez et al., 1995). Likewise, intercropping can augment AM abundance if the practice increases the density of plant roots (Sieverding, 1991).

Precropping can also augment AM fungus populations. Certain hosts, such as *Sorghum* spp., are especially effective at boosting fungus abundance for other crops and thereby increase subsequent productivity (Dodd et al., 1990; Johnson and Pfleger, 1992). This use of precropping is common in Cuban urban agriculture (pers. obs.). Urban plots begin with soil mixtures of compost and grain chaff, so inoculating and augmenting AM fungus populations with precrops is necessary. Similar approaches should work on farms with degraded soil communities. Moreover, with a better understanding of the ways different plant species promote particular species of fungi, precropping could be used strategically to shift the fungus population toward fungus species that are most effective with particular crops or those that have desired traits.

Increasing plant diversity through rotation, polycultures, and even mycorrhizal weeds can improve the community of AM fungi. Pure crop monocultures tend to promote AM fungi that are ineffective with that particular crop (Sieverding, 1989, 1991; Feldmann et al., 1998). In a negative feedback, some plants may promote fungus species that are least effective with themselves but most effective with other host species (Johnson et al., 1992a; Johnson and Pfleger, 1992; Bever, 1994).

Increased crop diversity would help alleviate these problems. In addition, mycorrhizal weeds can be beneficial if they promote effective species of fungi and thereby compensate for their competitive effects on crops. Incorporating mycorrhizal weeds into some tropical crops has been shown to improve AM colonization (citation in Janos, 1988), increase fungus diversity, and promote the most effective fungus species (Sieverding, 1991; Feldmann and Boyle, 1999). Each of these effects from the presence of weeds improved crop productivity compared to weed-free monocultures. Even if these diversity-mediated changes to the fungus community have little impact on crop yield, they complement the strong negative feedbacks that are generated by soil pathogens (Bever, Westover, and Antonovics, 1997; Mills and Bever, 1998). Such feedbacks explain why intercropping, rotation, and some weeds can prevent the yield declines associated with continuous monoculture.

It has been proposed that mycorrhizal weeds could be used to transfer nutrients and/or carbohydrates to crops (Bethlenfalvay et al., 1996; Jordan, Zhang, and Huerd, 2000). Weeds can capture nutrient pulses that are lost to crops, and then release those nutrients at optimal times through senescence, clipping, or grazing. Nutrient transfer from injured weeds to crops is most efficient via mycorrhizal networks that directly connect host weeds and crops (Bethlenfalvay et al., 1996). Even if AM do not normally transfer a significant amount of nutrients among interconnected hosts (e.g., Ikram et al., 1994), at least weeds could serve to reduce the carbon costs of establishing a functional mycelium.

Pesticides

The response of mycorrhizae to pesticides is highly variable, depending on the type of pesticide (Menge, 1982). Not surprisingly, soil fumigants are the class of pesticides most harmful to AM fungi. Although they are designed to eliminate pathogens, fumigants like methyl bromide will sterilize a soil and temporarily eliminate mycorrhizal fungi. As a consequence, fumigation causes stunting in many mycorrhizal-dependent crops, including avocado, citrus, cotton, peach, soy, white clover, and cassava (Menge, 1982; Howeler, Sieverding, and Saif, 1987; Sieverding, 1991).

This yield decline from fumigants is more likely to emerge in a mixed cropping system than in a continuous monoculture (Jawson et al., 1993). Fumigation will slightly increase yield in a maize monoculture because it eliminates the built-up populations of ineffective AM fungi and — more importantly — pathogens. In contrast, corn in rotation has a healthier soil community, and fumigation is seen to reduce P uptake and yield, presumably from a lack of AM fungi. Once again, the industrial model of production (monoculture) depends on chemical inputs such as fumigation, while the agroecological model can maximize productivity by rejecting those same inputs.

Fungicides are less harmful than soil fumigants, but their effects are more variable. AM response to fungicides depends on the class of fungicide, growth stage of the fungus, application rate, and species of fungus (Menge, 1982; Johnson and Pfleger, 1992). For example, systemic fungicides, aromatic hydrocarbons, and benzimidizole fungicides (e.g., Benomyl) can significantly reduce AM colonization and spore development. But other fungicides can enhance AM colonization by stimulating exudates from roots or by eliminating competition from root pathogens (Sieverding, 1991; Johnson and Pfleger, 1992).

Similarly, effects of herbicides, nematocides, and insecticides are also variable (Howeler, Sieverding, and Saif, 1987; Sieverding, 1991; Munyanziza, Kehri, and Bagyaraj, 1997). Herbicides can reduce abundance of mycorrhizal weeds and thereby reduce AM colonization in crops (Sieverding, 1991). Nematocides and insecticides can reduce populations of soil fungivores and thereby increase AM fungus populations (Kurle and Pfleger, 1994). It is generally difficult to predict the impact of these kinds of pesticides.

In contrast, biocontrol agents seem to be compatible with AM fungi. Soil microbes, such as rhizobacteria (*Pseudomonas*, *Bacillus*) and fungi (*Trichoderma*), are commonly used to control root pathogens, and they are not known to inhibit mycorrhizae (Azcón-Aguilar and Barea, 1996). In fact, these agents can increase AM colonization as a result of improved root growth and plant health (Sieverding, 1991). Because of the compatibility of biocontrol and mycorrhizal microbes, Cuban researchers have developed methods to simultaneously inoculate crops with both agents (pers. obs.; also Cupull et al., 1997).

Transgenic Crops

Use of transgenic crops has risen dramatically since the mid-1990s. Because transferred genes are functionally diverse, we should expect their impacts on soil organisms to be variable. In particular, crops with antimicrobial compounds should have different effects than crops that tolerate herbicides or contain pesticidal products (Donegan and Seidler, 1999; Miller, 1993).

The antimicrobial crops are likely to have the most detrimental effects on beneficial microbes like AM fungi, but they are the least studied so far. Genes for pathogenesis-related proteins are often inserted to resist disease. These genes express chitinases and β-1,3-glucanases throughout the plant — including the roots — where the enzymes degrade fungus cell walls. In one study of transgenic tobacco (Vierheilig et al., 1995), one variety expressing glucanase had altered fungus structures and significantly lower AM colonization. The effect from this particular transgene was completely unexpected, given the limited effects of the other transgenes in the study.

It is not known whether AM fungi are affected by crops with pesticidal transgenes (e.g., proteinase inhibitors and toxins from the bacterium *Bacillius thuringiensis*, or Bt), but their impacts on the soil community suggest a need for further investigation. Bt toxins from transgenic crops can persist in the soil for many months and accumulate if the crops are repeatedly planted (Donegan and Seidler, 1999; Saxena, Flores, and Stotzky, 1999). Residue from these crops can increase total fungus populations (Donegan and Seidler, 1999) but reduce fungus diversity (K. Donegan, pers. comm.). Given the tremendous area recently planted to transgenic crops over the span of a few years, a better understanding of their impacts on the soil community is urgently needed.

Management Recommendations

Based on the information in this section, the following practices are recommended to promote an abundant, effective, and diverse community of AM fungi:

- Reduce or eliminate tillage. If tillage is necessary, use zone tillage to confine the disturbed soil to the 20- to 30-cm strips where crops will be planted (Jordan, Zhang, and Huerd, 2000). The unplowed areas between strips will retain an intact fungus mycelium that should quickly recolonize the disturbed soil.
- Use cover crops, intercrops, continuous cropping, and perhaps mycorrhizal weeds to maintain a constant supply of live roots to the AM community. Avoid continuous monocultures because they promote ineffective fungus species for the monocultured crop. Use host species strategically to promote effective AM fungus species or isolates.
- Increase soil organic matter by eliminating tillage, using deep-rooted cover crops and compost, and by reducing burning (especially in dry tropical regions).
- When restoring soil structure is paramount, use crops that produce extensive root systems, exhibit high mycorrhizal colonization, and/or foster abundant AM fungus populations. Promote fungus species that produce extensive mycelium (these may need to be introduced via inoculation — see next section).
- Promote a soil community that makes nutrient cycling as efficient as possible. Reduce application of synthetic fertilizer to avoid selecting ineffective fungi. When necessary, use slow-release fertilizers — like rock phosphate or mature compost — to have minimal impacts on AM abundance. Compost is preferable because it can be produced on the farm and will increase soil organic matter. Rock phosphate is also preferable to more soluble fertilizers (e.g., triple-superphosphate) because it is less expensive and is domestically produced in tropical countries (Sieverding, 1991).
- For efficient use of soil P, use crop varieties that are most responsive to mycorrhizae. Avoid varieties that depend on fertilizers and pesticides for high productivity.
- Avoid fungicides that are harmful to AM fungi (Menge, 1982), especially soil fumigants.

MYCORRHIZAL FUNGI AS AN AGRICULTURAL INPUT: INOCULATION

In some situations, managing the indigenous mycorrhizal community may not be sufficient to take advantage of mycorrhizae in the agroecosystem. In degraded soils where the mycorrhizal community is almost eliminated, inoculation with effective isolates of fungi may be required. This section describes procedures and strategies for inoculating farms with effective fungi.

Procedures for Large-Scale Inoculation

Selecting Superior Species or Strains

The first step is to screen germplasm of fungus species to determine which are most effective. In order to screen as many species as possible, soil should be collected from a broad and diverse range of soils types within a region (Sieverding, 1991). The highest species richness in these soils will be found by looking for spores with two complementary strategies (Morton, Bentivenga, and Bever, 1995; Picone, 1999). AM fungus spores should be isolated directly from the field-collected soil and also from soil trap cultures with planted hosts. Each method can produce fungus spore populations that are not present in the other method. Using at least two plant species in the trap cultures can help increase the likelihood that all species present will produce spores (Dodd and Thomson, 1994). Once small populations of individual species are isolated, they can be further propagated and amplified by growing them with plant hosts.

The effectiveness of individual species then needs to be evaluated. Species can be initially screened in pot cultures in a greenhouse, but they should be evaluated under a range of conditions that local farms may experience (Abbott and Robson, 1982; Sieverding, 1991). For example, conditions might vary in soil type, pH, nutrient availability, moisture stress, heavy metal concentrations, presence of soil pathogens, or other critical factors on local farms. Likewise, experimental host plants should include the range of locally grown crops. Fungus species that are most effective in one set of soil conditions, or with particular crops, may be ineffective in another set of conditions, or with other crops. Fungus effectiveness should be evaluated for multiple functions, such as nutrient uptake in infertile soil, defense against pathogens, and ability to stabilize aggregates in degraded soil. Ideally, species also should be assessed in mixtures to test for complementary and synergistic effectiveness.

The most promising species and mixtures from greenhouse studies can then be evaluated in the field (Abbott and Robson, 1982; Sieverding, 1991). Field trials are critical to determine which isolates will continue to be effective in the presence of the indigenous soil community. Effective species are only useful if they persist and compete well against the indigenous community.

Generally, inoculated species decline over time (Dodd and Thomson, 1994) and require repeated applications in subsequent years. Benefits may persist into the second year after inoculation, but by the third year there is typically little or no effect, depending on the fungus species (Sieverding, 1991). However, benefits of inoculating coffee seedlings may persist into the fifth harvest (Siqueira et al., 1998). Such long-term persistence is an ideal trait for introduced species in order to reduce the need for further mycorrhizal inputs.

Producing and Applying Inoculum

Once effective, persistent species are selected, sufficient inoculum needs to be produced for a large-scale application. As with the soil trap cultures described above, living plant hosts are required to multiply fungus populations, both as spores and as colonized roots.

Sieverding (1991) recommends on-site production of inoculum to reduce costs of transport. Field soil, either sterile or fumigated, can be preinoculated with spores and roots from effective isolates. Soil can be in pots or large troughs to reduce the risk of contaminant fungi entering the cultures. Host plants should be chosen for extensive production of spores and colonized roots. Commonly planted hosts include sorghum (*Sorghum bicolor* and *Sorghum vulgare*), sudangrass (*S. bicolor* var. *sudanense*), and bahiagrass (*Paspalum notatum*), among others (Bagyaraj and Varma, 1995). In some cases, legumes may produce more spores than pasture grasses (Howeler, Sieverding, and Saif, 1987), but grasses produce more infected roots (Sieverding, 1991).

After several months, the soil cultures can be used to inoculate areas of the farm that need it. If kept dry, AM inoculum can remain viable for several years (pers. obs.; also Feldmann and Idczak, 1992; Bagyaraj and Varma, 1995).

Unfortunately, soil inoculum is awkward and ponderous to apply, requiring 3000–5000 liters/ha (Sieverding, 1991). These numbers are not unreasonable given that annual applications of organic fertilizer reach 20 to 30 tons/ha. Yet either fertilization strategy may only be practical for gardens and very small farms.

Alternatively, some researchers produce inoculum in soilless growth media, such as perlite, vermiculite, peat moss, and expanded clay (Sieverding, 1991; Bagyaraj and Varma, 1995). These media have several advantages. They are light (ca. 420 kg/m^3) and easy to transport. They contain extremely high propagule densities (30–50 propagules/cm^3), so only a tiny volume is needed (Feldmann and Idczak, 1992). The granular structure of some media allow them to be used with conventional fertilizing machinery and seed drills, applied at 25 to 75 kg/ha (Bagyaraj, 1992; Sieverding, 1991). In addition, these media help avoid soil pathogens and contaminants that are often present in soil inoculum.

Another practical alternative to soil media is coating the crop seeds with fungus inoculum before planting. Spores and fine roots can be cemented to seeds with adhesives, such as methyl cellulose (Bagyaraj, 1992). The amount of inoculum and inputs required is thus greatly reduced compared to other inoculation techniques (Fernandez et al., 1997; Tejada, Soto, and Guerrero, 1997). Seed coats are commonly used to inoculate crops with bacteria, but it may prove more difficult with AM fungi because their large spores are 40 to 800 μm in diameter. Perhaps this technique will be limited to large-seeded crops (Bagyaraj, 1992). On the other hand, Cuban farmers are using such seed coats successfully on a wide variety of crops, including rice, yucca, peanut, beans, maize, carrot, tomato, and garlic (Gomez et al., 1997).

In contrast to soilless media and seed coats, precropping is a low-tech alternative to transporting large amounts of fungus inoculum. Small amounts of effective fungi can be planted in the field around seeds of host species used for culturing the fungi. These hosts can be grown for 10 to 12 weeks prior to planting the economic crop of interest, and in this way effective fungi may be multiplied in the field with few inputs (Bagyaraj, 1992).

Transplanted crops provide an easy, low-tech method of inoculating farms with effective fungi. Many crops must be pregerminated before being transplanted into the field. These plants can be inoculated easily as seedlings, which then transport the fungi into the field at no added cost to the farmer. This inoculation strategy has been used effectively with chili, finger millet, tomato, tobacco, citrus, mango, coffee, ornamental flowers (asters, marigold), and trees used in agroforestry (Sieverding, 1991; Bagyaraj, 1992; Tejada, Soto, and Guerrero, 1997). The transplants themselves directly benefit from initial inoculation with effective fungi, and the field is prepared with fungus species for subsequent crops that may be seeded directly.

Results from Field-Scale Inoculation

Is AM inoculation in the field really an effective strategy for increasing crop yields? Most of the studies published from tropical research indicate that it is, even when inoculating live (unsterilized) soils. In one review, Sieverding (1991) lists a positive effect of large-scale inoculation in each of 15 trials with annual grains, including wheat, maize, sorghum, barley, soy, and beans. Sorghum productivity was increased 10 to 29% in a sandy loam in India (Singh and Tilak, 1992). Inoculation increased cassava yields by 23 to 25% in an infertile Colombian oxisol (Howeler, Sieverding, and Saif, 1987), and treatments were effective at least a year after planting into a low-nutrient Nigerian alfasol (Fagbola, Osonubi, and Mulongoy, 1998). Similar success stories have been reported from Cuba in soy, sorghum, onion (Fernandez et al., 1997), pepper (Pulido and Peralta, 1997a), tomato (Pulido and Peralta, 1997b), and potato (Ruiz et al., 1997). Likewise, tree crops have higher productivity and survival if they are inoculated in nurseries before being transplanted to the field. This has been demonstrated in many trees incorporated in agroforestry, including coffee (Siqueira et al., 1998), papaya, coconut, peijbaje (*Bactris gasipes*), achiote (*Bixa orellana*), cacao (*Theobroma grandiflorum*), brazil nut (*Bertholletia excelsia*), rubber (*Hevea braziliensis*), and mahogany (*Swietenia macrophylla* and *Carapa guianensis*) (Feldmann et al., 1995). In one exception, the productivity of soybean in Taiwan was improved by only four of twelve inoculations (Young, Juang, and Chao, 1988), but the fungus species used apparently had not been selected for high effectiveness.

Increased plant productivity from mycorrhizal inoculation permits reduced inputs of chemical fertilizers. For example, yield of cassava in nutrient-poor soils in Colombia is similar in inoculated fields compared to noninoculated fields that have twice as much phosphorus applied (Sieverding, 1991). Cuban research presents the best examples of reduced dependence on chemical fertilizers through AM inoculation (Rosset and Benjamin, 1993). In the early 1990s, Cuba was faced with a fertilizer shortage because of the loss of industrial inputs following the demise of the Soviet Union, as well as the U.S. embargo against trade with Cuba. To compensate for shortfall of synthetic fertilizers, they developed a unique, nationally promoted program for extensive use of mycorrhizae. Laboratories are now dispersed throughout the country to produce low-cost inoculum for farmers, and they have also developed the techniques for seed coats described above. From this effort, they have been able to reduce fertilizer use 12 to 50% in many crops, while maintaining constant yield (Fernandez et al., 1997; Ruiz et al., 1997). Inoculation of tomatoes, even on poor soils, can allow phosphorous inputs to be eliminated without a loss of yield (Domini, Cara, and Gomez, 1997).

Principles for Effective Inoculation

Despite the high success rate of inoculation reflected in publications, it should be emphasized that simply adding AM fungus inoculum to a soil — without regard for fungus effectiveness, crop host, and soil factors — is likely to be an exercise in frustration. Too many people consider beneficial microbes to be simple replacements for synthetic fertilizers. They often are treated as just another input. As biological agents, however, mycorrhizae are more complex than chemical inputs, so they are more difficult to effectively incorporate into an agricultural system. Below are some principles that can help ensure that time and money are not wasted when adding and/or managing these organisms in an agroecosystem.

AM Fungi Must Be Effective and Appropriate to the Particular Agricultural System

As should be obvious from the previous sections, whether mycorrhizae are going to help will depend on the effectiveness of the fungus species. Moreover, they must be adapted to the range of soil and environmental conditions on the farm. Too often, AM fungus species may be marketed as effective, but they have only been tested under experimental conditions that may not resemble conditions on one's farm. Commercially available inoculants should provide the buyer information on the conditions from which the fungus was isolated and tested, including soil type, climate, and crop systems. An effective fungus from a lowland, acidic, nutrient-poor oxisol may not be useful in an andisol in a cooler climate.

Host Crops Are Responsive to Inoculation at the Relevant Soil Fertility

Most crops appear to be facultatively mycorrhizal, and their responses to inoculation are greatest in soils with low to moderate levels of fertility. In soils with extremely low P, mycorrhizal colonization and functions are inhibited (Figure 4.5). Under those conditions, both crop and fungus are likely competing for the same labile pool of P. Crop response to inoculation is often negligible in soils with <3 mg P/kg, but the response is dramatic if small amounts of P are added (Howeler, Sieverding, and Saif, 1987; Diederichs, 1990; Habte and Fox, 1993; Vaast, Zasoski, and Bledsoe, 1996; Siqueira et al., 1998). As a consequence, restoration of severely eroded, low P soils may require additional P amendments in order for mycorrhizal inoculation to be worthwhile (Habte and Fox, 1993). At the other extreme, facultatively mycorrhizal crops (e.g., soybean, maize, sorghum, and some forage species) will not respond to inoculation in soils with high P concentrations (Figure 4.1; also Young, Juang, and Chao, 1988; Simpson and Daft, 1990b; Arias et al., 1991). In soils with excessive P, inoculation can even reduce yield of some crops due to the carbon cost of AM fungi (e.g., Peng et al., 1993).

The concentration of soil P at which an optimal response occurs will differ among plant species degree because mycorrhizal dependence varies among crops. The benefits of inoculating some facultative grasses can disappear with soil P concentrations that exceed only 5 ppm, while legumes, cassava, and coffee are highly responsive even in soil with 40 to 100 ppm (Arias et al., 1991; Sieverding, 1991; Vaast, Zasoski, and Bledsoe, 1996). Grass roots tend to be fine, and therefore more efficient at extracting nutrients than roots of many other crops. Regardless of soil fertility, nonmycorrhizal crops, such as spinach, crucifers, beets, and amaranth, will not benefit from inoculation (e.g., Khasa, Furlan, and Fortin, 1992) and might even be inhibited (Francis and Read, 1995). These interactions between crop species and soil fertility will influence whether or not inoculation is worthwhile.

Interactions between fungus species and soil fertility will also affect success. For example, isolates of *Glomus manihotis* (= *G. clarum*) can be very effective over a wide range of soil P, whereas other fungi may provide benefits only at intermediate or high P availability (Figure 4.3; also Howeler, Sieverding, and Saif, 1987; Sieverding, 1991; Vaast, Zasoski, and Bledsoe, 1996). For inoculation to improve nutrient uptake and crop productivity, soil fertility must be appropriate to the fungus species used.

The Native Fungus Community Is Depauperate or Ineffective

Mycorrhizal inoculation is most likely to increase crop productivity in soils where the native AM fungus community is deficient. Common soil treatments in industrial agriculture, such as tillage, chemical fertilization, and fumigation, can severely reduce mycorrhizal abundance. Following those conditions, field-scale inoculation has proved to be quite effective with many crops (e.g., Khasa, Furlan, and Fortin, 1992). Soils with poor root colonization or few spores will have the highest crop response to inoculation relative to soils where AM fungi are more abundant (Howeler, Sieverding, and Saif, 1987; Sieverding, 1991; Habte and Fox, 1993).

In addition to fungus abundance, the quality of the native community can influence the outcome of inoculation. Even when roots are well colonized in the native soil, crops will respond best to inoculation if the indigenous species are ineffective (Diederichs, 1990). Likewise, when very effective species are dominant in the soil, inoculation will have no effect (Howeler, Sieverding, and Saif, 1987).

Caveats for Commercial-Scale Inoculation

It is important to emphasize that mycorrhizae are not simply biological fertilizers that should replace synthetic fertilizers. As noted above, their biology and ecology are far more complex than the biology behind synthetic fertilizers, so treating mycorrhizae as replacements for synthetics is a recipe for failure. More importantly, mycorrhizae must not become yet another input for farmers to buy. Researchers and proponents of sustainable agriculture must think beyond the input substitution mentality (Rosset and Altieri, 1997). Organic inputs may be ecologically sensible replacements for chemical ones, but most do not alleviate the economic debts that burden farmers. Those debts make much of modern agriculture socially unsustainable, even in cases where it may be ecologically sustainable.

Therefore, strategic management of indigenous fungi should supercede large-scale inoculation efforts. Of course, inoculation will be necessary to initiate an effective community in severely degraded and eroded soils. But more research should emphasize the long-term establishment and subsequent management of such inoculants in order to reduce the need for repeated application. Moreover, research should better address ways that cultural practices can augment populations of effective species that already exist among the indigenous soil community. For example, even in soils that have been degraded by industrial agriculture for decades, several years with unplowed, organic, perennial plant crops can restore the majority of AM fungus species from native grassland communities (Picone, unpubl. data).

The commercial production of inoculum also raises two more concerns. First, the most effective local species can be easily overlooked because of the ease of buying commercial inoculum. Although it can be a tedious process, evaluating local species for effectiveness and managing them strategically will produce communities adapted to the range of soils and climate of interest. In contrast, buying the most effective isolates from commercial dealers may produce communities that fail in certain local conditions or in stressful seasons. In addition, some commercial isolates are likely selected for their ease of mass production, not necessarily for their effectiveness. Let the buyer beware — or better yet, use inoculum that is evaluated and produced locally.

Second, the international trade in mycorrhizal fungus isolates does not yet seem to consider the risk of invasive fungus species. For example, effective isolates of *Glomus manihotis* from Colombia have been used in field trials in Cuba, Nicaragua, and Costa Rica (pers. obs.). Granted, the risk is low because these are mutualist microbes, and introduced isolates seem to disappear over time (Sieverding, 1991; Dodd and Thomson, 1994). It may be possible, however, to introduce a species that outcompetes vital members of the native fungus community. Given our ignorance of microbial interactions in the soil, it seems prudent to avoid transferring inoculants among disparate regions.

CONCLUSION

Arbuscular mycorrhizae function in many ways that can help us design sustainable systems of agriculture. To use these microbes successfully, however, will require research into (1) the ways agricultural practices select particular fungus species and traits and (2) how those species and traits affect agronomic functions, such as nutrient uptake, pest resistance, and soil aggregation.

Despite our ignorance of the details required to select and manage optimal fungus communities, AM fungi can be incorporated into sustainable agriculture by simply following basic practices recommended for other reasons. Whether or not reduced use of chemical fertilizers selects effective AM fungi, it provides other benefits to the environment and human health. Although we do not yet understand whether no-till systems promote effective species, reduced tillage is recommended for increasing the total abundance of AM inoculum, for improving soil structure, for increasing soil organic matter, and for reducing dependence on fossil fuels. Perhaps diverse crop systems provide only a subtle growth benefit by increasing populations of effective AM species, but such systems provide many other benefits via pest control and economic security. Fortunately, a healthy soil community may emerge by piggybacking on other recommended practices.

In order for sustainable agriculture research to get the funding and interest required to precisely use soil microbes such as AM fungi, two related shifts in perspective will be required. First, the soil must be considered as an ecosystem, not simply a substrate. If the farm is approached as a factory, then the soil is just a chemical substrate, or a means of transferring chemical fertilizers to plant roots. If the farm and its soil are approached as ecosystems, however, then we will begin to recognize the interrelationships among soil biota and can take advantage of them. Second, society must move away from the high-energy, extractive economy and move toward the high-information, renewable economy. Many of the functions performed by microbes in a soil, such as nutrient cycling and pest control, have been replaced by energy-intensive industrial inputs. Fossil energy has superseded ecological information. As fossil energy becomes increasingly scarce, ecological information must be elucidated and developed to replace it. A detailed knowledge of mycorrhizal ecology should be one component of this inevitable future transition.

ACKNOWLEDGMENTS

This chapter was greatly improved by the thorough and thoughtful reviews of Dave Janos, Jim Bever, and Michelle Schroeder. Support was provided by The Land Institute, Salina, Kansas (www.landinstitute.org).

REFERENCES

Abbott, L.K., Comparative anatomy of vesicular-arbuscular mycorrhizas formed in subterranean clover, *Aust. J. Bot.,* 30:485–499, 1982.

Abbott, L.K. and Robson, A.D., The role of vesicular-arbuscular mycorrhizal fungi in agriculture and selection of fungi for inoculation, *Aust. J. Agric. Res.,* 33:389–408, 1982.

Abbott, L.K. and Robson, A.D., , Factors influencing the occurrence of vesicular arbuscular mycorrhizas, *Agric. Ecosyst. Environ.,* 35:121–150, 1991.

Abbott, L.K., Robson, A.D., and De Boer, G., The effect of phosphorus on the formation of hyphae in soil by the vesicular-arbuscular mycorrhizal fungus *Glomus fasciculatum, New Phytol.,* 97:437–446, 1984.

Abdalla, M.E. and Abdel-Fattah, G.M., Influence of the endomycorrhizal fungus *Glomus mosseae* on the development of peanut pod rot disease in Egypt, *Mycorrhiza,* 10:29–35, 2000.

Allen, E.B. and Allen, M.F., Competition between plants of different successional stages: mycorrhizae as regulators, *Can. J. Bot.,* 62:2625–2629, 1984.

Allen, E.B. et al., Disturbance and seasonal dynamics of mycorrhizae in a tropical deciduous forest in Mexico, *Biotropica,* 30:261–274, 1998.

Allen, M.F., Allen, E.B., and Friese, C.F., Responses of the non-mycotrophic plant *Salsola kali* to invasion by vesicular-arbuscular mycorrhizal fungi, *New Phytol.,* 111:45–49, 1989.

Arias, I. et al., Growth-responses of mycorrhizal and nonmycorrhizal tropical forage species to different levels of soil phosphate, *Plant Soil,* 132:253–260, 1991.

Augé, R.M., Water relations, drought and vesicular-arbuscular mycorrhizal symbiosis, *Mycorrhiza,* 11:3–43, 2001.

Azcón-Aguilar, C. and Barea, J.M., Arbuscular mycorrhizas and biological control of soil-borne plant pathogens — an overview of the mechanisms involved, *Mycorrhiza,* 6:457–464, 1996.

Bagyaraj, D.J., Vesicular-arbuscular mycorrhiza: application in agriculture, in *Methods in Microbiology,* Norris, J.R., Read, D.J., and Varma, A.K., Eds., Academic Press, New York, 1992, p. 359–373.

Bagyaraj, D.J. and Varma, A., Interaction between arbuscular mycorrhizal fungi and plants — their importance in sustainable agriculture in arid and semiarid tropics, *Adv. Microb. Ecol.,* 14:119–142, 1995.

Beare, M.H. and Bruce, R.R., A comparison of methods for measuring water-stable aggregates: implications for determining environmental effects on soil structure, *Geoderma,* 56:87–104, 1993.

Bellgard, S.E., Soil disturbance and infection of *Trifolium repens* roots by vesicular-arbuscular mycorrhizal fungi, *Mycorrhiza,* 3:25–29, 1993.

Bentivenga, S.P., Ecology and evolution of arbuscular mycorrhizal fungi, *McIlvainea,* 13:30–39, 1998.

Bethlenfalvay, G.J. et al., Mycorrhizae, biocides, and biocontrol. 2. Mycorrhizal fungi enhance weed control and crop growth in a soybean-cocklebur association treated with the herbicide bentazon, *Appl. Soil Ecol.,* 3:205–214, 1996.

Bever, J.D., Feedback between plants and their soil communities in an old field community, *Ecology,* 75:1965–1977, 1994.

Bever, J.D., Westover, K.M., and Antonovics, J., Incorporating the soil community into plant population dynamics: the utility of the feedback approach, *J. Ecol.,* 85:561–573, 1997.

Boddington, C.L. and Dodd, J.C., A comparison of the development and metabolic activity of mycorrhizas formed by arbuscular mycorrhizal fungi from different genera on two tropical forage legumes, *Mycorrhiza,* 8:149–157, 1998.

Boddington, C.L. and Dodd, J.C., Evidence that differences in phosphate metabolism in mycorrhizas formed by species of *Glomus* and *Gigaspora* might be related to their life-cycle strategies, *New Phytol.,* 142:531–538, 1999.

Boddington, C.L. and Dodd, J.C., The effect of agricultural practices on the development of indigenous arbuscular mycorrhizal fungi. I. Field studies in an Indonesian ultisol, *Plant Soil,* 218:137–144, 2000a.

Boddington, C.L. and Dodd, J.C., The effect of agricultural practices on the development of indigenous arbuscular mycorrhizal fungi. II. Studies in experimental microcosms, *Plant Soil,* 218:145–157, 2000b.

Brundrett, M., Abbott, L., and Jasper, D., Glomalean mycorrhizal fungi from tropical Australia. I. Comparison of the effectiveness and specificity of different isolation procedures, *Mycorrhiza,* 8:305–314, 1999a.

Brundrett, M.C., Jasper, D.A., and Ashwath, N., Glomalean mycorrhizal fungi from tropical Australia. II. The effect of nutrient levels and host species on the isolation of fungi, *Mycorrhiza,* 8:315–321, 1999b.

Carpenter, F.L., Mayorga, S.P., and Quintero, E.G., Land-use and erosion of a Costa Rican ultisol affect soil chemistry, mycorrhizal fungi, and early regeneration, *Forest Ecol. Manage.,* 144:1–17, 2001.

Clark, R.B., Zeto, S.K., and Zobel, R.W., Arbuscular mycorrhizal fungal isolate effectiveness on growth and root colonization of *Panicum virgatum* in acidic soil, *Soil Biol. Biochem.,* 31:1757–1763, 1999.

Cox, T.S., Picone, C., and Jackson, W., Research priorities in natural systems agriculture, *J. Crop Prod.,* in press.

Cuenca, G., De Andrade, Z., and Escalante, G., Diversity of Glomalean spores from natural, disturbed and revegetated communities growing on nutrient-poor tropical soils, *Soil Biol. Biochem.,* 30:711–719, 1998.

Cuenca, G. and Meneses, E., Diversity patterns of arbuscular mycorrhizal fungi associated with cacao in Venezuela, *Plant Soil,* 183:315–322, 1996.

Cupull, R. et al., Utilización de *Trichoderma*, Micorriza y *Azotobacter* en la producción de posturas de cafeto, *III Encuentro Nacional de Agricultura Orgánica,* Asociación Cubana de Agricultura Orgánica (ACAO), Villa Clara, Cuba, 1997, p. 80.

Diederichs, C., Improved growth of *Cajanus cajans* (L) Millsp. in an unsterile tropical soil by 3 mycorrhizal fungi, *Plant Soil,* 123:261–266, 1990.

Diederichs, C., Influence of different P sources on the efficiency of several tropical endomycorrhizal fungi in promoting the growth of *Zea mays* L., *Fert. Res.,* 30:39–46, 1991.

Diederichs, C. and Moawad, A.M., The potential of VA mycorrhizae for plant nutrition in the tropics, *Angew. Bot.,* 67:91–96, 1993.

Dodd, J.C. et al., The management of populations of vesicular-arbuscular mycorrhizal fungi in acid-infertile soils of a savanna ecosystem. 2. The effects of precrops on the spore populations of native and introduced VAM-fungi, *Plant Soil,* 122:241–247, 1990.

Dodd, J.C. et al., Mycelium of arbuscular mycorrhizal fungi (AMF) from different genera: form, function and detection, *Plant Soil,* 226:131–151, 2000.

Dodd, J.C. and Thomson, B.D., The screening and selection of inoculant arbuscular-mycorrhizal and ectomycorrhizal fungi, *Plant Soil,* 159:149–158, 1994.

Domini, M.E., Lara, D., and Gomez, R., Reducción del uso del fertilizante químico en el cultivo del tomate (*Lycopersicon esculentum*, Mil) mediante la aplicación de biofertilizantes utilizando la tecnología de recubrimento de semillas, *III Encuentro Nacional de Agricultura Orgánica,* Asociación Cubana de Agricultura Orgánica (ACAO), Villa Clara, Cuba, 1997, p. 81–82.

Donegan, K., personal communication, 2001.

Donegan, K.K. and Seidler, R.J., Effects of transgenic plants on soil and plant microorganisms, in *Recent Research Development in Microbiology,* Pandalai, S.G., Ed., Research Signpost, Trivandrum, India, 1999, p. 415–424.

Egerton-Warburton, L.M. and Allen, E.B., Shifts in arbuscular mycorrhizal communities along an anthropogenic nitrogen deposition gradient, *Ecol. Appl.,* 10:484–496, 2000.

Elsen, A., Declerck, S., and Waele, D.D., Effects of *Glomus intraradices* on the reproduction of the burrowing nematode (*Radopholus similis*) in dixenic culture, *Mycorrhiza,* 11:49–51, 2001.

Entry, J.A. et al., Influence of compaction from wheel traffic and tillage on arbuscular mycorrhizae infection and nutrient uptake by *Zea mays, Plant Soil,* 180:139–146, 1996.

Evans, D.G. and Miller, M.H., Vesicular-arbuscular mycorrhizas and the soil-disturbance-induced reduction of nutrient absorption in maize. I. Causal relations, *New Phytol.,* 110:67–74, 1988.

Fagbola, O., Osonubi, O., and Mulongoy, K., Growth of cassava cultivar TMS 30572 as affected by alley-cropping and mycorrhizal inoculation, *Biol. Fertil. Soils,* 27:9–14, 1998.

Feldmann, F. and Boyle, C., Weed-mediated stability of arbuscular mycorrhizal effectiveness in maize monocultures, *J. Appl. Bot. Angew. Bot.,* 73:1–5, 1999.

Feldmann, F. and Idczak, E., Inoculum production of vesicular-arbuscular mycorrhizal fungi for use in tropical nurseries, *Methods Microbiol.,* 24:339–357, 1992.

Feldmann, F. et al., Recultivation of degraded, fallow lying areas in central Amazonia with equilibrated polycultures — response of useful plants to inoculation with VA-mycorrhizal fungi, *J. Appl. Bot. Angew. Bot.,* 69:111–118, 1995.

Feldmann, F. et al., The strain-inherent variability of arbuscular mycorrhizal effectiveness. I. Development of a test system using *Petroselinum crispum* Hoffm. as host, *Symbiosis,* 25:115–129, 1998.

Fernandez, F. et al., Tecnología de recubrimiento de semillas con biofertilizante micorrizógeno, alternativa sostenible de bajo costo, *III Encuentro Nacional de Agricultura Orgánica,* Asociación Cubana de Agricultura Orgánica (ACAO), Villa Clara, Cuba, 1997, p. 76–77.

Fontenla, S., Garcia-Romera, I., and Ocampo, J.A., Negative influence of non-host plants on the colonization of *Pisum sativum* by the arbuscular mycorrhizal fungus Glomus mosseae, *Soil Biol. Biochem.,* 31:1591–1597, 1999.

Francis, R. and Read, D.J., The contributions of mycorrhizal fungi to the determination of plant community structure, *Plant Soil,* 159:11–25, 1994.

Francis, R. and Read, D.J., Mutualism and antagonism in the mycorrhizal symbiosis, with special reference to impacts on plant community structure, *Can. J. Bot.,* 73:S1301–S1309, 1995.

Galvez, L. et al., An overwintering cover crop increases inoculum of VAM fungi in agricultural soil, *Am. J. Alternative Agric.,* 10:152–156, 1995.

Giovannetti, M. and Sbrana, C., Meeting a non-host: the behaviour of AM fungi, *Mycorrhiza,* 8:123–130, 1998.

Gomez, R. et al., La biofertilización de los cultivos de importancia económica como parte integral de la agricultura sostenible en las condiciones tropicales de Cuba, *III Encuentro Nacional de Agricultura Orgánica,* Asociación Cubana de Agricultura Orgánica (ACAO), Villa Clara, Cuba, 1997, p. 75.

Grime, J.P. et al., Floristic diversity in a model system using experimental microcosms, *Nature,* 328:420–422, 1987.

Habte, M. and Fox, R.L., Effectiveness of VAM fungi in nonsterile soils before and after optimization of P in soil solution, *Plant Soil,* 151:219–226, 1993.

Hamel, C. et al., Composition of the vesicular-arbuscular mycorrhizal fungi population in an old meadow as affected by pH, phosphorus and soil disturbance, *Agric. Ecosyst. Environ.,* 49:223–231, 1994.

Harinikumar, K.M. and Bagyaraj, D.J., Effect of crop-rotation on native vesicular arbuscular mycorrhizal propagules in soil, *Plant Soil,* 110:77–80, 1988.

Hartnett, D.C. et al., Mycorrhizal influence on intra- and interspecific neighbor interactions among co-occurring prairie grasses, *J. Ecol.,* 81:787–795, 1993.

Heggo, A., Angle, J.S., and Chaney, R.L., Effects of vesicular-arbuscular mycorrhizal fungi on heavy metal uptake by soybeans, *Soil Biol. Biochem.,* 22:865–869, 1990.

Hetrick, B.A.D. and Wilson, G.W.T., Effects of mycorrhizal fungus species and metalaxyl application on microbial suppression of mycorrhizal symbiosis, *Mycologia,* 83:97–102, 1991.

Hetrick, B.A.D., Wilson, G.W.T., and Cox, T.S., Mycorrhizal dependence of modern wheat cultivars and ancestors — a synthesis, *Can. J. Bot.,* 71:512–518, 1993.

Howeler, R.H., Sieverding, E., and Saif, S., Practical aspects of mycorrhizal technology in some tropical crops and pastures, *Plant Soil,* 100:249–283, 1987.

Ikram, A., Jensen, E.S., and Jakobsen, I., No significant transfer of N and P from *Pueraria phaseloides* to *Hevea brasiliensis* via hyphal links of arbuscular mycorrhiza, *Soil Biol. and Biochem.*, 26:1541–1547, 1994.

Jaizme-Vega, M.C. and Azcon, R., Responses of some tropical and subtropical cultures to endomycorrhizal fungi, *Mycorrhiza,* 5:213–217, 1995.

Jaizme-Vega, M.C. and Pinochet, J., Growth response of banana to three mycorrhizal fungi in *Pratylenchus goodeyi* infested soil, *Nematropica,* 27:69–76, 1997.

Jakobsen, I., Abbott, L.K., and Robson, A.D., External hyphae of vesicular-arbuscular mycorrhizal fungi associated with *Trifolium subterraneum* L. 1. Spread of hyphae and phosphorus inflow into roots, *New Phytol.,* 120:371–380, 1992.

Janos, D.P., Vesicular-arbuscular mycorrhizae affect lowland tropical rain forest plant growth, *Ecology,* 61:151–162, 1980.

Janos, D.P., VA mycorrhizas in humid tropical ecosystems, in *Ecophysiology of VA Mycorrhizal Plants,* Safir, G.R., Ed., CRC Press, Boca Raton, FL, 1987, p. 107–134.

Janos, D.P., Mycorrhiza applications in tropical forestry: are temperate zone approaches appropriate?, in *Trees and Mycorrhiza,* Ng, F.S.P., Ed., Forest Research Institute, Kuala Lumpur, Malaysia, 1988, p. 133–188.

Janos, D.P., Mycorrhizas, succession, and the rehabilitation of deforested lands in the humid tropics, in *Fungi and Environmental Change,* Frankland, J. C., Magan, N., and Gadd, G.M., Eds., *British Mycological Society Symposium,* Cambridge University Press, U.K., 1996.

Jastrow, J.D., Miller, R.M., and Lussenhop, J., Contributions of interacting biological mechanisms to soil aggregate stabilization in restored prairie, *Soil Biol. Biochem.,* 30:905–916, 1998.

Jawson, M.D. et al., Soil fumigation within monoculture and rotations — response of corn and mycorrhizae, *Agron. J.,* 85:1174–1180, 1993.

Johnson, N.C., Can fertilization of soil select less mutualistic mycorrhizae?, *Ecol. Appl.,* 3:749–757, 1993.

Johnson, N.C. et al., Mycorrhizae — possible explanation for yield decline with continuous corn and soybean, *Agron. J.,* 84:387–390, 1992a.

Johnson, N.C., Graham, J.H., and Smith, F.A., Functioning of mycorrhizal associations along the mutualism — parasitism continuum, *New Phytol.,* 135:575–586, 1997.

Johnson, N.C. and Pfleger, F.L., VA mycorrhizae and cultural stresses, in *Mycorrhizae in Sustainable Agriculture,* Bethlenfalvay, G.J. and Linderman, R.G., Eds., American Society of Agronomy, 1992, p. 71–99.

Johnson, N.C., Tilman, D., and Wedin, D., Plant and soil controls on mycorrhizal fungal communities, *Ecology,* 73:2034–2042, 1992b.

Johnson, N.C. and Wedin, D.A., Soil carbon, nutrients, and mycorrhizae during conversion of dry tropical forest to grassland, *Ecol. Appl.,* 7:171–182, 1997.

Jordan, N.R., Zhang, J., and Huerd, S., Arbuscular-mycorrhizal fungi: potential roles in weed management, *Weed Res.,* 40:397–410, 2000.

Kabir, Z. et al., Seasonal changes of arbuscular mycorrhizal fungi as affected by tillage practices and fertilization: hyphal density and mycorrhizal root colonization, *Plant Soil,* 192:285–293, 1997.

Kabir, Z. et al., Vertical distribution of arbuscular mycorrhizal fungi under corn (*Zea mays* L.) in no-till and conventional tillage systems, *Mycorrhiza,* 8:53–55, 1998.

Kaldorf, M. et al., Selective element deposits in maize colonized by a heavy metal tolerance conferring arbuscular mycorrhizal fungus, *J. Plant Physiol.,* 154:718–728, 1999.

Khalil, S., Loynachan, T.E., and Tabatabai, M.A., Mycorrhizal dependency and nutrient-uptake by improved and unimproved corn and soybean cultivars, *Agron. J.,* 86:949–958, 1994.

Khasa, P., Furlan, V., and Fortin, J.A., Response of some tropical plant-species to endomycorrhizal fungi under field conditions, *Trop. Agric.,* 69:279–283, 1992.

Koslowsky, S.D. and Boerner, R.E.J., Interactive effects of aluminum, phosphorous, and mycorrhizae on growth and nutrient uptake of *Panicum virgatum* L. (Poaceae), *Environ. Pollut.,* 61:107–125, 1989.

Kurle, J.E. and Pfleger, F.L., The effects of cultural practices and pesticides on VAM fungi, in *Mycorrhizae and Plant Health,* Pfleger, F.L. and Linderman, R.G., Eds., American Phytopathological Society, St. Paul, MN, 1994, p. 101–131.

Linderman, R.G., The role of VAM fungi in biocontrol, in *Mycorrhizae and Plant Health,* Pfleger, F.L. and Linderman, R.G., Eds., American Phytopathological Society, St. Paul, MN, 1994, p. 1–25.

Lopez-Aguillon, R. and Mosse, B., Experiments on competitiveness of three endomycorrhizal fungi, *Plant Soil,* 97:155–170, 1987.

Louis, I. and Lim, G., Differential response in growth and mycorrhizal colonization of soybean to inoculation with 2 isolates of *Glomus clarum* in soils of different P availability, *Plant Soil,* 112:37–43, 1988.

Manske, G.G.B., Genetic-analysis of the efficiency of VA mycorrhiza with spring wheat, *Agric. Ecosyst. Environ.,* 29:273–280, 1990.

Marler, M.J., Zabinski, C.A., and Callaway, R.M., Mycorrhizae indirectly enhance competitive effects of an invasive forb on a native bunchgrass, *Ecology,* 80:1180–1186, 1999.

Marschner, H. and Dell, B., Nutrient-uptake in mycorrhizal symbiosis, *Plant Soil,* 159:89–102, 1994.

McGonigle, T.P. and Miller, M.H., Mycorrhizal development and phosphorus absorption in maize under conventional and reduced tillage, *Soil Sci. Soc. Am. J.,* 57:1002–1006, 1993.

Menge, J.A., Effects of soil fumigants and fungicides on vesicular-arbuscular fungi, *Phytopathology,* 72:1125–1132, 1982.

Miller, R., Nontarget and ecological effects of transgenically altered disease resistance in crops — possible effects on the mycorrhizal symbiosis, *Mol. Ecol.,* 2:327–335, 1993.

Miller, R. and Jastrow, J., Hierarchy of root and mycorrhizal fungal interactions with soil aggregation, *Soil Biol. Biochem.,* 22:579–584, 1990.

Miller, R. and Jastrow, J., The role of mycorrhizal fungi in soil conservation, in *Mycorrhizae in Sustainable Agriculture,* Bethlenfalvay, G.J. and Linderman, R.G., Eds., American Society of Agronomy, Madison, WI, 1992, p. 29–44.

Miller, R., Reinhardt, D.R., and Jastrow, J.D., External hyphal production of vesicular-arbuscular mycorrhizal fungi in pasture and tallgrass prairie communities, *Oecologia,* 103:17–23, 1995.

Mills, K.E. and Bever, J.D., Maintenance of diversity within plant communities: soil pathogens as agents of negative feedback, *Ecology,* 79:1595–1601, 1998.

Molina, R., Massicotte, H., and Trappe, J.M., Specificity phenomena in mycorrhizal symbiosis: community-ecological consequences and practical implications, in *Mycorrhizal Functioning,* Allen, M.F., Ed., Chapman & Hall, New York, 1992, p. 357–423.

Morton, J.B. and Benny, G.L., Revised classification of arbuscular mycorrhizal fungi (Zygomycetes) — a new order, Glomales, 2 new suborders, Glomineae and Gigasporineae, and 2 new families, Acaulosporaceae and Gigasporaceae, with an emendation of Glomaceae, *Mycotaxon,* 37:471–491, 1990.

Morton, J.B., Bentivenga, S.P., and Bever, J.D., Discovery, measurement, and interpretation of diversity in arbuscular endomycorrhizal fungi (Glomales, Zygomycetes), *Can. J. Bot.,* 73:S25–S32, 1995.

Munyanziza, E., Kehri, H.K., and Bagyaraj, D.J., Agricultural intensification, soil biodiversity and agro-ecosystem function in the tropics: the role of mycorrhiza in crops and trees, *Appl. Soil Ecol.,* 6:77–85, 1997.

Noyd, R.K., Pfleger, F.L., and Norland, M.R., Field responses to added organic matter, arbuscular mycorrhizal fungi, and fertilizer in reclamation of taconite iron ore tailing, *Plant Soil,* 179:89–97, 1996.

Oades, J.M. and Waters, A.G., Aggregate hierarchy in soils, *Aust. J. Soil Res.,* 29:815–828, 1991.

Pearson, J.N., Abbott, L.K., and Jasper, D.A., Mediation of competition between 2 colonizing VA mycorrhizal fungi by the host plant, *New Phytol.,* 123:93–98, 1993.

Pearson, J.N., Abbott, L.K., and Jasper, D.A., Phosphorus, soluble carbohydrates and the competition between 2 arbuscular mycorrhizal fungi colonizing subterranean clover, *New Phytol.,* 127:101–106, 1994.

Pearson, J.N. and Jakobsen, I., The relative contribution of hyphae and roots to phosphorus uptake by arbuscular mycorrhizal plants, measured by dual labeling with P-32 and P-33, *New Phytol.,* 124:489–494, 1993.

Peng, S.B. et al., Growth depression in mycorrhizal citrus at high-phosphorus supply — analysis of carbon costs, *Plant Physiol.,* 101:1063–1071, 1993.

Picone, C., unpublished data.

Picone, C., Comparative Ecology of Arbuscular Mycorrhizal Fungi in Lowland Tropical Forest and Pasture, Ph.D. thesis, University of Michigan, Ann Arbor, MI, 1999.

Picone, C., Diversity and abundance of arbuscular-mycorrhizal fungus spores in tropical forest and pasture, *Biotropica,* 32:734–750, 2000.

Pulido, L. and Peralta, H., Producción de pimiento con el uso de MVA: Comportamiento en las etapas de semillero y plantación, *III Encuentro Nacional de Agricultura Orgánica,* Asociación Cubana de Agricultura Orgánica (ACAO), Villa Clara, Cuba, 1997a, p. 77.

Pulido, L. and Peralta, H., Uso de micorrizas arbusculares y solubilizadores de fósforo para la producción de tomate, *III Encuentro Nacional de Agricultura Orgánica,* Asociación Cubana de Agricultura Orgánica (ACAO), Villa Clara, Cuba, 1997b, p. 77–78.

Redecker, D., Kodner, R., and Graham, L.E., Glomalean fungi from the Ordovician, *Science,* 289:1920–1921, 2000a.

Redecker, D., Morton, J.B., and Bruns, T.D., Ancestral lineages of arbuscular mycorrhizal fungi (Glomales), *Mol. Phylogenet. Evol.,* 14:276–284, 2000b.

Redecker, D., Morton, J.B., and Bruns, T.D., Molecular phylogeny of the arbuscular mycorrhizal fungi *Glomus sinuosum* and *Sclerocystis coremioides, Mycologia,* 92:282–285, 2000c.

Rivas-Platero, G.G. and Andrade, J.C., Interacción de hongos endomicorrizicos con *Meloidogyne exigua* en café, *Manejo Integrado de Plagas (Costa Rica),* 49:68–72, 1998.

Rose, S.L. and Paranka, J.E., Root and VAM distribution in tropical agricultural and forest soils, in *Mycorrhizae in the Next Decade: Practical Applications and Research Priorities,* Sylvia, D.M., Hung, L.L., and Graham, J.H., Eds., paper presented at the 7th North American Conference on Mycorrhizae, Institute of Food and Agricultural Science, University of Florida, Gainesville, 1987, p. 56.

Rosset, P. and Altieri, M.A., Agroecology versus input substitution: a fundamental contradiction of sustainable agriculture, *Soc. Nat. Res.,* 10:283–295, 1997.

Rosset, P. and Benjamin, M., Eds., *Greening of the Revolution: Cuba's Experiment with Organic Agriculture,* Global Exchange, San Francisco, CA, 1993.

Ruiz, L.A. et al., Empleo de las micorrizas, la fosforina, el azotobacter y dosis crecientes de NPK en minituberculos de papa para semilla, *III Encuentro Nacional de Agricultura Orgánica,* Asociación Cubana de Agricultura Orgánica (ACAO), Villa Clara, Cuba, 1997, p. 72–73.

Saif, S.R., Growth-responses of tropical forage plant-species to vesicular-arbuscular mycorrhizae. 1. Growth, mineral uptake and mycorrhizal dependency, *Plant Soil,* 97:25–35, 1987.

Salinas, J.G., Sanz, J.I., and Sieverding, E., Importance of VA mycorrhizae for phosphorus supply to pasture plants in tropical oxisols, *Plant Soil,* 84:347–360, 1985.

Sanchez, P.A., *Properties and Management of Soils in the Tropics,* John Wiley & Sons, New York, 1976.

Sanders, I.R. and Koide, R.T., Nutrient acquisition and community structure in co-occurring mycotrophic and non-mycotrophic old-field annuals, *Funct. Ecol.,* 8:77–84, 1994.

Saxena, D., Flores, S., and Stotzky, G., Insecticidal toxin in root exudates from *Bt* corn, *Nature,* 402:480, 1999.

Schenck, N.C. and Perez, Y., *Manual for the Identification of VA Mycorrhizal Fungi,* Synergistic Publications, Florida, 1990.

Schreiner, R.P. and Bethlenfalvay, G.J., Mycorrhizal interactions in sustainable agriculture, *Crit. Rev. Biotechnol.,* 15:271–285, 1995.

Schreiner, R.P. and Koide, R.T., Antifungal compounds from the roots of mycotrophic and non- mycotrophic plant-species, *New Phytol.,* 123:99–105, 1993.

Sieverding, E., Ecology of VAM fungi in tropical agrosystems, *Agric. Ecosyst. Environ.,* 29:369–390, 1989.

Sieverding, E., *Vesicular-Arbuscular Mycorrhiza Management in Tropical Agrosystems,* Deutsche Gesellschaft für Technische Zusammenarbeit (GTZ), Eschborn, Germany, 1991.

Sieverding, E. and Leihner, D., Influence of crop rotation and intercropping of cassava with legumes on VA mycorrhizal symbiosis of cassava, *Plant Soil,* 80:143–146, 1984.

Simpson, D. and Daft, M.J., Interactions between water-stress and different mycorrhizal inocula on plant-growth and mycorrhizal development in maize and sorghum, *Plant Soil,* 121:179–186, 1990a.

Simpson, D. and Daft, M.J., Spore production and mycorrhizal development in various tropical crop hosts infected with *Glomus clarum, Plant Soil,* 121:171–178, 1990b.

Singh, M. and Tilak, K., Inoculation of sorghum (*Sorghum bicolor*) with *Glomus versiforme* under field conditions, *Trop. Agric.,* 69:323–326, 1992.

Siqueira, J.O., Colozzi-Fiho, A., and Oliveira, E., Ocorrencia de micorrizas vesicular-arbusculares em agro e ecosistemas naturais do estado de Minas Gerais, *Pesquisas agropecuarias Brasileiras,* 24:1499–1506, 1989.

Siqueira, J.O. et al., Arbuscular mycorrhizal inoculation and superphosphate application influence plant development and yield of coffee in Brazil, *Mycorrhiza,* 7:293–300, 1998.

Smith, S.E. and Read, D.J., *Mycorrhizal Symbiosis,* Academic Press, New York, 1997.

Soule, J.D. and Piper, J.K., *Farming in Nature's Image: An Ecological Approach to Agriculture,* Island Press, Washington, D.C., 1992.

Struble, J.E. and Skipper, H.D., Vesicular-arbuscular mycorrhizal fungal spore production as influenced by plant species, *Plant Soil,* 109:277–280, 1988.

Sylvia, D. and Williams, S., Vesicular-arbuscular mycorrhizae and environmental stresses, in *Mycorrhizae in Sustainable Agriculture,* Bethlenfalvay, G.J. and Linderman, R.G., Eds., American Society of Agronomy, Madison, WI, 1992, p. 101–124.

Tejada, T., Soto, F., and Guerrero, G., La peletización como método de aplicación de micorrizas en cafeto, *III Encuentro Nacional de Agricultura Orgánica,* Asociación Cubana de Agricultura Orgánica (ACAO), Villa Clara, Cuba, 1997, p. 82.

Tilman, D. et al., Forecasting agriculturally driven global environmental change, *Science,* 292:281–284, 2001.

Tisdall, J.M. and Oades, J., The management of ryegrass to stabilize aggregates of a red-brown earth, *Aust. J. Soil Res.,* 18:415–422, 1980.

Trappe, J.T., Phylogenetic and ecologic aspects of mycotrophy in the angiosperms from an evolutionary standpoint, in *Ecophysiology of VA Mycorrhizal Plants,* Safir, G.R., Ed., CRC Press, Boca Raton, FL, 1987, p. 5–25.

Vaast, P., Zasoski, R.J., and Bledsoe, C.S., Effects of vesicular-arbuscular mycorrhizal inoculation at different soil P availabilities on growth and nutrient uptake of *in vitro* propagated coffee (*Coffea arabica* L.) plants, *Mycorrhiza,* 6:493–497, 1996.

van der Heijden, M.G.A. et al., Different arbuscular mycorrhizal fungal species are potential determinants of plant community structure, *Ecology,* 79:2082–2091, 1998a.

van der Heijden, M.G.A. et al., Mycorrhizal fungal diversity determines plant biodiversity, ecosystem variability and productivity, *Nature,* 396:69–72, 1998b.

Vandermeer, J., *The Ecology of Intercropping,* Cambridge University Press, Cambridge, U.K., 1989.

Vierheilig, H. et al., Colonization of transgenic tobacco constitutively expressing pathogenesis-related proteins by the vesicular-arbuscular mycorrhizal fungus *Glomus mosseae, Appl. Environ. Microbiol.,* 61:3031–3034, 1995.

Wacker, T.L., Safir, G.R., and Stephenson, S., Evidence for succession of mycorrhizal fungi in Michigan asparagus fields, *Acta Hort.,* 271:273–278, 1990.

Wilson, J. et al., Long-term changes in vesicular-arbuscular mycorrhizal spore populations in *Terminalia* plantations in Côte d'Ivoire, in *Mycorrhizas in Ecosystems,* Read, D.H. et al., Eds., CAB International, Egham, Surrey, U.K., 1992, p. 268–275.

Wright, S.F. and Upadhyaya, A., A survey of soils for aggregate stability and glomalin, a glycoprotein produced by hyphae of arbuscular mycorrhizal fungi, *Plant Soil,* 198:97–107, 1998.

Wright, S.F., Upadhyaya, A., and Buyer, J.S., Comparison of N-linked oligosaccharides of glomalin from arbuscular mycorrhizal fungi and soils by capillary electrophoresis, *Soil Biol. Biochem.,* 30:1853–1857, 1998.

Young, C.C., Juang, T.C., and Chao, C.C., Effects of rhizobium and vesicular-arbuscular mycorrhiza inoculations on nodulation, symbiotic nitrogen-fixation and soybean yield in subtropical-tropical fields, *Biol. Fertil. Soils,* 6:165–169, 1988.

Zhu, Y. et al., Genetic diversity and disease control in rice, *Nature,* 406:718–722, 2000.

CHAPTER 5

Technological Change and Biodiversity in the Rubber Agroecosystem of Sumatra

Laxman Joshi, Gede Wibawa, Hendrien Beukema, Sandy Williams, and Meine van Noordwijk

CONTENTS

INTRODUCTION: DOMESTICATING TREES OR THE FOREST?

Large areas of the humid tropics have land-use patterns that do not fit into a simple culture/nature or agriculture/forest dichotomy and thus the term *deforestation* refers

0-8493-1581-6/03/$0.00+$1.50

to a gradual loss of forest functions, rather than an abrupt change. The definition of *agroforest* by de Foresta and Michon (1996), an intermediate stage between natural forest and agricultural plantations, captures the mixed heritage of the wild and the domesticated aspect of these systems. Outside perspectives on these systems have focused on either side of the coin: poorly managed, low productivity, because too wild or interesting biodiversity but not like a real forest, because too domesticated. Yet, these land-use systems should be understood from a farmer's/manager's perspective if we want to understand what scenarios exist for their future development. Can farmers increase productivity (and/or profitability) while maintaining the current biodiversity of the system? Or will any intensification beyond current practices lead to a further decrease of biodiversity values, which in the past were largely derived from the natural forest context of the system? This chapter discusses these perspectives on the basis of ongoing research by ICRAF and partners in Jambi, one of the main rubber-producing provinces in Sumatra (Indonesia).

A conceptual scheme for the analysis of complex agroecosystems such as rubber agroforests (Figure 5.1) should consider interactions between farmer management decisions (the human part) and a considerable wild, spontaneous, or natural component in the agroecosystem. Both the planned/planted and the spontaneous components can be harvested and contribute to farm profitability, but the nonharvested components contribute to long-term sustainability and environmental functions for outside stakeholders. While traditionally agricultural research has focused on the upper part of the diagram (the planted and harvested part), a more complete understanding is desirable.

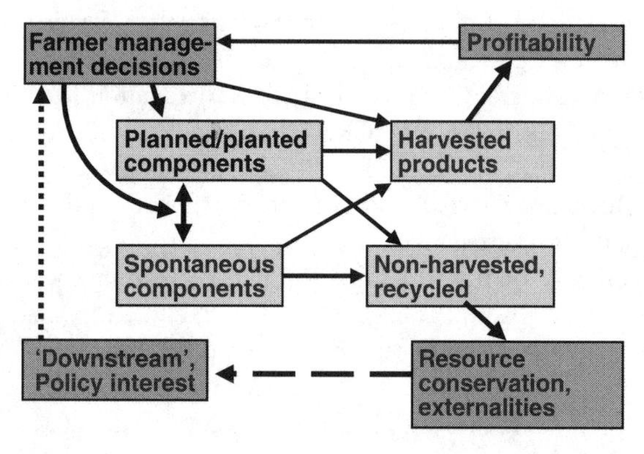

Figure 5.1 Conceptual scheme for analyzing complex agroecosystems in which farmer management decisions interact with a considerable spontaneous or natural component in the agroecosystem, and where both the planned/planted and the spontaneous components can be harvested and contribute to farm profitability, while the non-harvested components contribute to long-term sustainability and environmental service functions for outside stakeholders. (Modified from Swift and Ingram, 1996; Vandermeer et al., 1998.)

Complex agroforests (de Foresta and Michon, 1997, de Foresta et al., 2000) can be derived from forest in essentially two ways:

- By gradually modifying a forest through interplanting of desirable local (such as cinnamon, tea, fruit trees) or introduced (such as coffee) forest species
- By modifying forest succession in the fallow vegetation after a slash-and-burn land clearing and food-crop production episode, using local (benzoe) or introduced (rubber) trees

Both methods can be repeated (or exchanged) in subsequent management for rejuvenation of the agroforest, as we will see.

Agroforests represent an important stage in the domestication of forest resources (Wiersum, 1997a) or an alternative pathway for domesticating the forest rather than the trees as such (Michon and de Foresta, 1997, 1999). Domestication involves both the biological resource (and an increasing human control over reproduction and gene flow into subsequent generations) and the land used (with increasing private control over what starts of as open-access resources, Figure 5.2). Wiersum (1997a,b) identified three thresholds in the process of domestication: *controlled utilization* (separating the open access from the controlled harvesting regime), *purposeful regeneration* (separating the dependence on natural regeneration from the interventions that generally require control over subsequent utilization), and *domestication* (into horticultural or plantation style production system). Agroforests contain trees planted, seeded, or otherwise regenerated by the farmer, as well as trees established spontaneously, but tolerated. Human control over the genetic makeup of the trees, however, is generally limited and there is thus scope for further *domestication.*

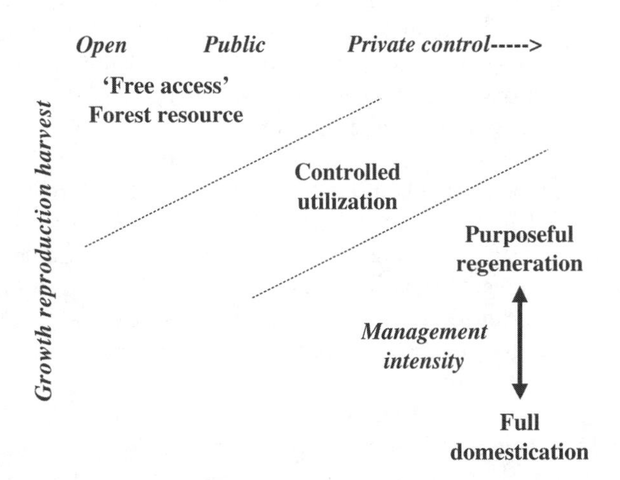

Figure 5.2 Stages in domestication of forest resources, on the basis of the type of control (tenure) of land and the type of control over reproduction and growth of the plants involved. (Modified from Wiersum, 1997b.)

While there has been a tradition of trading various types of resin and latex collected from the forest, the introduction a century ago of *Hevea brasiliensis* or "para" from the Amazone (Para) to Southeast Asia formed the basis of a large-scale spontaneous adoption of new agroforestry practices at a scale not easily matched elsewhere (Van Gelder, 1950; Webster and Baulkwill, 1989). An estimated 7 million people in Sumatra and Kalimantan islands currently make their living from rubber-based agroforests from an area of 2.5 million hectares. Smallholder rubber constitutes 84% of the total Indonesian rubber area; and 76% of the total rubber production volume (Ditjenbun, 1998). Rubber is a major export commodity supporting the Indonesian economy. Around 70% of farmers in Jambi province are engaged in smallholder rubber production and derive, on average, nearly 70% of household income from rubber (Wibawa et al., 2002).

This chapter discusses the origins of the rubber agroforest, their current value for biodiversity conservation, and the search for technological improvements that suit farmers' priorities but also maintain the environmental service functions that current systems provide.

RUBBER AGROFORESTS ARE HISTORICALLY DERIVED FROM CROP-FALLOW ROTATIONS

Fallow rotation systems, in the definition of Ruthenberg (1976), are an intermediate stage between "shifting cultivation or long rotation fallow systems" (where land is cropped for less than one third of the time, $R < 0.33$) and continuous cropping (where land is cropped more than two thirds of the time, $R > 0.67$). Ruthenberg's (1976) R value is the fraction of time (or land area) used for annual food crops as part of the total cropping cycle (area). The equivalence of time and area only applies in steady-state conditions of land-use intensity. Although crop yields per unit cropped field are directly related to the soil fertility of the plot, and hence to the preceding length of the fallow period, shortening the fallow period can generally increase total yields per unit land. According to a simple model formulation (Trenbath, 1989; van Noordwijk, 1999), the maximum sustainable returns to land can be expected where soil fertility can recover during a fallow to just over half its maximum value. More intensive land use (higher R values) are only possible if the soil restoring functions of a fallow can be obtained in less time, by a so-called improved (more effective) fallow, or that these functions have to be fully integrated into the cropping system. In practice, there is a danger of overintensification leading to degradation (Trenbath, 1989; van Noordwijk et al., 1997). *Imperata* fallows may on the one hand prevent complete soil degradation, but they are not productive, do little to restore soil fertility, and may lead to land abandonment. In the lowland peneplains of Sumatra, however, para rubber arrived in time to provide an alternative method of intensifying land use by increasing the forest productivity of the fallow and increasing the cycle length.

Fallows in various stages of their succession to secondary forests can be used as grazing land and for producing firewood, honey, thatching material, etc. The first step in intensification of farmer management of fallow lands is usually the retention or promotion of certain plant species that appear in the fallow and are considered to be of value for one of the several functions of a fallow: its role with respect to a future crop, or its role as a direct resource. During further intensification, however, choices among the multiple functions may be necessary, as more effective fallows will tend to become shorter in duration while many elements for a more productive fallow impose an increased duration on the system.

The transition from swidden-fallow systems into more intensive land-use systems can essentially follow three routes (Cairns and Garrity, 1999; van Noordwijk, 1999) that focus on three elements of the system: food crops, tree crops, or fodder supply/pasture systems. Efforts to increase the harvestable output per unit area can be achieved in food-crop-based systems by reducing the length of the fallow period and the age at which secondary forests are reopened by slash-and-burn methods. Agroforest development emphasizes the harvestable part of the forest fallow and will lead to a reduction of annual crop intensity as the economic lifespan of the trees dominates decisions on cycle length. Specialization on fodder supply or pasture systems is relatively unimportant in the humid forest zone of Indonesia, but it is a dominant pattern in Latin America. In the Indonesian context, beginning at the start of the 20th century, the swidden lands were gradually transformed by slash-and-burn farmers into rubber agroforests.

A major incentive in this process was the local rule system that essentially allowed private ownership claims over formerly communal land resources to be established by planting trees (Gouyon, de Foresta, and Levang, 1993). This ownership claim strictly applies to only the trees planted, and, for example, durian fruits in such a garden are still treated as a village-level resource. However, in practice, planting rubber trees, even with a low rate of success in tree establishment and regardless of the genetic quality, yields full control over the land, including the right to sell (Suyanto, Tomich, and Otsuka, 2001; Suyanto and Otsuka, 2001). In Sumatra's lowland peneplains, nearly all shifting cultivation has now been replaced by rubber-based agroforestry (van Noordwijk et al., 1995, 1998), but small reserves for bush fallow rotations are maintained in some villages as an option for poor farmers to grow food crops. In addition to providing cash income for the farmer, jungle rubber agroforests also provide a range of nonrubber products and other environmental benefits.

JUNGLE RUBBER AGROFORESTRY SYSTEMS IN JAMBI

At the turn of the new millennium, smallholder rubber production systems in Indonesia still spanned a wide range of intensities of management. Despite decades of government efforts, only about 15% smallholder rubber farmers have adopted the improved monoculture plantation (Ditjenbun, 1998) with selected (domesticated) tree germ plasm of higher (up to fivefold in on-station experiments) latex production per tree. A vast majority of rubber producing areas in Indonesia, located mainly in North Sumatra, Jambi, South Sumatra, Riau, and West, South, and Central Kalimantan provinces, are still in the form of jungle rubber agroforests with varying levels of dominance of native nonrubber flora. The majority of the rubber area is still in the form of complex multistrata agroforests (de Foresta and Michon, 1993, 1996).

Around 70% of farmers in Jambi province are directly involved in smallholder rubber production and derive on average nearly 70% of household income from rubber (Table 5.1). Rubber agroforests have been primarily established by slash-and-burn techniques on logged-over forest land or land under some form of secondary forest, previously used for food crop/fallow rotations.

Mostly established in the 1940s to 1960s, the existing rubber agroforests in Jambi are old with very low latex production potential (Hadi, Manurung, and Purnama, 1997), essentially still based on rubber germ plasm that came directly from Brazil and became naturalized in Indonesia, spreading by seed. Latex productivity per unit land from these jungle rubber agroforests is very low, at about 600 kg dry rubber/ha/year, less than half that of estate plantations (Wibawa et al., 1998). Returns to labor, however, are similar if land is not valued in the profitability assessment (Tomich et al., 2001) and can only be surpassed by collection of nontimber forest products (with low returns per hectare), illegal logging, or (at least before the economic crisis of 1997) oil palm production. Rubber is a major livelihood provider, but many of the rubber gardens are getting old and productivity per hectare declines. Occasionally, trees that according to the villagers are 100 years old and that survived from the earliest plantings, around 1920, close to the river, are still being tapped.

Many farmers rejuvenate their rubber agroforest only after production from the old rubber becomes very low by slash-and-burn to start a new cycle of jungle rubber system. The system is also known as cyclical rubber agroforestry system, or CRAS, (Figure 5.3), using either locally obtained rubber seedlings or improved clonal planting material. In the first year or two, farmers often plant upland food crops

Table 5.1 Household Annual Income and Expenditure Figures for Villages in the Lowland Peneplain of Jambi (Sumatra, Indonesia)

Source/Expense	Indonesian Rupiah* in 2000	Percentage of Total
Income Sources		
Rubber	4819	69
Nonrubber farm	1424	20
Off farm	768	11
Subtotal	*7011*	*100*
Expenses		
Consumption (mainly food)	4344	68
Education	46	1
Miscellaneous	2028	31
Subtotal	*6418*	*100*

Note: *1 U.S. dollar = 7500 Indonesian rupiah approximately.

Source: From Wibawa et al., *ASA Proceedings*, in press. With permission.

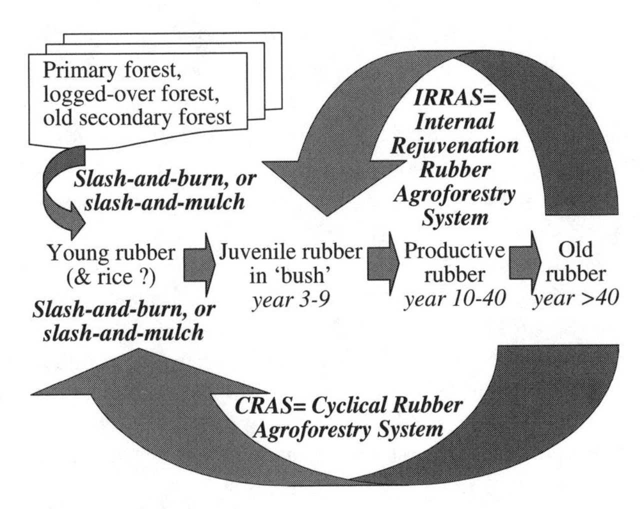

Figure 5.3 Schematic representation of cyclical and internal-rejuvenation forms of rubber agroforests.

such as rice, maize, soybean, mungbean, pineapple, or banana, while estates plant leguminous cover crops during the establishment of young plants. Small-scale rubber producers are often reluctant to rejuvenate their rubber agroforest, primarily because of the following:

- Potential loss of income during replacement/establishment of rubber trees, especially for heavily rubber dependent households
- Limited financial capital (particularly money and labor) for replacing old rubber trees with new ones, and for clonal material, high management costs (input material)
- High risk of pig and monkey damage on young rubber plants; this forms a major constraint in establishment of rubber gardens in Jambi — in up to half of the newly established plots, insufficient rubber trees survive to tappable age

Rubber trees spread by seed, and in the more extensively managed rubber gardens, spontaneous rubber seedlings are common. Some of these may grow to reach tappable size, but techniques of assisted natural regeneration are required to promote this. Recent observations in the smallholder jungle rubber system in the Jambi region in Indonesia indicate that many farmers practice a technique of rubber tree rejuvenation in order to fill in gaps or replace unproductive trees with productive rubber seedlings in rubber gardens. This is a strategy to cope with the decreased or declining productive rubber tree population without a need for the drastic slash-and-burn of the plot. Locally known as *sisipan* (literally meaning planting new plants between old plants), new rubber seedlings are transplanted over a number of years within gaps in the forest to replace dead, dying, unproductive, and unwanted trees. A permanent forest cover is maintained, but we cannot as yet expect the agroforest to be permanent; hence, the system is called the internal-rejuvenation rubber agroforestry system (IRRAS). The system can be recognized from its range of development stages and forms of rubber trees.

Thus, two methods exist for rejuvenating the stand (Figure 5.3): slash-and-burn followed by a replant, depending on natural regeneration, or the technique of *sisipan* for gap enrichment planting. The *sisipan* technique is emerging as an important, farmer-developed solution to investment constraints associated with slash-and-burn. Gap-level enrichment planting most often leads to a permanent cover rubber agroforestry in contrast to the (supposedly) more common cyclical system involving slash-and-burn.

The low-input internal-rejuvenation technique deserves full evaluation of its development prospects and environmental aspects. The sustainability of farmers' *sisipan* method and its viability as an alternative to slash-and-burn in the jungle rubber agroforestry system can be debated. These, including possible interventions that could assist in promoting this interesting technique, are discussed in this chapter. But first we will review data on the biodiversity of rubber agroforests as systems intermediate between natural forest and monocultural rubber plantation.

BIODIVERSITY ASSESSMENTS: SPECIES USED, SPECIES TOLERATED

Biodiversity in jungle rubber gardens is a result of farmers' management decisions that (implicitly) determine the structure and composition of the vegetation, providing a habitat for birds, mammals, insects, and other organisms. Weeding is usually restricted to the first few years after slash-and-burn when rice and annual crops are grown with the newly planted rubber. Thereafter, the farmer relies on the quick growth of bushy and woody vegetation to shade out harmful weeds like *Imperata cylindrica* (Bagnall-Oakeley et al., 1997). Perennial species are managed by farmers through planting and through positive and negative selection of spontaneous seedlings. Apart from rubber, a few other perennial species such as fruit trees are planted, usually in small numbers and around the temporary dwelling farmers may construct to live on site during the first year. In addition, many tree species that establish spontaneously are allowed to grow with the rubber as far as they are considered useful. Those tree species are mainly used for timber and fuelwood and for constructing fences around new rubber plots. Spontaneous seedlings of desired species, including rubber seedlings, are protected or even transplanted to a more suitable spot in the garden. Slashing or ring barking removes unwanted species. Thus, the perennial framework of jungle rubber is created by steering the secondary forest succession in addition to planting. This leads to a diversified tree stand dominated by rubber, similar to a secondary forest in structure (Gouyon, de Foresta, and Levang, 1993). In addition, there are numerous species, especially undergrowth species and epiphytes, that are not of direct use to the farmer but are not considered harmful either. They are left to grow as most farmers find that slashing of undergrowth or removal of epiphytes does not pay in terms of higher output. Keeping undergrowth may even be beneficial: rubber seedlings are hidden from pigs (the main vertebrate pests); the microclimate on the ground is kept moist and cool, which is conducive for latex flow when tapping; and (at least in the farmer's perception) soil moisture is kept, which allows for continued tapping during periods of drought.

However, benefits for the farmer result from selected species and from the vegetation structure as such, not from species richness.

BIODIVERSITY CONSERVATION: RUBBER AGROFORESTS AS LAST RESERVOIR OF LOWLAND FOREST SPECIES

Since the early 1970s, forests in the Sumatran lowlands are being rapidly transformed by large-scale logging and estate development (oil palm, trees for pulp-and-paper factories), turning the extremely species-rich lowland rainforest into large, monotonous monoculture plantations. In terms of forest biodiversity, not much can be expected from such plantations, while on the other hand strict conservation of sufficiently large areas of protected lowland rainforest has not been a realistic option in the process of rapid land-use change. The ongoing development is changing the role of rubber agroforests in the landscape: from adding anthropogenic vegetation types to the overall natural forest diversity, rubber agroforests are probably becoming the most important forest-like vegetation that we can find covering substantially large areas in the lowlands. It has become a major reservoir of forest species itself and provides connectivity between forest remnants for animals that need larger ranges than the forest remnants provide.

While ecologists are aware that jungle rubber cannot replace natural forest in terms of conservation value, the question of whether such a production system could contribute to the conservation of forest species in a generally impoverished landscape is very relevant. However, jungle rubber farmers are not interested in biodiversity in the sense that conservationists are. They make a living by selectively using species richness and ecosystem functions and base their management decisions on maximizing profitability and minimizing ecological and economical risks. Michon and de Foresta (1990) were the first to draw attention to this issue, including the need for researchers to take both the farmer's perspective and the ecologist's perspective into account. They started the discussion on complex agroforestry systems and the conservation of biological diversity in Indonesia and pleaded for "assessment of existing and potential capacity of agricultural ecosystems to preserve biological diversity."

As part of a research program on complex agroforestry systems, researchers from Orstom and Biotrop started working on biodiversity in rubber systems in the Sumatra lowlands (de Foresta and Michon, 1994). Vegetation profiles were drawn of four jungle rubber plots in the Jambi province (Kheowongsri, 1990) and one in the South Sumatra province (de Foresta, 1997), including lists of tree species and analysis of structure. In addition, a 100-m transect line was sampled for all plant species in a natural forest, a jungle rubber garden in Jambi, and a rubber plantation in South Sumatra (Figure 5.4A). Bird species (Thiollay, 1995) and soil fauna were compared between natural forest and jungle rubber, and an inventory was done to document the presence of mammal species in jungle rubber. In an overview paper presenting the results, Michon and de Foresta (1999) conclude that different groups are affected differently by human interference. Levels of soil fauna diversity are quite similar between forest and agroforest, while bird diversity in the agroforest is

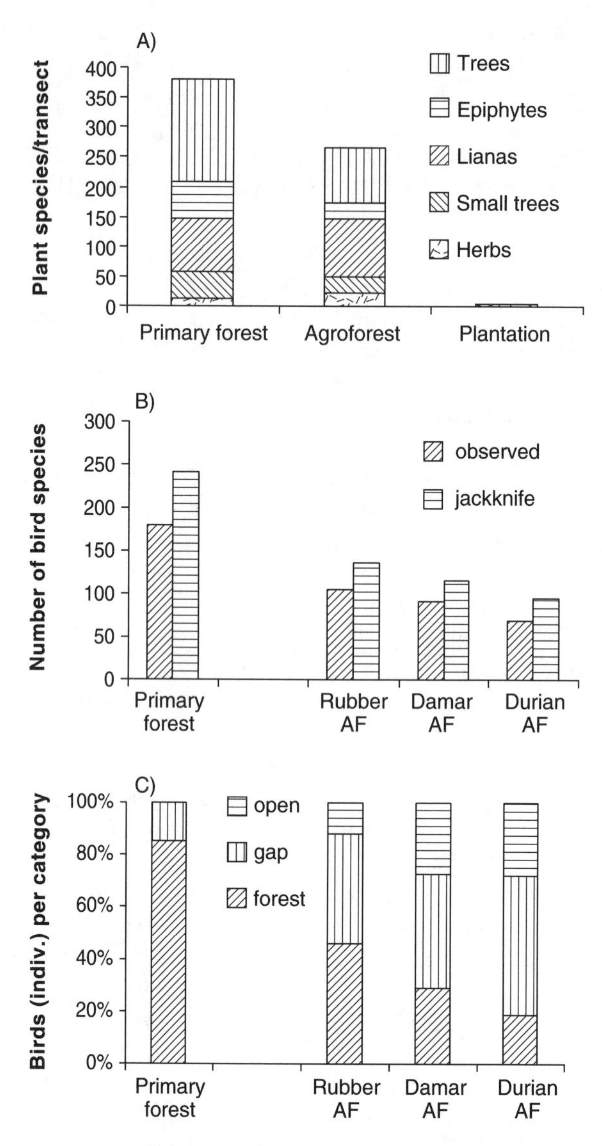

Figure 5.4 Comparisons of plot-level richness of plant and bird species between (A) natural forest, rubber agroforest, and rubber plantation for higher plants (Michon and de Foresta, 1999), (B) and (C) natural forest and three types of agroforest for birds based on total species richness (directly observed or extrapolated by the jackknife method) and contribution of forest, gap, and open land bird species. (For C, from Thiollay, J.-M., *Conserv. Biol.*, 9:335–353, 1995. With permission.)

reduced to about 60% of that in primary forest (Figure 5.4B), with a shift from typical forest birds (including ground dwellers) to birds of more open vegetation (Figure 5.4C). Danielsen and Heegaard (1994, 2000) confirmed the results of Thiollay (1995) that different groups of birds were affected differently by changes in vegetation structure, floristic richness, and the associated variety of food resources. Some groups were drastically reduced while others were thriving in agroforests.

Almost all forest mammals were found to be present in the agroforest, but population densities were not studied yet, and occasional recordings of rhinoceros or elephant do not indicate that agroforests are in themselves a suitable habitat for charismatic megafauna. For vegetation, Michon and de Foresta (1995) concluded that overall diversity is reduced to approximately 50% in the agroforest and 0.5% in plantations. These statements on relative diversity, however, apply to plot-level assessments only and cannot be extrapolated to larger scales until we have data on the scaling relations beyond the plot for forests as well as agroforests. Another multitaxa study (including plants, birds, mammals, canopy insects, and soil fauna) was reported by Gillison et al. (1999) and covered a wider range of land-use types, from forest to *Imperata* grassland, with similar results for the relative diversity of agroforest. From these studies it is clear that jungle rubber is an interesting system, potentially combining biodiversity conservation and sustainable production, but some questions remain. Apart from signaling changes in overall species richness, understanding the ecological significance of differences in species composition between forest, jungle rubber, and rubber plantations is necessary to be able to judge the value of jungle rubber for the conservation of forest species. Another problem to be solved is the problem of scale. Results from studies based on few plots or relatively small plots in a limited area cannot be safely extrapolated, because some land-use types are more repetitive in species composition than others (alpha versus beta diversity).

Studying terrestrial pteridophytes, Beukema and van Noordwijk (in press) found that average plot-level species richness was not significantly different among forest, jungle rubber, and rubber plantations; however, at the landscape level, the species-area curve for jungle rubber had a significantly higher slope parameter, indicating a higher beta diversity. When pteridophytes were grouped according to their ecological requirements, the species-area curves based on forest species alone were far apart, showing that jungle rubber supports intermediate numbers of forest species as compared to natural forest (much higher) and rubber plantations (much lower).

We can conclude from all these studies that jungle rubber is indeed diverse, but also that it is different from forest both as a habitat that has more gaps and open spaces, and in scaling relations. The percentages of forest species conserved in complex agroforestry systems such as jungle rubber are not easily estimated from the relative richness at plot level, because they depend on taxonomic or functional group and on the scale of evaluation.

Biodiversity studies in jungle rubber have been integrated with socioeconomic and agronomic studies from the beginning (Gouyon, de Foresta, and Levang, 1993). To optimally use limited research capacity, further biodiversity studies should ideally be targeted at taxonomic groups that are either of direct interest to farmers, such as timber trees and other secondary products (Hardiwinoto et al., 1999; Philippe, 2000), or that are important to ecosystem functioning (soil fauna, pollinators, seed dispersers). There is also an important role for biological research in studying effects of the secondary forest component such as competition for light and nutrients (Williams, 2000) or the ecology of vertebrate consumers of rubber seeds and seedlings (pigs) or young leaves (monkey) (Gauthier, 1998) and fungal diseases of rubber. Direct conflict between rubber farmers and top carnivores (such as the tiger) have been largely settled to the detriment of the latter ("people have eaten the tiger," as a villager expressed, where eating implies the concept of getting benefit from the sale of skins and bones).

We will now return to the question of whether productivity of rubber agroforests can be increased while conserving complexity and biodiversity values.

FACTORS INFLUENCING FARMERS' CHOICE OF RUBBER REJUVENATION METHOD

There is limited access to new land for clearance, and intensification within the existing rubber domain is the main option available. Four major classes of factors seem to be important to farmers when selecting a method for rubber agroforest rejuvenation: economic resources (including labor), forest resources, land resources, and knowledge of and confidence in the *sisipan* method (Figure 5.5). Lack of income during the rubber establishment period (5 to 8 years after planting), aversion to risk by vertebrate pests, and inability of farmers to finance large costs of clearing and rubber establishment are the primary factors promoting farmers' choice for the *sisipan* strategy. Other influencing factors include availability of new land for clearance, household income and assets, alternative sources of household income, and knowledge about the *sisipan* technique.

The *sisipan* method is not a new invention, but it has escaped researcher attention until recently. In Rantau Pandan and Muara Buat villages in Jambi, some farmers estimate that around 75% of the farmers practice *sisipan,* although most also slash and burn, but not on the same plots.

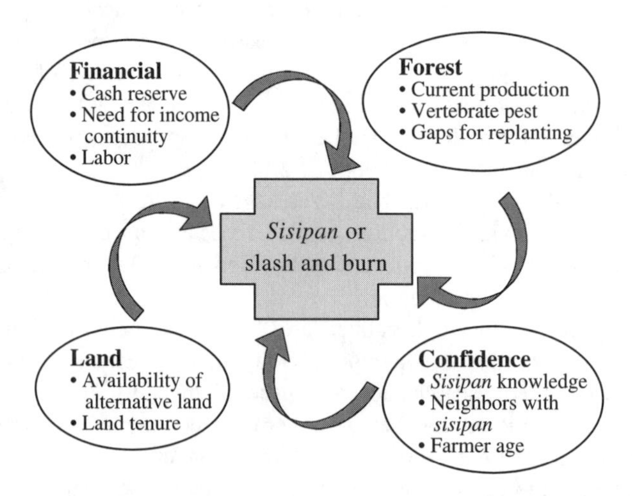

Figure 5.5 Determinants influencing farmers' choice between *sisipan* and slash-and-burn to rejuvenate rubber forests.

The practice of *sisipan* for gap rejuvenation of the jungle rubber agroforestry system is of much interest among the development professionals for the following reasons:

- It is environmentally friendly because there is no need to slash and burn the old *kebun* (garden), avoiding the smoke problems and greenhouse gas emissions of slash-and-burn fires and maintaining a higher time-averaged carbon stock (estimated difference around 20 Mg C/ha).
- It mimics the natural continuity of forest flora and fauna (biodiversity) and, by reducing the scale of management decisions from the field to the gap or tree level, provides more opportunity for maintaining valuable nonrubber trees.
- It does not require large capital investment because the work can be spread over a number of years.
- There is no break in farmer's income because the existing production trees continue to be harvested while new trees mature.
- It is a farmer-initiated strategy that is appropriate to the farmers' socioeconomic and biophysical conditions.

ECONOMIC EVALUATIONS OF IRRAS VS. CRAS

Although attractive from an environmental perspective and requiring less intensive management and less initial capital investment, the *sisipan* method has a principal drawback: its low productivity and a long establishment period for rubber trees compared to the slash-and-burn method. Latex production from farmers' jungle rubber system is around 590 kg dry rubber per hectare, while for private rubber estates it is 1065 kg/ha and for government estates it is 1310 kg/ha (Penot, 1995). Farmers in general do not use the higher yielding domesticated planting material in jungle rubber agroforests.

Economic evaluation carried out on previous survey data (Wibawa et. al., in press), however, indicated a higher profitability of the internal-rejuvenation system than that of the cyclical system primarily because the former is almost free of cost, while the second requires a substantial financial investment for establishing new trees. *Sisipan* activities are carried out while the farmer is in the plot tapping trees and are not regarded as requiring additional labor. However, both systems are economically feasible even at an interest rate of 20% (Wibawa et al., in press). Assumptions made for the analysis included 208 days of tapping in a year, at 4.2 kg dry rubber/ha/tapping day, or an average of 870 kg dry rubber/ha/year and the absence of a waiting period in the *sisipan* approach; planting of new seedlings can be spread over a number of years. Economic indicators such as net present value (NPV), internal rate of return (IRR), and cost benefit analysis (CBA) reflected the higher profitability with the internal-rejuvenation system. Even at a lower yield of 680 kg dry rubber/ha/year, the *sisipan* system was still viable while CRAS became uneconomical.

FARMER KNOWLEDGE

Research and development professionals have only recently focused on *sisipan* methods in Jambi. The practice most probably also exists in other rubber-growing provinces in Indonesia. Hence, scientific understanding of the ecological factors and processes, particularly for growth of rubber seedlings within the mature rubber system, hardly exists whereby improvements can be developed. On the other hand, farmers who are practicing *sisipan* have observed and understood the ecological processes occurring in the system. A knowledge-based systems approach (Sinclair and Walker, 1999; Walker and Sinclair, 1998) was adopted to understand farmer knowledge and perception related to rubber seedling growth in the jungle rubber agroforests. Thirty farmers in five villages (Rantau Pandan, Muara Buat, Sepunggur, Lubuk, and Muara Kuamang) in Jambi Province were consulted for investigating local knowledge. Examples of ecological relationships between various components that occur as understood and known by local farmers are reported in the following sections.

Factors Influencing Survival and Growth of Rubber Seedlings

Planting materials of rubber for *sisipan* come from the following:

- Clonal seedlings: 1-year-old seedlings uprooted from rubber plantations and usually available at local markets. These are believed to have a higher latex yield potential than the local seedlings.
- Seedlings raised by farmers in nursery from seed collected from existing forests, either local or from clone plantations. These may be raised in polybags or in seedling beds from which seedlings will be uprooted when ready, kept in running water for a few weeks before planting in the field.
- Naturally regenerated seedlings or wildlings growing in the rubber agroforest and translocated to another site without any prior treatment.

In a mature rubber agroforest, seeding and seed germination is generally not a problem where underground vegetation is not very dense. However, even in the presence of *in situ* wildlings, farmers still prefer to plant seedlings brought from outside, trusting that these seedlings have a higher genetic potential for latex production. High-yielding grafted plants are rarely used in the *sisipan* system.

Many development professionals hold the view that rubber for latex production should be grown in plantations and intensively managed. This is the general message conveyed by development agents to rubber farmers. It is now known from experimental trials that clonal rubber plants can also be successfully grown under less intensive management, *viz.* less weeding intensity and less fertilizer application (Williams, 2000). It is also possible to grow rubber seedlings inside existing stands of mature rubber trees. However, rubber is still a poor competitor when compared to other natural flora, and this is well understood by rubber farmers. Young rubber plants, whether natural or planted, require deliberate management if they are to grow into productive trees. Local knowledge and perception of gap, both at the canopy level for light infiltration and at the ground level for nutrient and moisture, is quite robust. In a jungle rubber context, gap is a concept of farmers that reflects sufficient space for seedling growth. Loosely, this is space of 6 m or more between two live rubber tree trees. Gaps can develop naturally through natural death or deliberate removal of trees and other vegetation. Farmers often create gaps by selectively killing, through ring barking, of undesired trees, including unproductive and old rubber trees, and/or debranching of existing trees to increase light infiltration through the canopy. At the ground level, light weeding is carried out to reduce competition from weeds.

On the one hand, overly large gaps or intensive removal of vegetation encourages weed dominance; on the other hand, rubber seedling growth is very slow if the size of the gap is too small. Rubber seedlings can tolerate a reasonable amount of shading in their establishment years. However, for continued growth, gradual opening of canopy and underground vegetation is essential. Hence, a gradual opening with careful monitoring of both rubber seedlings and weed growth in the gaps is essential. Insufficient light infiltration leads to increased height of rubber seedlings with little or no growth in diameter and branching. Too much light encourages weedy vegetation (Figure 5.6).

Farmers have knowledge about the importance of gap on the ground in addition to space in the canopy for survival and growth of rubber seedlings; hence the practice of spot-weeding around rubber seedlings, which is essential until these seedlings have developed a sound root system. This is particularly important with transplanted seedlings whose root systems are always drastically cut back during uprooting the plant and root trimming prior to transplanting. Seedlings grown in polybags suffer less from this transplanting stress. Farmers report that this intensity of stress depends on the intact root system and the size of the seedling. The larger the seedling size (girth), the greater the stress to the seedling and the lower its chances of survival.

Weeding and Pig Damage

Vertebrate pests, especially wild pigs (*Sus scrofa*), are the biggest constraint in the rubber production system in Jambi, surpassing other management practices (Williams et. al., 2001). Although wild pigs do not eat seedlings (some farmers perceive that pigs like to chew the sweet root collar of these seedlings), the seedlings are often uprooted and broken off when pigs dig soil in search of soil insects and rubber seed. Farmers are quite adamant that the decreasing size of natural forests has led to a decline in populations of tiger, which hunted pigs, and consequently allowed the pig population to increase. Another factor is what seems to be a change in the feeding habits of wild pigs, a shift toward rubber seeds and other agricultural crops from their fast-disappearing feed in natural forests. Village communities in Jambi are predominantly Muslim, and pigs are generally not hunted. Hence, the pig problem is of a higher severity in Jambi than in other pig-consuming provinces such as North Sumatra of West Kalimantan.

The relationship between weeding and pig damage to seedlings was well articulated by farmers. Seedlings in a clearly weeded plot are highly prone to pig damage due to increased visibility and access to seedlings as well as easier digging of soil. However, farmers also are aware that high weed biomass in their rubber gardens

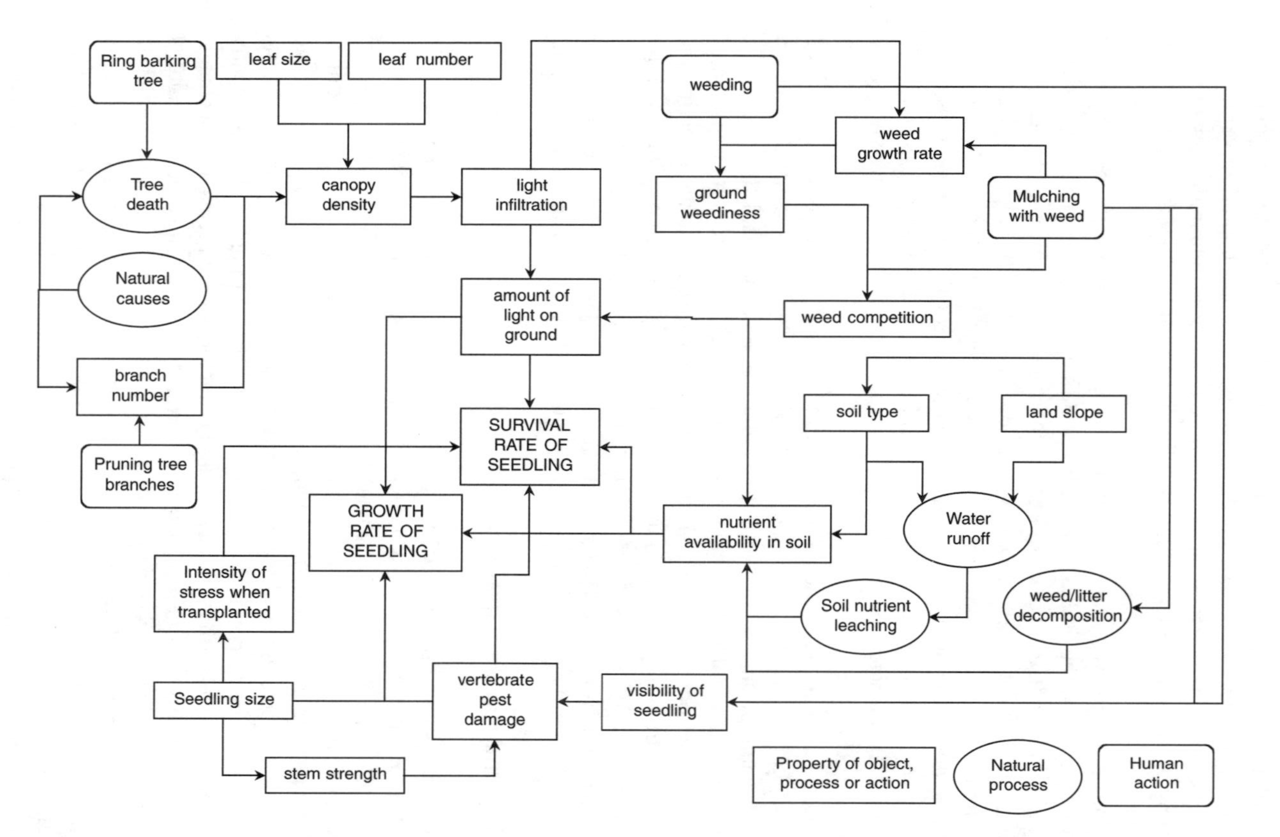

Figure 5.6 Diagrammatic representation of farmers' ecological knowledge about factors influencing seedling survival and growth in *sisipan* system. Arrows indicate source node causing an effect on target node.

provides hiding places and nests for pigs. Annual crops, normally available in the first 2 to 3 years after planting rubber, become attractions for wild animals. Farmers must stay on guard to scare away the animals during this crucial period for a reasonable chance of successful establishment of rubber plants. The conversion of sites recently cleared and planted with rubber to *alang-alang* (*Imperata cylindrica*) encroachment is not uncommon, mostly because of high seedling mortality due to damage from pig and other vertebrate pests.

One of the strategies to minimize seedling damage is to use large-sized seedlings (often over 5 cm in diameter). But this strategy has other drawbacks. Farmers are well aware that survival and growth of seedlings are lower due to damage during uprooting and preparation for transplanting. This increases transplanting stress and decreases their chances of survival.

In the *sisipan* system, farmers weed around seedlings but leave the weed litter in order to physically hide the seedlings. In addition to physical protection of seedlings, weed litter upon decomposition provides mulch, a source of soil nutrients and moisture (Figure 5.7). Farmers have observed that rubber seedlings are susceptible to weed competition in the first 3 years after planting, after which they are able to outgrow weeds and their crown is dense enough to retard further weed growth.

Local knowledge reflects ecological knowledge that farmers have acquired and put into practice. However, the dilemmas that farmers often face, such as weeding method and intensity, fertilizer application, selecting planting material, and tolerating nonrubber vegetation in the system, pose considerable but researchable constraints.

INTRODUCING GENETICALLY IMPROVED CLONAL RUBBER TO THE JUNGLE RUBBER SYSTEM

A logical approach to enhance the productivity of the jungle rubber system is to incorporate high-yielding clonal material into the system whereby both production and environmental functions can be optimized. There is a great need to improve the productivity of rubber agroforests, with moderate changes in management, if they are not to disappear from the landscape under pressure from monocultural oil palm systems that may be more risky, but that are more profitable in the short term, especially in terms of income per unit area of land (Tomich et al., 1998). Genetically improved planting material will contribute to improved yields per unit area and to closing the technology gap (Kumar and Nair, 1997) between smallholders and plantations.

This gap has developed since the 1920s, due to great advances in the plantation rubber sector, with the breeding of higher yielding, genetically improved material, and the development of the technique of grafting buds of this onto well-developed rootstock stumps to produce clones. These clones are capable of yielding two to three (or even up to five) times more than the unselected material (regenerated seedlings collected from existing agroforests) being used by smallholders in the jungle rubber system. Smallholder farmers would benefit greatly from the increased yields possible from this improved genetic material if it could be integrated into their agroforests (van Noordwijk et al., 1995).

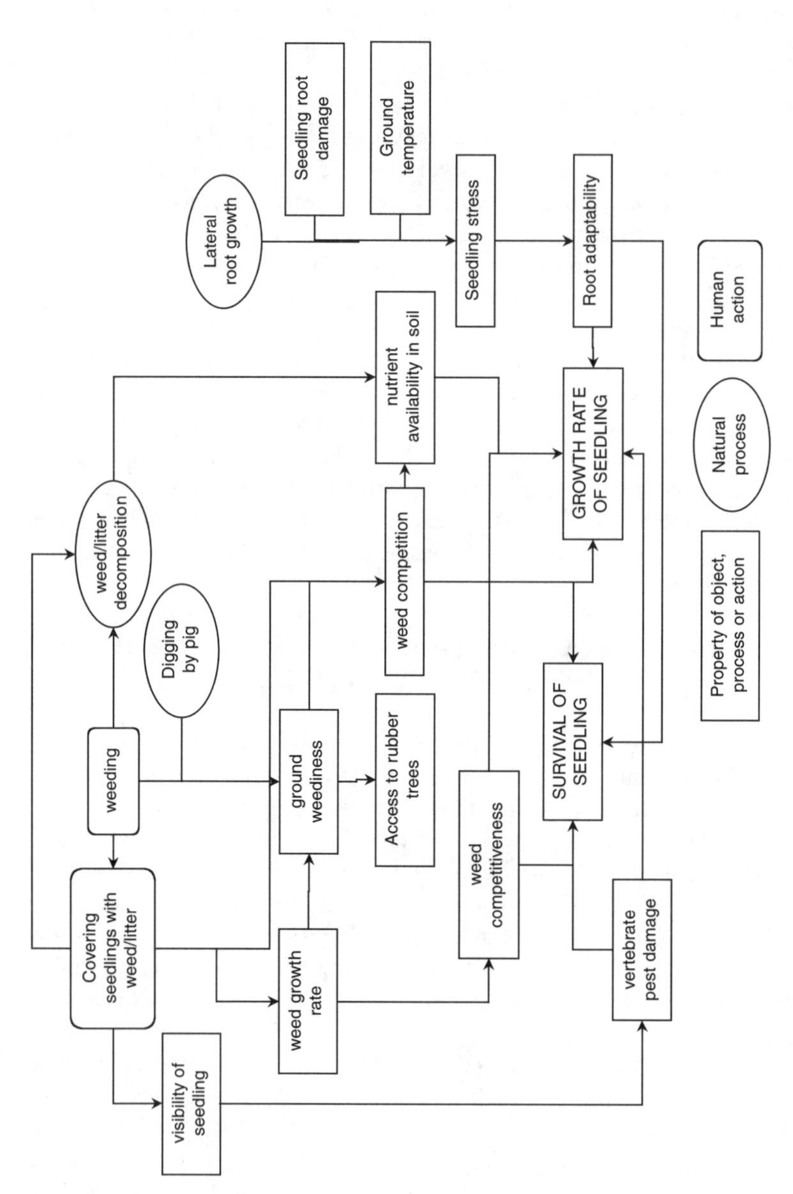

Figure 5.7 Farmers' knowledge about interaction between weed, weeding, and seedling performance in *sisipan* system. Arrows indicate source node causing an effect on target node.

Therefore, an improved rubber agroforestry system, modeled on the traditional jungle rubber system in its cyclical form, was designed to maximize the following:

- Productivity (by introducing new technology in the form of rubber clones, which give higher yields per tree and require less labor for tapping)
- Biodiversity (by keeping the spontaneous secondary forest component of the jungle rubber system — this yields local benefits to the farmer in terms of potentially harvestable products (Figure 5.1), and global benefits if agroforests function as a reservoir in areas where primary forest has been lost)
- Affordability and adoptability (by keeping management and input levels to a minimum, and so within the reach of smallholder farmers)

The system comprises rows of clonal rubber trees, planted with a spacing of 3 m within the rows, and an interrow area, 6-m wide, where secondary forest species are allowed to regenerate (Penot et al., 1994). It falls under the cyclical classification and is established at the level of an entire field, after the slashing and burning of secondary forest or old jungle rubber vegetation.

This system (RAS 1) is currently being tested within a network of on-farm trials set up by the ICRAF/CIRAD/GAPKINDO Smallholder Rubber Agroforestry Project (SRAP) in two provinces in Indonesia (Penot, 1995). Results from the establishment phase of one trial, in Jambi province, highlighted many issues relevant to the introduction of high-value planting material into complex agroforestry systems.

The objective of the experiment reported here was to test a range of low-input management practices that were designed to ensure survival and growth of the clones in a highly competitive multispecies environment. Interactions between the effects of secondary forest species and farmer management practices on the establishment of clonal rubber were studied in order to do the following:

- Assess the effect of four weeding treatments on the growth of clonal rubber
- Identify and quantify constraints to, and factors affecting, clonal rubber growth under on-farm conditions

The trial involved clonal rubber trees grown by farmers in a total of 20 plots, in five replicate experimental blocks, spread across four farms (Williams et al., 2001). The amount of labor invested in strip weeding the rubber tree rows was significantly ($p < 0.001$) and positively correlated with rubber growth (Figure 5.8). However, unexpectedly, breakage of rubber tree stems by vertebrate pests (banded leaf monkeys, *Presbytis melalophos nobilis,* and wild pigs, *Sus scrofa*) was the most important influence on establishment. Pest damage was significantly ($p < 0.001$) but negatively correlated with rubber tree size, explaining almost 70% of the variation in rubber growth in the trial (Figure 5.8). In three farms, none of the trees escaped damage. These results were confirmed in a multiple linear regression analysis, where weeding effort and pest damage in combination explained 80% of the variation in tree diameter in the experiment (Williams, 2000).

In one trial, farmers generally decided to completely cut back the diverse vegetation between rows of rubber trees, including potentially valuable trees, rather than weeding within the rows and selectively pruning trees in the interrow. Farmers

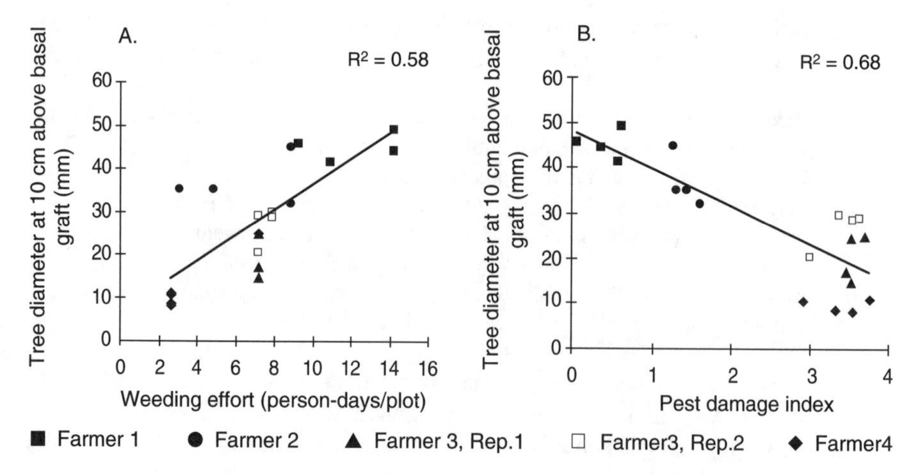

Figure 5.8 Linear regression between (A) mean rubber tree diameter per plot (21 months after planting) and weeding effort (person-days/plot), and (B) mean rubber tree diameter per plot and pest damage index*, for 20 plot means in an agroforestry trial in Jambi province, Indonesia. *The number of stem breaks sustained over 21 months for each tree, totalled for all trees in the plot, then divided by the total number of trees. (From Williams et al., 2001.)

thought that the interrow vegetation would harbor vertebrate pests and compete with the clonal rubber, and they had access to fruits, firewood, and other nontimber forest products on other land. Thus, contrary to expectations, when offered clonal germ plasm, farmers opted to use plantation methods to protect what they considered a valuable asset suited to monoculture, rather than maintain the traditional multispecies strategy they use with local germ plasm.

In other trials based on this system (RAS 1) in the province of Kalimantan, however, farmers from a different ethnic group were happy to keep the diverse secondary vegetation in the interrow area (Penot, pers. comm.). In this area there was no pressure from pests, so there was less of a risk of the trees failing to become established. Thus, different outcomes of the intensification trials were observed in different circumstances; and in the forest margins in Jambi, the major constraint to establishment of clonal rubber was not, in fact, competition from secondary forest species, but vertebrate pest damage.

Because we now understand that the resource-poor smallholders prefer *sisipan*-style management rather than cyclical rubber agroforestry, new experiments were started to test the option of grafting clonal buds directly onto *in situ* local rubber seedlings. Initial results are variable but encouraging. Further work is being carried out to explore and adapt the technology through farmer-led experimentation, combined with researcher-led on-farm trials.

CONCLUSIONS AND DIRECTIONS FOR RESEARCH

The property of combining biodiversity conservation and economic profitability is a strength of jungle rubber, but it is also a weakness because the balance between the two is easily disrupted. Jungle rubber is not a forest type, it is a plantation, even though it does not look like one. It is owned and managed by a farmer who makes rational decisions based on socioeconomic circumstances and information available. Jungle rubber is a low-input, medium-output system that depends on relatively large areas of available land to make a living for a family, even though its returns to labor are attractive when compared to other types of labor available in Indonesia.

Conversion of natural forests leads to a lack of habitat for natural populations that are probably needed to maintain the present biodiversity levels in jungle rubber (Beukema, in prep.). The extent to which current diversity levels in agroforests are a transient phenomenon, based on influx of seed and animals from surrounding forests in the past, and the extent to which they form a habitat ensuring effective reproduction and maintenance of populations, are as yet unresolved questions.

As the scaling results for ferns indicate, much of the richness of agroforests as a category is related to the diversity between plots and farms. Landscape-level diversity of rubber agroforestry systems probably depends on the diversity in management intensity. This, again, may be a transient phenomenon reflecting a past in which there was room for expansion into forest lands. Over time, a gradual process of intensification on the existing rubber land may reduce the between-farm diversity, possibly accelerated by the impact of extension of specific rubber-based technologies. On the other hand, diversification of the economic basis of rubber agroforests, with value accruing from rubberwood and other timber trees (some of them high-value but relatively slow-growing species) and fruits (with improved road access to urban markets, which would make fruits like *duku* [*Lansium domesticum*] from Jambi marketable in Jakarta), could provide an incentive for maintaining diversity in farming styles.

Overall sustainability (i.e., probability for persistence as well as continued scope for adaptation) of complex agroforestry systems should be the focus of further research to develop alternatives to mainstream intensification of tropical agroecosystems, based on the interaction of ecology and economy.

It had been assumed that jungle rubber agroforestry systems are essentially cyclical and old stands are rejuvenated through slash-and-burn methods at the start (or end) of each cycle. Research and extension activities have been designed and implemented accordingly. The significance of the farmer-developed *sisipan* or enrichment planting method is much higher than initially perceived. An increasing proportion of smallholder rubber cultivators are actively adopting the *sisipan* method, primarily due to prevailing financial constraints and decreasing new land availability. This also has a positive impact on the environment and on biodiversity and carbon stocks. Preliminary economic analysis of a pilot study indicated that the *sisipan* method is perhaps more rewarding than the slash-and-burn approach due to low capital investment at a household level, although this is likely to be different when returns are calculated per unit area of land.

Ongoing work on local knowledge indicates that farmers have a fairly good understanding of the ecology in the *sisipan* method. Because little is known in the scientific community about this approach, farmers' knowledge can play a significant role in the direction and type of research and development program. Local ecological knowledge can partly fill in the large gaps that still exist in the professional understanding about the jungle rubber system, at least within the foreseeable future. Preliminary analysis of local ecological knowledge also reveals gaps and constraints, as well as windows for improvements.

Another improvement strategy investigated through rubber agroforestry research under the SRAP project revealed the technical possibility for establishing rubber plantations under less-intensive management. However, farmers' behavior of intolerance of any weeds and other vegetation when clonal germ plasm is used reflects a constraint in their knowledge system. Studies of gap manipulation, pest management, incorporating planting material with higher latex productivity, and the economic and environmental consequences of cultivating nonrubber species could substantially increase our understanding of this promising method.

The *sisipan* method offers much potential, but much more still remains to be researched and understood despite substantial qualitative ecological knowledge among farmers. Introduced less than a century ago, rubber is relatively new, and rubber production systems, including jungle rubber agroforests, are constantly being modified as economic factors, forest and land factors, as well as social environments change. The short-term financial gain and farmers' current perception of a need for intensive management for high production may act against this potential method of *sisipan.* Although we use the term "permanent" to describe the system, the history of rubber is too short to actually evaluate the permanence of the system. Again, the continued dynamism of jungle rubber system would mean that any suggestion of permanence would remain very much in doubt.

Although very promising, the *sisipan* technique largely remains invisible to most rubber research and development professionals, and many do not believe that *sisipan* is a possible alternative to monocropping as a viable economic activity despite its widespread existence. Nonetheless, farmers practicing *sisipan* do not speak highly of this, perhaps a reflection of the predominance of monocropping technology in extension messages. Research on appropriate mixtures of rubber and nonrubber for optimum productivity and biodiversity while maintaining farmers' affordability and inherent flexibility needs to be initiated soon in order to enhance the knowledge base and confidence in the *sisipan* strategy. Enhancing the productivity of rubber production systems through *sisipan* without compromising environmental and other socioeconomic advantages remains a challenge to the research community.

ACKNOWLEDGMENTS

The initial study on internal-rejuvenation rubber agroforests was funded by ACIAR as part of the alternatives to slash and burn program; research on farmer ecological knowledge is supported by University of Wales, Bangor, U.K., under research funded by the U.K. Department for International Development. The ICRAF/CIRAD/GAP-KINDO Smallholder Rubber Agroforestry Project was funded by USAID and GAP-KINDO, and Sandy Williams' Ph.D. research was funded by a 2-year, study-abroad studentship from the Leverhulme Trust and a scholarship from the School of Agricultural and Forest Sciences, University of Wales, Bangor, U.K.

REFERENCES

Bagnall-Oakeley, H. et al., *Imperata* management strategies used in smallholder rubber-based farming systems, *Agrofor. Syst.*, 36:83–104, 1997.

Beukema, H. and van Noordwijk, Terrestrial pteridophytes as indicators of a forest-like environment in rubber production systems in the lowlands of Jambi, Sumatra, *Agriculture, Ecosystems and Environment*, in press.

Cairns, M. and Garrity, D.P., Improving shifting cultivation in Southeast Asia by building on indigenous fallow management strategies, *Agrofor. Syst.*, 47:37–48, 1999.

Danielsen, F. and Heegaard, M., The impact of logging and forest conversion on lowland forest birds and other wildlife in Seberida, Riau province, Sumatra, in *Rain Forest and Resource Management,* Sandbukt, Ø. and Wiriadinata, H., Eds., paper presented at the Proceedings of the NORINDRA seminar, Jakarta, Indonesia, Indonesian Institute of Sciences (LIPI), 1994, p. 59–60.

Danielsen, F. and Heegaard, M., The biodiversity value of "intermediate systems" of forest management as an alternative to logged forests and plantations: birds in the traditional agroforests of the Sumatran lowlands, paper presented at the Workshop on Cultivating (in) Tropical Forests: The Evolution and Sustainability of Intermediate Systems between Extractivism and Plantations, Lofoten, Norway, 2000.

de Foresta, H., Smallholder rubber plantations viewed through forest ecologist glasses — an example from south Sumatra, paper presented at Smallholder Rubber Agroforestry Project Workshop, ICRAF, Bogor, Indonesia, 1997.

de Foresta, H. and Michon, G., Creation and management of rural agroforests in Indonesia: potential applications in Africa, in *Tropical Forests, People and Food. Biocultural Interactions and Applications to Development,* Hladik, C.M. et al., Eds., UNESCO MAB Series, 13, UNESCO and Parthenon Publishing Group, 1993, p. 709–724.

de Foresta, H. and Michon, G., Orstom-Biotrop Cooperation Final Report (September 1989–June 1994), Orstom, Bogor, Indonesia, 1994.

de Foresta, H. and Michon, G., Tree improvement research for agroforestry: a note of caution, *Agrofor. Forum,* 7:8–11, 1996.

de Foresta, H. and Michon, G., The agroforest alternative to *Imperata* grasslands: when smallholder agriculture and forestry reach sustainability, *Agrofor. Syst.*, 36:105–120, 1997.

de Foresta, H. et al., *Ketika Kebun Berupa Hutan-Agroforest Khas Indonesia-Sebuah Sumbangan Masyarakat*, International Centre for Research in Agroforestry, ICRAF, Bogar, Indonesia, 2000, p. 249.

Ditjenbun, Statistik Perkebunan Indonesia 1997–1999: Karet, Departemen Kehutanan Dan Perkebunan, Jakarta, 1998.

Gauthier, R.C.T., Policies, Livelihood and Environmental Change at the Forest Margin in North Lampung, Indonesia: a Coevolutionary Analysis, Ph.D. dissertation, Wye College, University of London, U.K., 1998.

Gouyon, A., de Foresta, H., and Levang, P., Does "jungle rubber" deserve its name? An analysis of rubber agroforestry systems in southeast Sumatra, *Agrofor. Syst.,* 22:181–206, 1993.

Hadi, P.U., Manurung, V.T., and Purnama, B.M., General socio-economic features of the slash-and-burn cultivator community in North Lampung and Bungo Tebo, in *Alternatives to Slash and Burn Research in Indonesia*, van Noordwijk, M. et al., Eds., ASB-Indonesia Report 6, Bogor, Indonesia, 1997.

Hardiwinoto, S. et al., Stand structure and species composition of rubber agroforests in tropical ecosystems of Jambi, Sumatra, *Report of Joint Research between ICRAF and University Gajah Mada*, Yogyakarta, 1999.

Kheowongsri, P., Les jardins à Hévéas des contreforts orientaux de Bukit Barisan, Sumatra, Indonésie, thesis, Academie De Montpellier, Université des Sciences et Techniques du Languedoc, 1990.

Kumar, A.K.K. and Nair, P.K., Transfer of technology — Indian experience, paper presented at the ANRPC Seminar on Modernising the Rubber Smallholding Sector, Padang, Indonesia, 1997.

Michon, G. and de Foresta, H., Complex agroforestry systems and the conservation of biological diversity, agroforests in indonesia: the link between two worlds, paper presented at the Proceedings of the International Conference on Tropical Biodiversity "In Harmony with Nature," Kheong, Y.S. and Win, L.S., Eds., Kuala Lumpur, Malaysia, 1990, p. 457–473.

Michon, G. and de Foresta, H., Agroforests: pre-domestication of forest trees or true domestication of forest ecosystems, *Neth. J. Agric. Sci.,* 45:451–462, 1997.

Michon, G. and de Foresta, H., Agro-forests: incorporating a forest vision in agroforestry, in *Agroforestry in Sustainable Agricultural Systems,* Buck, L.E., Lassoie, J.P., and Fernandes, E.C.M., Eds., CRC Press, Boca Raton, FL, 1999, p. 381–406.

Penot, E., Taking the "jungle" out of the rubber, improving rubber in Indonesian agroforestry systems, *Agrofor. Today,* 7:11–13, 1995.

Penot, E., de Foresta, H., and Garrity, D., Rubber agroforestry systems: methodology for on-farm experimentation, Internal ICRAF Document, Bogor, Indonesia, 1994.

Philippe, L., draft paper, Assessment of the Potential of Agroforests to Conserve Valuable Timber Species, ICRAF, Muara Bungo, Jambi, 2000.

Ruthenberg, H., *Farming Systems in the Tropics,* 2nd ed., Oxford University Press, Oxford, 1976.

Sinclair, F.L. and Walker, D.H., A utilitarian approach to the incorporation of local knowledge in agroforestry research and extension, in *Agroforestry in Sustainable Agricultural Systems,* Buck, L.E., Lassoie, J.P., and Fernandes, E.C.M., Eds., CRC Press, Boca Raton, FL, 1999, p. 245–275.

Suyanto, S. and Otsuka, K., From deforestation to development of agroforests in customary land tenure areas of Sumatra, *Asian Econ. J.,* 15:1–17, 2001.

Suyanto, S., Tomich, T.P., and Otsuka, K., Land tenure and farm management: the case of smallholder rubber production in customary land areas of Sumatra, *Agrofor. Syst.,* 52:145–160, 2001.

Swift, M.J. and Ingram, J.S.I., Effects of Global Change on Multispecies Agroecosystems, Implementation Plan for GCTE Activity 3.4, GCTE Focus 3 Office, Wallingford, U.K., 1996, p. 56.

Thiollay, J.-M., The role of traditional agroforests in the conservation of rain forest bird diversity in Sumatra, *Conserv. Biol.,* 9:335–353, 1995.

Tomich, T. et al., Agricultural development with rainforest conservation: methods for seeking best bet alternatives to slash-and-burn, with applications to Brazil and Indonesia, *J. Agric. Econ.,* 19:159–174, 1998.

Tomich, T. et al., Agricultural intensification, deforestation, and the environment: assessing tradeoffs in Sumatra, Indonesia, in Tradeoffs or Synergies? Agricultural Intensification, Economic Development and the Environment, CAB-International, Lee, D.R. and Barrett, C.B., Eds., Wallingford, U.K., 2001, p. 221–244.

Trenbath, B.R., The use of mathematical models in the development of shifting cultivation, in *Mineral Nutrients in Tropical Forest and Savanna Ecosystems*, Proctor, J., Ed., Blackwell, Oxford, 1989, p. 353–369.

Vandermeer, J. et al., Global change and multi-species agroecosystems: concepts and issues, *Agric. Ecosyst. Environ.,* 67:1–22, 1998.

van Gelder, A., Bevolkingsrubbercultuur, in *De Landbouw in de Indische Archipel,* van Hall, C.J.J. and van de Koppel, C., Eds., W. van Hoeve's Gravenhage, the Netherlands, 1950, p. 427–475.

van Noordwijk, M., Productivity of intensified crop fallow rotations in the Trenbath model, *Agrofor. Syst.,* 47:223–237, 1999.

van Noordwijk, M. et al., Alternatives to Slash-and-Burn in Indonesia, Summary Report of Phase 1, ASB-Indonesia Report 4, ICRAF, Bogor, Indonesia, 1995.

van Noordwijk, M. et al., Sustainable food-crop based production systems, as alternative to *Imperata* grasslands? *Agrofor. Syst.*, 36:55–82, 1997.

van Noordwijk, M. et al., Forest soils under alternatives to slash-and-burn agriculture in Sumatra, Indonesia, in *Soils of Tropical Forest Ecosystems: Characteristics, Ecology and Management,* Schulte, A. and Ruhiyat, D., Eds., Springer-Verlag, Berlin, 1998, p. 175–185.

Walker, D.H. and Sinclair, F.L., Acquiring qualitative knowledge about complex agro-ecosystems. 2. formal representation, *Agric. Syst.,* 56:365–386, 1998.

Webster, C.C. and Baulkwill, W.J., Eds., *Rubber,* Longman, Singapore, 1989.

Wibawa, G., Hendratno, S., and Anwar, C., Study mekanisme persaingan untuk penyusunan paket teknologi pola tanam berbasis karet menurut pendekatan ekoregional, paper presented in Sembawa Research Station Annual Meeting, 1998.

Wibawa, G., Hendratno, S., and van Noordwijk, M., Permanent smallholder rubber agroforestry systems in Sumatra, Indonesia: environmental benefits and reasons for smallholder interest, Sanchez, P. et al., Eds., American Society of Agronomy, Madison, WI, 2002.

Wiersum, K.F., Indigenous exploitation and management of tropical forest resources: an evolutionary continuum in forest-people interactions, *Agric. Ecosyst. Environ.,* 63:1–16, 1997a.

Wiersum, K.F., From natural forest to tree crops, co-domestication of forests and tree species, an overview, *Neth. J. Agric. Sci.,* 45:425–438, 1997b.

Williams, S.E., Interactions between Components of Rubber Agroforestry Systems in Indonesia, Ph.D. thesis, University of Wales, Bangor, U.K., 2000.

Williams, S.E. et al., On-farm evaluation of the establishment of clonal planting stock in multistrata rubber agroforests, *Agrofor. Syst.,* 53(2):227–237, 2001.

Williams, S.E., Gillison, A.N., and van Noordwijk, M., Boodiversity: issues relevant to integrated natural resource management in the humid tropics. ASB LN 5, in: Towards integrated natural resource management in forest margins of the humid tropics local action and global concerns, van Noordwijk, M., Williams, S.E., and Verbist, B., Eds., ASB-Lecture Notes 1–12. International Centre for Research in Agroforestry (ICRAF), Bogor, Indonesia, 2001. Also available from: http://www.icraf.cgiar.org/ sea/Training/Materials/ASB-TM/ASB-ICRAFSEA-LN.htm

CHAPTER 6

The Coffee Agroecosystem in the Neotropics: Combining Ecological and Economic Goals

Ivette Perfecto and Inge Armbrecht

CONTENTS

0-8493-1581-6/03/$0.00+$1.50

THE SOCIOPOLITICAL AND ECONOMIC LANDSCAPE OF COFFEE

In Latin America, a region of rich and diverse natural resources and intensifying anthropogenic pressures upon them, policy makers, economists, and conservationists struggle to balance economic development with environmental conservation. The interest in combining conservation and development has resulted in more attention being paid to managed agroecosystems, in particular those that incorporate high levels of planned biodiversity (Vandermeer and Perfecto, 1997). Among the agroecosystems that have received considerable attention recently is the coffee agroforest. It has been argued that coffee production in Latin America, if managed with a diverse canopy of shade trees, presents the opportunity to generate economic benefits, conserve biodiversity, and enhance the livelihood of small producers (Perfecto et al., 1996; Rice and Ward, 1996). This chapter examines the agroecology of the shade coffee agroecosystem, focusing on its biodiversity and the potential that this system presents for combining economic and conservation goals in Latin America.

Economic Importance of Coffee

Coffee, along with petroleum and cotton, is one of the world's most traded commodities (McLean, 1997; International Coffee Council, 2001). Approximately 34% of the world's coffee production and 30% of the world's coffee area is based in northern Latin America, an area that extends from Mexico to Colombia and includes the Caribbean (Rice, 1999). As early as the mid-1800s, coffee had been economically linked to the countries of the region, becoming one of their main export crops. Until the mid-1980s, when production declined due to the civil war and adverse policies, coffee accounted for more than 50% of total exports in El Salvador (Consejo Salvadoreño del Café, 1997). In Mexico over the past few decades, coffee has become one of the most important exports, generating 36% of the agricultural export value (Nolasco, 1985; Nestel, 1995); and in Peru, coffee is the single most important export crop in terms of value (Greenberg and Rice, 2000). Furthermore, the coffee produced in this region belongs to varieties of *Coffea arabica*, which produces a higher-quality coffee and demands higher prices in the international market than varieties *of C. robusta* grown in Brazil and in lower elevations in the region. In Colombia, coffee constitutes around 66% of permanent crops in the country (Rice and Ward, 1996) and traditionally has been the dominant agricultural activity of the country, with 20% of the value of agricultural production (Sanint, 1994).

The 2001 Coffee Crisis

The economic importance of coffee for northern Latin America transcends figures of export value. The great demand for labor that is generated from this commodity ensures that a large sector of the agricultural labor force is involved in coffee (Rice and Ward, 1996). Until the most recent coffee crisis, this crop was an important and reliable source of income for many small producers in Latin America. This began to change with a remarkable drop in price as a consequence of overproduction on a global scale. By the end of 2001, coffee prices had reached a 30-year low. In just 4 years, from 1997 to 2001, coffee prices went from $3.00/lb to $0.42/lb (De Palma, 2001), causing widespread poverty, desperation, and conversion of coffee farms to other types of agriculture. By the harvest season of 2001, many coffee producers were abandoning their farms, setting up shanty towns near large cities, waiting in line for food handouts, and in the case of Mexican and Central American producers, trying to make their way north to find jobs in the U.S. (Oxfam, 2002).

To a large degree, the coffee crisis stems from an excess of coffee production. In the past 5 years, coffee demands have remained constant, but in the same time period production has increased by nearly 7%. Much of the overproduction stems from a general intensification of coffee production over the past 30 years. Though coffee is traditionally grown as an understory crop under a diverse shade canopy, many producers have opted for higher-yielding varieties that are grown on farms with little or no shade and high chemical input levels, largely to boost per-farm productivity. As a result, coffee yields in Central America were at an all-time high. Furthermore, increases in coffee production in Vietnam flooded the world market with cheap coffee. In the 1990s, Vietnam was producing little coffee, but then a massive project funded by the World Bank and the Asian Development Fund promoted intensive coffee production. By 2001, production levels had skyrocketed, placing Vietnam in second place among world coffee-producing countries, second only to Brazil (Oxfam, 2002).

The consequences of the coffee crisis are manifold. Rural poverty and unemployment have increased astronomically in coffee-growing regions, and coffee farmers and workers in many areas are faced with poverty and hunger. Reports from Guatemala claimed 40% rural unemployment in 2001; in Nicaragua, thousands of jobless workers set up camps along the highways, begging for food (Jordan, 2001; González, 2001), and in Colombia, more than 2 million people were displaced from several regions including the coffee-growing regions (Human Rights Watch, 2001).* Furthermore, many small coffee producers chose to either abandon their largely shade-grown coffee farms or convert them to subsistence crops or cattle pasture. In South America, many farmers turned to growing more lucrative crops such as coca (Wilson, 2001). By 2001, in Peru, 10,000 of the 180,000 small coffee producers had already converted to coca production (Human Rights Watch, 2001).

The environmental and political ramifications of such land conversions are many. It is within this sociopolitical and economic landscape that we discuss the agroecology of coffee production in northern Latin America and explore the possibility of

* These problems, although not a direct result of the coffee crisis, have been accentuated by the crisis.

combining economic goals with conservation and social justice goals for coffee producers in the region.

Ecological Importance of Coffee

Globally, coffee is cultivated on 26,000 square miles, which is equivalent to a strip 1 mile wide around the equator. In northern Latin America, coffee farms cover 3.1 million hectares of land (FAO, 1997). However, the ecological importance of coffee is not as much with the extension of land that its covers, but on the particular locations where coffee is grown. In Latin America, coffee is important in countries that have been identified as megadiverse, such as Colombia, Brazil, and Mexico (Mittermeier et al., 1998). *Coffea arabica* is grown primarily in mid-elevation mountain ranges and volcanic slopes where deforestation has been particularly high. The northern Latin American region has three of the five countries with the highest rates of deforestation in the world (FAO, 2001). In some countries of the region, traditional coffee plantations are among the few remaining forested areas, especially in the medium-to-high elevation ranges. An extreme example of the ecological importance of coffee can be found in El Salvador, one of the most deforested countries of this hemisphere. El Salvador has lost more than 90% of its original forests; however, 92% of its coffee is shade grown (Rice and Ward, 1996). Shaded coffee has been estimated to represent about 80% of the nation's remaining forested areas (Panayotou, Faris, and Restrepo, 1997; Monro et al., 2001). High levels of biodiversity and endemism also characterize tropical mid-elevation areas. In Mexico, the main coffee-growing areas coincide with areas designated by the national biodiversity agency (CONABIO) as priority areas for conservation because of the high numbers of endemic species they contain (Moguel and Toledo, 1999).

BIODIVERSITY CONSERVATION IN SHADE COFFEE

Coffee is produced under a wide range of cultivation technologies. However, the traditional and, until the late 1970s, most common way of producing coffee was under the diverse canopy of shade trees (Perfecto et al., 1996). In some cases, farmers would cut the original vegetation and establish agroforestry systems of shade trees, fruit and timber trees, and coffee shrubs. But the most traditional way of establishing a coffee plantation was by removing the understory of a forest, leaving most of the original trees intact, and replacing the understory with coffee plants (Perfecto et al., 1996; Moguel and Toledo, 1999) (Figure 6.1). This rustic coffee represents an agroforestry system that maintains many of the environmental functions of an undisturbed forest (Rice, 1990; Fournier, 1995; Perfecto et al., 1996; Moguel and Toledo, 1999). Other management systems consist of planted shade trees with varying degrees of floristic diversity, height, and density of shade trees (Figure 6.2). The most technified plantations are coffee monocultures, also called sun coffee (Figure 6.3), where newer varieties of coffee replace the older varieties and agrochemicals are used to replace the functions of shade trees such as weed suppression and nitrogen fixation.

Figure 6.1 A rustic coffee plantation in Chiapas, Mexico.

In recent years, conservationists have focused their attention on shaded coffee as an agroecosystem where biodiversity can be conserved (Perfecto et al., 1996; Moguel and Toledo, 1999; Botero and Baker, 2001). This interest arises from many studies conducted over the past 20 years that demonstrate that shaded coffee plantations contain high levels of biodiversity, sometimes comparable to those in forests. These studies have also demonstrated the significant ecological role of shaded coffee in the region. From their erosion-suppression qualities (Rice, 1990), to their importance as habitat and refuge for biodiversity (Perfecto et al., 1996; Moguel and Toledo, 1999) and for carbon sequestration (Fournier, 1995; Márquez-Barrientos, 1997; DeJong et al., 1995, 1997), shaded coffee, and in particular rustic coffee, has been demonstrated to behave in a similar fashion to natural forests.

Birds and Other Vertebrates

Regional large-scale and detailed local surveys of birds in the Caribbean, Mexico, Central America, and northern South America revealed that coffee plantations support high diversity and densities of birds, and in particular some species that depend on closed canopy forest (Aguilar-Ortíz, 1982; Robbins et al., 1992; Wunderle and Wide, 1993; Vennini, 1994; Wunderle and Latta, 1994, 1996; Greenberg, Bichier, and Sterling, 1997b; Johnson, 2000). Coffee plantations have also been cited as an important habitat for migratory birds, which can be found in coffee agroforests in higher densities than in natural forests (Borrero, 1986; Greenberg, Bichier, and

Figure 6.2 A coffee plantation with a shade canopy dominated by *Inga* sp. in Chiapas, Mexico.

Figure 6.3 A sun (unshaded) coffee plantation in Chiapas, Mexico.

Sterling, 1997b). Shade coffee plantations may serve as dry-season refugia for birds at a time when energetic demands are high and other habitats are food poor (Wunderle and Latta, 1994, Johnson, 2000). Certain tree species that are used as shade trees can provide important nectar and insect resources to birds in coffee plantations. For example, it has been documented that trees in the genus *Inga* support large numbers of arthropods and that birds tend to be in higher abundances in areas dominated by this shade tree (Johnson, 2000). Wunderle and Latta (1998) also described how birds in Dominican coffee plantations dominated by *Inga* and *Citrus* spp. foraged primarily in the shade tree canopy. *Inga* also provides abundant nectar resource for nectivores (Koptur, 1994; Celedonio-Hurtado, Aluja, and Liedo, 1995; Greenberg et al., 1997a).

A large percentage of the birds found in coffee plantations are canopy omnivores and partial nectivores (Wunderle and Latta, 1996; Greenberg, Bichier, and Sterling, 1997b). Although some studies have found similar levels of bird species richness in shaded plantations when compared to adjacent forests (Aguilar-Ortíz, 1982; Corredor, 1989; Greenberg, Bichier, and Sterling, 1997b; Dietsch, 2000), the species composition tends to be different. According to Greenberg et al. (1997b), many forest-edge and second-growth species contribute significantly to the high diversity of birds in coffee plantations. Being more generalists than residents, migrants seem to fare better in coffee plantations. Forest residents that have very specific foraging and nesting requirements may be more affected by the habitat modifications that take place even in rustic plantations. In addition, larger resident birds may be more susceptible to hunting pressure in coffee plantations than in isolated large tracts of forests. However, in areas where forests have been highly fragmented or depleted, coffee agroforests seem to offer an adequate habitat for the conservation of many bird species. It is because of this high potential that the Smithsonian Migratory Bird Center, as well as many conservation organizations, has taken special interest in the conservation of shade coffee plantations in northern Latin America, especially along the main migration routes.

Other vertebrates have not received as much attention as birds from the scientific community, and therefore many of the accounts are anecdotal. However, the few studies that have been published suggest that shaded plantations, especially the rustic systems (which preserve most of the canopy species from the original forest), support a diverse medium- and small-sized mammalian fauna (Gallina, Mandujano, and González-Romero, 1992, 1996; Estrada, Coates-Estrada, and Merrit, 1993; Estrada, 1994). Estrada, Coates-Estrada, and Merrit (1993) reported a high diversity and abundance of bats in shaded coffee as compared to other agricultural habitats. The majority of the bats found in coffee plantations are partially frugivores and nectivores, deriving most of their diet from the fruits and flowers produced by shade trees (Estrada et al., 1993). Likewise, nonflying mammals have been reported to be richer in species and biomass in coffee plantations than in other agricultural habitats (Estrada, 1994; Gallina, Mandujano, and González-Romero, 1992, 1996; Horváth, March, and Wolf, unpubl. data). Nonflying mammals are primarily omnivores, but Gallina et al. (1992) reported that some specialized mammals, such as small cats and otters, have been observed in coffee agroforests in Veracruz, Mexico.

There are also accounts of regular observations of howler monkeys in the same region (Estrada, 1994). Although no large mammals such as deer and large cats have been officially recorded in coffee, some rare and threatened species such as the chupamiel (*Tamandua mexicana*), the nutria (*Lutra longicaudis*), and the vizt-lacuache (*Coenduc mexicanus*) can be observed in diverse coffee agroforests (Moguel and Toledo, 1999). The diversity and richness of small- and medium-sized mammals have been found to be associated with horizontal plant diversity and vertical foliage diversity (Estrada, 1994), as well as with the vegetation structure of coffee plantations (Gallina et al., 1992).

A limiting factor for mammals in coffee agroforests could be the availability of food (seeds, fruits, insects) throughout the year, which suggests that shaded plantations dominated by one or a few shade tree species might not be sufficiently diverse to provide the ample and continuous food resources needed to maintain a diverse mammalian community. Although most studies have found coffee agroforests to fare better than other agricultural habitats with respect to mammals, they have also found lower mammal diversity in coffee agroforests than in closed-canopy forests (Estrada et al., 1993; Estrada, 1994; Horváth, March, and Wolf, unpubl. data). However, in a study comparing a forest fragment with coffee plantations under different shade levels, Witt (2001) reported higher species richness and densities of small rodents in the agroforests than in the forest fragment. This study reported a total of three small rodent species in the forest fragment and five in the more diverse coffee plantations, which included the three found in the forests (Witt, 2001). This study suggests that, in the absence of a large reserve or continuous original forests, which is the case in most of the midelevation regions in northern Latin America where coffee is grown, coffee agroforests could provide a matrix of suitable habitat for medium- and small-sized mammals, if not for permanent colonization, at least as a safe travel route from one forest fragment to another (Witt, 2001).

Studies documenting populations of amphibians and reptiles in coffee agroforests are even scarcer than those for mammals, and results are contradictory. Although Lenart et al. (1997) documented that all five species of *Norops* (formerly *Anolis*) lizards reported locally in a region of the Dominican Republic were also found in three-tiered coffee plantations, Seib (1986) and Rendón-Rojas (1994) documented much lower numbers of reptiles and amphibians in coffee plantations than in natural forests. Komar and Domínguez (2001) sampled 24 coffee plantations in El Salvador but did not find enough amphibians and reptiles to quantify the potential benefits of certifying high-shade plantations for these groups. Seib (1986) reported that mixed shade plantations supported approximately 50% of snakes found in the original forest in Guatemala, and Rendón-Rojas (1994) reported only 16 species of reptiles (11) and amphibians (5) in coffee plantations in the state of Oaxaca, Mexico, compared to 77 and 94 species reported for undisturbed forests in Los Tuxtla (Pérez-Higereda et al., 1987) and the Lacandon forest (Lazcano-Barrero et al., 1992), respectively. Unfortunately, none of these studies involved extensive surveys comparable to those that have been undertaken in forest reserves, and therefore it is hard to draw conclusions about the role of agroforests in maintaining populations of reptiles, amphibians, and mammals.

Arthropods

In one of the earliest studies of arthropod diversity in coffee plantations, Morón and López-Méndez (1985) reported a total of 27,000 individuals of ground scavengers representing 78 families in a mixed shaded coffee plantation in Chiapas, Mexico. Ibarra-Núñez (1990) also reported a high abundance and diversity of arthropods on coffee bushes in the same plantation: almost 40,000 individuals representing 258 families and 609 morphospecies, with the Diptera (22%), Hymenoptera (21.8%), Coleoptera (13.3%), Homoptera (11.5%), and spiders (10.7%) being the most diverse taxa. A more detailed analysis of three families of web spiders yielded a total of 87 species, with 6 genera and 32 species representing new records for Chiapas and 3 genera and 11 species creating new records for Mexico (Ibarra-Núñez and Garcia Ballinas, 1998). Species richness in this plantation registered 31% of that reported for the entire state of Chiapas and 14% of that reported for all the country.

The potential of shaded coffee plantations to harbor high arthropod biodiversity was highlighted by the study of Perfecto et al. (1997) in Heredia, Costa Rica. Using the same methodology pioneered by Erwin and Scott (1980), Perfecto and colleagues fogged the canopy of shade trees in a traditional coffee plantation. In the canopy of a single *Erythrina poeppigeana*, they recorded 30 species of ants, 103 species of other Hymenoptera, and 126 species of beetles. In a second tree in the same plantation, they recorded 27 species of ants, 67 of other hymenopterans, and 110 species of beetles. Furthermore, the overlap of species between these two trees was only 14% for the beetles and 18% for the ants. This level of species richness is within the same order of magnitude as those reported for canopy arthropods in tropical forests (Erwin and Scott, 1980; Adis et al., 1984; Wilson 1987).

Other studies have also found the diversity of arthropods in coffee plantations to be similar to that of adjacent forests. For example, in Colombia, studies comparing soil arthropods (Sadeghian, 2000) in general and coprophagous beetles (Scarabinae) in particular (Molina, 2000) concluded that the two most diverse habitats were the forest and the shaded coffee plantation. Similarly, in a study with fruit-feeding butterflies in Chiapas, Mexico, Mas (1999) found no significant differences in species richness between a forest fragment and an adjacent rustic coffee plantation. Estrada et al. (1998) used rarefaction analysis and sampled different agricultural habitats and native forests to conclude that the forest had the highest diversity of dung beetles but that a cacao/coffee mixed shade plantation was the next most diverse habitat.

Although these studies underscore the importance of the shade coffee agroecosystem in the conservation of arthropod diversity, a few studies have reported significant differences in species composition and richness between native forest and coffee plantations. In a study in Las Cruces, Costa Rica, Ricketts et al. (2001) found a decline in species richness as well as in the number of unique species of moths between a forest reserve and both shade and sun coffee plantations. They concluded that distance from the forest rather than habitat type was the most important factor determining moth species richness. It is important to point out that the shaded coffee plantations that were sampled in this study were monospecific stands of shade trees of either *Erythrina* sp. or *Inga* sp. and therefore represent the less diverse side of the coffee management spectrum.

THE COFFEE TECHNIFICATION PROCESS

The loss of forest cover in Latin America, a genuine ecological crisis, is in part due to agrodeforestation in the coffee sector (Perfecto et al., 1996). Attempts to modernize coffee plantations in Latin America started in the 1950s (Rice, 1990), but it was not until the arrival in Brazil of the coffee leaf rust (*Hemeleia vastatrix*) in 1970 that the so-called technification programs really took hold. Countries in Central America and the Caribbean, encouraged by more than $81 million from USAID, began to implement programs aimed at converting coffee production from the low-input, low-intensity, and low-productivity shaded system to the highly technified unshaded system (Rice and Ward, 1996).

A recent study suggests that approximately 67% of all coffee production in northern Latin America has been affected by the technification trend in one way or another (Rice, 1999). Countries differ in the degree of coffee technification, ranging from less than 20% in El Salvador and Venezuela, to up to 69% in Colombia. But technification pressures persist in most countries, and unless better alternatives are offered to producers, this process may eventually eliminate most shaded plantations from the Latin American landscape, perhaps with dramatic social and environmental consequences for the region.*

The technification process includes a reduction or elimination of most planned biodiversity (i.e., the species that are intentionally incorporated into the agroecosystem). In the shaded coffee plantations, the planned biodiversity includes coffee plus all the shade, fruit, and timber trees. The most extreme technification results in the complete elimination of all trees except for the coffee bushes, essentially creating a monoculture (also called sun coffee or unshaded coffee) (Figure 6.3). However, this is only one component of the technification process, which frequently involves planting high-yielding varieties of coffee at a higher density, plus the application of agrochemicals such as fertilizers, herbicides, fungicides, and insecticides.

CONSEQUENCES OF TECHNIFICATION FOR BIODIVERSITY

The reduction or elimination of shade trees can have a devastating effect on biodiversity. Studies comparing sun coffee with shade coffee or with coffee with different levels of shade have shown that the technification of this agroecosystem results in a loss of biodiversity for most organisms.

Impact of Coffee Technification on Birds

The possibility that deforestation in the American tropics was responsible for the decline of several species of neotropical migratory birds (Askins, Lynch, and

* The most recent coffee crisis had surprising consequences. In the early 1990s when prices fell in the international market, large producers simply let their farms idle for awhile, awaiting better times (Perfecto, pers. obs. in Costa Rica). The small producers who had diverse farms with many fruit trees were able to gain some income from the noncoffee harvest from their farms. However, this recent crisis has resulted in coffee producers opting out of coffee altogether and transforming their plantations to other land uses such as cattle or corn milpas (Perfecto, pers. obs. in Mexico; Armbrecht, pers. obs. in Colombia).

Greenberg, 1990) focused attention on the coffee agroecosystem. Given that coffee agroforests had been recognized as important habitat and refugia for both migrant and resident birds (see previous section), the conversion of these diverse plantations to sun coffee could have detrimental effects on bird conservation. However, few studies have examined these effects on birds. In one of the earliest published studies, Borrero (1986) documented a dramatic decline in bird diversity in sun plantations in Colombia. In the Dominican Republic, Wunderle and Latta (1996) documented a shift from forest to shrubby second growth bird species when comparing monogeneric shade plantations with sun coffee. However, in Guatemala, Greenberg et al. (1997a) documented relatively low bird diversity in shaded coffee plantations dominated by either *Inga* or *Gliricidia* species, and both of these types of plantations had similar species richness as the sun plantation, although the *Inga* coffee plantation had slightly higher species richness than the others. Comparing bird species richness found in this study with a previous study in a traditional farm in Chiapas (Greenberg et al., 1997b), the authors estimated that sun coffee plantations support approximately half of the species diversity and density that traditional plantations do and suggested that coffee could only be important for bird conservation if a tall, taxonomically and structurally diverse canopy is maintained. Along those lines, Komar and Domínguez (2001) sampled 24 plantations with varying degrees of shade and structural diversity in El Salvador and found that 16 species of residents were negatively affected by intensification. Analyzing resident species in more detail, they reported that of 13 measured habitat variables, shade cover was the one that better predicted species richness and abundance of resident species. Based on these results, they developed a model that established 44% shade cover and 15 species of shade trees per 0.5 hectares as a threshold for the conservation of species that are sensible to perturbation. These results deviate somewhat from what is required by the Bird Friendly® and the Eco-OK™ coffee certification programs (discussed below) — 40% shade cover and 10 species of shade trees per hectare. With respect to the decline of bird diversity and abundance along the intensification gradient, the density of emergent trees (>5 m above the canopy) also appears to be important for resident species (Greenberg et al., 1997a; Komar and Domínguez, 2001).

Neotropical migrants do not seem to be as affected by coffee intensification as residents. Since they are largely omnivores and have more generalized habitat requirements than most resident birds, vegetation changes associated with the intensification of coffee are less likely to affect them, especially when the transformation does not imply a complete removal of the canopy. Dietsch and Mas (2001) found that resident birds have a stronger forest association than migrants in Chiapas, Mexico, and that rustic coffee plantations provide the strongest conservation benefit for forest-associated birds. The most likely candidates to be affected by technification are the largely nectivorous Baltimore oriole, the Tennessee warbler, and the Cape May warbler, all three of which have experienced sharp population declines since 1980 when the technification process intensified (Perfecto et al., 1996).

Arthropods and Coffee Technification

Among arthropods, generalist ground-foraging ants have received considerable attention because they are easy to sample and occupy the same habitat in forests, shaded plantations, and sun plantations. Most studies with this group show a significant decrease in diversity along an intensification gradient. These studies are summarized in Table 6.1.

Perfecto and Vandermeer (1994) reported a 39% decline in ant species richness when comparing a traditional coffee plantation with a plantation with only *E. poeppigeana* as shade, and further 65% decline when comparing the monospecific shaded coffee plantation with a sun plantation. Perfecto and Snelling (1995) also

Table 6.1 Studies that Compare Ant Species Richness in Coffee Plantations with Different Levels of Intensification

Country	Group/Theme	Intensification Effect	Reference
Colombia	Leaf litter ants	Yes	Armbrecht and Perfecto, 2002
Colombia	Ground ants	Yes	Sadeghian, 2000
Colombia	Leaf litter ants	Yes	Sossa and Fernández, 2000[a]
Costa Rica	Ground ants (baits)	Yes	Benitez and Perfecto, 1990
Costa Rica	Ground ants	Yes	Perfecto and Snelling, 1995
	Ants foraging in coffee (baits)	No	
Costa Rica	Ground ants (baits)	Yes	Perfecto and Vandermeer, 1994
Costa Rica	Competitive relations	Yes	Perfecto and Vandermeer, 1996
Costa Rica	Arboreal ants (canopy fogging)	Yes	Perfecto et al., 1997
Mexico	Ground ants nesting in twigs	Yes	Armbrecht and Perfecto, in press
Mexico	Ants in coffee plants (D-vac)	Yes	Ibarra-Núñez et al., 1995[b]
Mexico	Foraging dynamics	Yes	Nestel and Dickschen, 1990[c]
Mexico	Ground ants (baits)	Yes	Perfecto and Vandermeer, 2002
Mexico	Ground ants (baits and pitfall traps)	No	Ramos, 2001
Panama	Army ants (*Eciton* and *Labidus*)	Yes	Roberts et al., 2000
Puerto Rico	Direct observations	Yes	Torres, 1984[d]

[a] Although the abundance of ants was lower in the shaded plantations, species richness is higher than in the unshaded plantations.

[b] This study found a higher species richness in an organic coffee plantation with higher shade than in a technified conventional farm dominated by *Inga*. However, the difference is not tested statistically.

[c] This study shows a much higher ant foraging activity in the sun coffee plantation as compared to the shaded plantations, mainly due to the dominance of *Solenopsis geminata* in plantations with higher sun exposure.

[d] Table 1 of this study shows similar ant richness in coffee plantations and forest plots.

reported that ground-foraging ants were positively and significantly correlated with the floristic and structural diversity of coffee farms along an intensification gradient. The only exception to this pattern was reported by Ramos (2001) for ground-dwelling ants in Mexico. In this study, no significant differences were found when comparing ant diversity in forests, multispecies shaded coffee, and coffee shaded with only *Inga*. However, the author points out that a qualitative analysis revealed that each habitat appears to have a different ant assemblage and suggests that forests and coffee plantations under different management contribute to ant diversity at the landscape level. In situations like these, it is important to examine the overlap of species within habitat and between habitats. In a similar study, Perfecto and Snelling (1995) reported much higher species similarity indices among coffee monocultures than among coffee agroforests, demonstrating that, at a landscape level, the agroforests contributed significantly more to species diversity than the coffee monocultures, even though the differences in species diversity locally were not very high.

The few studies that have sampled arboreal ants foraging in the coffee layer showed mixed results (see Table 6.1). While Perfecto and Snelling (1995) found no significant difference in ant diversity in the coffee layer between shaded and unshaded plantations, Ibarra-Núñez et al. (1995) and Perfecto et al. (1997) reported a higher ant species richness in coffee bushes in diverse plantations compared to more technified plantations. These conflicting results could be a consequence of the methods used for sampling the ant community. Perfecto and Snelling (1995) used tuna baiting, which tends to capture the generalist subcommunity of ants, while Ibarra-Núñez et al. (1995) and Perfecto et al. (1997) used D-vac sucking and insecticidal fogging of entire plants, respectively. It can thus be argued that, when a more complete sample of the ant assemblage is taken, a significant difference in ant species richness is detected between shaded and unshaded plantations. Canopy ants, those that nest and forage in the canopy of shade trees, have been less studied than ground-foraging ants or ants that forage in the coffee bushes, and the only study published to date that compares canopy ants along a coffee intensification gradient shows an even more accentuated reduction of species richness than that documented for ground or leaf litter ants (Perfecto et al., 1997).

Army ants also seem to be affected by the elimination of shade trees. A study in Panama reported that two species of army ants commonly found in forests were also present and abundant in shaded coffee plantations but not in unshaded plantations (Roberts et al., 2000). This study also found no difference in the number of swarms for these two species between forest and shade coffee plantations either near or far from the forest.

Several direct and indirect mechanisms have been proposed for the observed reduction of ant species richness along the technification gradient. Among the direct mechanisms are the loss of nesting sites for canopy and trunk nesting species (Perfecto and Vandermeer, 1994; Roberts et al., 2000) and changes in microclimatic conditions (Torres, 1984; Perfecto and Vandermeer, 1996). Indirect mechanisms include changes in the type of resources available for ants, which could alter the competitive interactions in the ant community (Perfecto, 1994; Perfecto and Snelling, 1995; Perfecto and Vandermeer, 1996).

Other arthropods also show declines in species richness with intensification (Table 6.2). Comparing shaded and unshaded plantations in Mexico, Nestel, Dickschen, and Altieri (1993) reported a reduction in species richness for soil macrocoleopterans. Similar results were reported by Estrada et al. (1998) for dung beetles. In both of these cases, the persistence of medium-sized diurnal mammals and the presence of decomposing fruits in the shaded plantations were given as possible reasons for the higher diversity in the shaded plantations.

Species richness of phytophagous insects has also been shown to decline with intensification. Mas (1999) reported a significant decline in fruit-feeding butterflies along a coffee intensification gradient. In this study, only the more rustic plantation was able to maintain high species richness, which suggests that this group of butterflies is very sensitive to the disturbances caused by intensification (such as reduction in canopy cover). In a study comparing sun coffee with shade coffee plantations in Costa Rica, Rojas et al. (1999) report lower homopteran species richness in the sun coffee system.

Not all arthropods appear to respond in the same fashion. As mentioned above, in the study with moths in Costa Rica, Ricketts et al. (2001) found no significant differences in moth richness between monospecific shade and sun plantations. A study in Colombia reported no difference between shaded and unshaded plantations for hymenopterans other than ants (Sossa and Fernandez, 2000). A study comparing

Table 6.2 Studies that Compare Arthropod Species Richness in Coffee Plantations with Different Levels of Intensification

Country	Group/Theme	Intensification Effect	Reference
Colombia	Mesoarhtropods (ground)	Yes	Sadeghian, 2000
Colombia	Scarabaeinae (ground)	Yes	Molina, 2000
Costa Rica	Coleoptera and Hymenoptera (shade)	Yes	Perfecto et al., 1996
	Coleoptera and Hymenoptera (coffee)	Yes	
Costa Rica	Arthropods (coffee)	Mixed	Perfecto and Snelling, 1995
Costa Rica	Moths (light traps)	No	Ricketts et al., 2001
Jamaica	Arthropods (in general)	Yes	Johnson, 2000[a]
Mexico	Scarabaeinae (ground)	Yes	Estrada et al., 1998
Mexico	Homoptera (coffee, D-vac)	Yes	Ibarra-Núñez et al., 1995
Mexico	Butterflies (traps: coffee and canopy levels)	Yes	Mas, 1999
Mexico	Macrocoleopterans (ground)	Yes	Nestel et al., 1993
Mexico	Coffee leaf miner (coffee) (abundance)	No	Nestel et al., 1994
Mexico	Spiders (coffee)	Yes	Pinkus-Rendón, 2000[b]

[a] This study did not examine coffee plantations with different intensification levels, but rather areas that were dominated by different species of shade trees (*Inga vera* versus *Pseudoalbizia berteroana*).

[b] This study showed the reverse pattern: higher density and diversity of spiders in the more technified plantations.

spiders in two farms in the Soconusco region of Mexico suggests no differences in spider species richness between an organic farm with diverse shade and a technified conventional farm with shade dominated by *Inga* (Ibarra-Núñez et al., 1995), while another study in the same plantation found significantly higher spider diversity in the technified farm as compared to the organic and more shaded farm (Pinkus-Rendón, 2000). Yet unpublished data from a study in Costa Rica by one of the authors (Perfecto) suggest that spider diversity is higher in shade coffee than in sun coffee. Fogging and sampling ten coffee bushes each in a sun and a shade plantation, Perfecto et al. (1996) found a total of 29 and 44 spider morphospecies, respectively.

Researchers comparing arthropods in shaded and unshaded plantations often make the decision of focusing on one compound of the community. Because it is not practical to sample all components of the shaded plantation (ground, leaf litter, coffee bushes, canopy of shade trees), most studies focus on the components that are more accessible (i.e., leaf litter or coffee bushes). This makes the interpretations of results of biodiversity studies difficult. Particularly problematic are organisms that can inhabit different levels in trees and shrubs, such as spiders and arboreal ants. Without sampling the canopy of the shade trees along with the coffee bushes, it is difficult to make a generalization about diversity in coffee plantations with different levels of shade. Arboreal spiders could be using the coffee bushes in a plantation with few or no trees, while in a shaded plantation most of these species could be found on the canopy of trees and not on the coffee layer. The canopy of *Inga* can be particularly attractive for insect predators such as spiders because the canopy of shade trees has a much higher abundance of insects than the coffee bushes (Johnson, 2000). However, the high density of insects in the canopy can also attract birds, which can prey on the spiders (these types of trophic interactions will be discussed in the next section).

Although the majority of these studies show a reduction in arthropod biodiversity with coffee intensification (Tables 6.1 and 6.2), most of the studies consist of comparisons between two systems, usually shade and sun plantations. The few studies that have examined a gradient of shade suggest that the particular level of shade is important (Perfecto and Vandermeer, 1994; Perfecto and Snelling, 1995; Perfecto et al., 1997; Mas, 1999). As was discussed above, some species of shade trees, like *Inga*, support higher diversity and abundance of arthropods than others (Johnson, 2000). Furthermore, there is no reason to think that the trajectories of species decline should be the same for different taxa. Although comparative studies that include different taxa within the same sites are rare, our knowledge of the natural history of different groups suggests that some taxa are more susceptible to technification than others. This is evident within birds, where residents have been shown to be more susceptible to intensification than migrants (Greenberg et al., 1997*a*). Preliminary data from a study in Chiapas (Perfecto et al., in press) also suggest that ants and butterflies follow a very different pattern of richness decline along an intensification gradient (Figure 6.4). These differences make it difficult to establish criteria for the certification of shade coffee for conservation purposes. It is important to note that the approach taken here emphasizes species richness without concern for the identity of those species. For conservation purposes it will be important to identify forest species or species that are sensitive to perturbations.

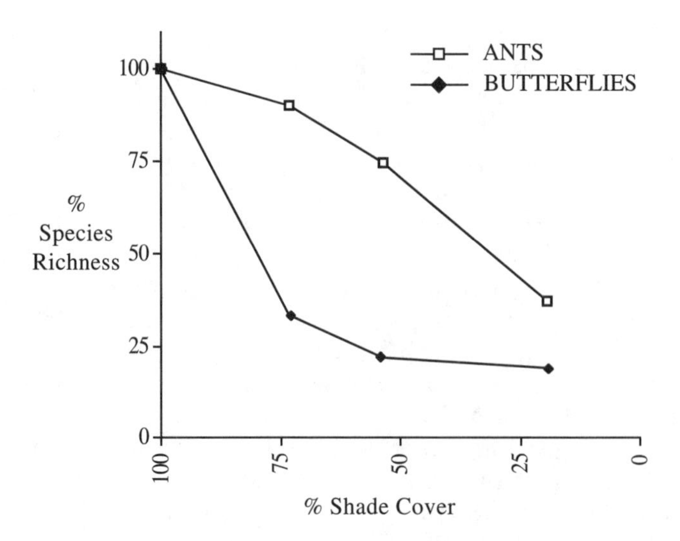

Figure 6.4 Percent of species richness (based on species richness in forest plots) of ants and butterflies in relationship to the percentage of shade cover in coffee plantations in the Soconusco region of Chiapas, Mexico. Data of species richness for 100% shade cover represents richness in forest plots. (Modified from Perfecto et al., in press.)

THE FUNCTION OF BIODIVERSITY IN THE REGULATION OF HERBIVORES

Based on ecological theory, we propose that the high diversity of organisms found in coffee plantations can play an important role in the functioning of that agroecosystem. Recent debate on the role of biodiversity in ecosystem function suggests that we should be cautious in making blanket statements about the subject (Huston, 1997; Huston et al., 2000; Loreau and Hector, 2001). One of the ecosystem functions that has been assumed to be enhanced by biodiversity is the regulation or control of insect pests (Altieri, 1993; Vandermeer and Perfecto, 1998). However, the relationship between pest control and biodiversity is a complicated one and should be examined more carefully on a system-by-system basis. In this section we will focus on the role of biodiversity in the regulation of insect herbivores in coffee because it is an area that is beginning to receive significant empirical attention. It also has obvious practical implications.

Coffee in the Western Hemisphere does not have a high incidence of insect pests. However, up to 200 species of herbivores have been reported to feed on coffee (Le-Pelley, 1973). In a baseline study of the arthropod community in a shaded coffee plantation of the Soconusco region in southern Mexico, Ibarra-Núñez (1990) reported that 37.5% of the individuals and 25% of the species collected were phytophagous. However, despite the fact that more than a third of the individuals collected by Ibarra-Núñez (1990) were phytophagous, only a few species are considered pests in coffee throughout Latin America. Among these are *Hypothenemus hampei* (Ferrari), the coffee berry borer, *Leucoptera coffeella* (Guer-Men), the coffee leaf miner, several coccids and pseudococcids (*Planococcus citri*, Rissi, *Pseudococcus jongispinus* Torgioni-Tozzeti), several shoot borers (*Plagiohammus macuosos* Bates, *P. mexicanus, P. spinipensis*), and the red mite (*Oligonychus coffeae* Nietar). It has been suggested that the structurally complex and floristically diverse traditional coffee plantations support a high density and diversity of predators and parasitoids, which are ultimately responsible for the reduced number of insect pests in traditional plantations (Ibarra-Núñez, 1990; Perfecto and Castiñeiras, 1998). Ibarra-Núñez's study (1990) reported that 42% of the species and 25% of the individuals collected were predators or parasitoids. It has been suggested that generalist or polyphagous predators, like birds, ants, and spiders, are better at preventing pest outbreak than at suppressing outbreaks once they have occurred (Holmes, 1990; Riechert et al., 1999). We suspect that as diverse shaded coffee plantations, like those sampled by Ibarra-Núñez (1990), are transformed to less diverse or sun plantations, the diversity of generalist predators will decline, releasing herbivores from the predation pressures that presumably maintain them below pest threshold population densities.

Impact of Birds on Coffee Arthropods

As discussed above, shaded coffee plantations support among the highest densities and species richness of birds of any habitats, either natural or anthropogenic, in northern Latin America (Aguilar-Ortíz, 1982; Wunderle and Wide, 1993; Greenberg, Bichier, and Sterling, 1997*b*). Most species of birds are either insectivores or omnivores — with arthropods comprising the majority of their diet. Experimental exclosure studies over the past 20 years have demonstrated that birds often remove a large portion of the standing crop of arthropod populations — particularly large herbivorous arthropods (Holmes et al., 1979; Gradwohl and Greenberg, 1982; Moore and Yong, 1991; Bock et al., 1992; Marquis and Whelan, 1994). Other studies have further demonstrated a reduction in herbivore damage in the presence of insectivorous birds (Atlegrim, 1989; Marquis and Whelan, 1994), which resulted in an increased growth rate of study plants. However, very few studies have examined the impact of insectivorous birds on the arthropod community in coffee plantations. In a study in coffee plantations in Jamaica, Johnson (2000) reported that coffee with *Inga* as the primary shade trees had higher abundances of arthropods and birds than coffee dominated by another shade tree species, and suggested that bird communities in coffee respond to spatial variation in arthropod availability. The only bird exclosure study conducted in coffee plantations so far showed a 64 to 80% reduction in arthropods greater than 5 mm in length (Greenberg et al., 2000). These data suggest that the effect of birds is quite generalized across ecological and taxonomic groups of arthropods. Furthermore, there was a small but significant increase in herbivore damage within the exclosures. The sample size for this study was small and the time frame of the experiment short, yet interesting significant results were obtained. Because overall bird density and diversity decline with the intensification of coffee plantations, it is reasonable to suggest that their ability to regulate insect herbivores will also be reduced with intensification.

Unlike temperate systems, which have often been the focus of exclosure studies of bird insectivory, the tropical coffee agroecosystem experiences herbivory and insectivory throughout the year. During the north temperate winter (which encompasses both the dry season and the season of coffee harvest in Chiapas) insectivorous bird populations may double with the influx of migrants from the North. During the north temperate summer, coffee plants are engaged in their peak vegetative growth and insectivorous bird populations are smaller and engaged in breeding activities. Based on these differences it is reasonable to expect seasonal variation in the impact of avian insectivory on arthropod abundance and herbivore damage. During the winter months, a relatively high density of birds and low abundance of arthropods may result in large proportional reductions in arthropod abundance (as was demonstrated in the study of Greenberg, Bichier, and Cruz Angon, 2000). During the summer, birds are less common and arthropods are more abundant, so birds may have a lower impact. However, because breeding birds usually rely heavily upon large arthropods to raise young (Greenberg, 1995), and because this is the period of greatest leaf production, we would expect the greatest absolute reduction in herbivory to occur during this period.

Preliminary results of bird exclosure experiments in Chiapas, Mexico, show that birds significantly reduce the number of arthropods larger than 5 cm, but this difference does not appear to be stronger for the winter months (unpubl. data) (Figure 6.5). Birds also significantly reduce herbivory in coffee plants (unpubl. data). A recent study where lepidopteran larvae were used to simulate a pest outbreak inside and outside bird exclosures demonstrated that birds rapidly remove caterpillars from the coffee layer. But even more significant, the rate of removal was significantly faster in the diverse shaded plantations than in the more technified plantation (Perfecto et al., unpubl. data). These results suggest that diversely shaded coffee

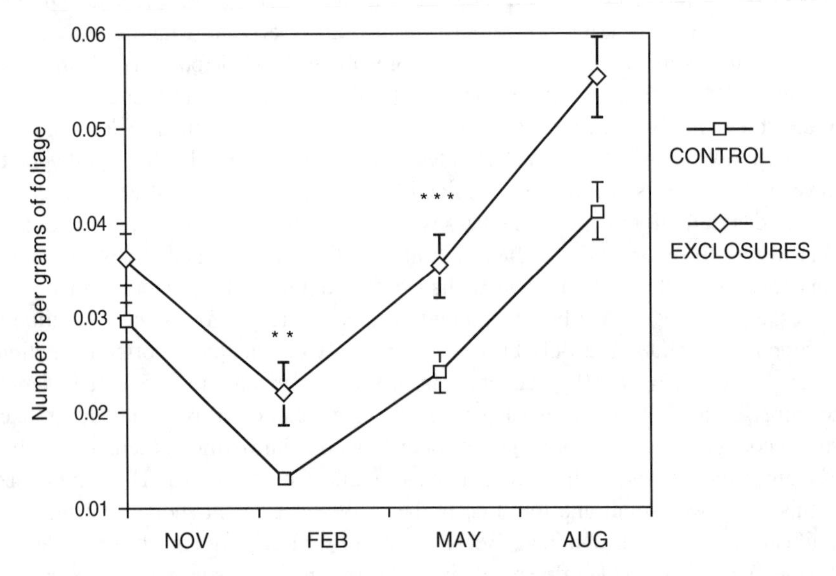

Figure 6.5 Numbers of arthropods >5 cm per gram of foliage of coffee in controls (n = 32) and bird exclosures (n = 32). ** = significance level of <0.01, *** = significance level of <0.001. Data from coffee farms in the Soconusco region of Chiapas.

plantations could also be more resistant to pest outbreaks than sun coffee because of the higher diversity and density of insectivorous birds that are present in these plantations. In a diverse coffee plantation, each herbivore species has to deal with a broader range of natural enemies than in the sun plantations, and therefore the probability that a particular herbivore would reach high population levels and become a pest is lower.

Impact of Ants on Other Coffee Arthropods

Ants are among the most important generalist predators in tropical ecosystems, both managed and natural. They are also numerically and ecologically dominant in these regions. For example, recent measurements suggest that ants and termites compose one third of the entire animal biomass of the Amazonian upland rainforest (Wilson, 1987). In Guianan cacao plantations, ants constitute 89% of the total insect numbers (Leston, 1973) and up to 70% of the arthropod biomass (Majer, 1976). In Ibarra-Núñez's study (1990), ants represented 12.2% of the total arthropods. Although not as diverse as other taxa, ants can have a high diversity in tropical agroecosystems and are almost always the most numerous of the arthropod taxa.

There are many instances of ants being used to limit pests, both in the tropics and in temperate ecosystems (see reviews in Way and Khoo, 1992; Perfecto and Castiñeiras, 1998). A study conducted in a shade coffee plantation in Chiapas, Mexico, demonstrated the potential of two Ponerinae species in controlling phytophagous insects (López Méndez, 1989). More recently, Ibarra-Núñez et al. (2001) conducted a prey analysis study for *Ectatomma ruidum* and *E. tuberculatum,* two common ant species in coffee plantations in Chiapas, and found that 17.8% and 11.3% of the prey items of these two species, respectively, were herbivores of coffee. Furthermore, in Colombia, leaf litter and soil ants were recently discovered preying on the coffee berry borer, the main coffee insect pest in all of Latin America (Vélez et al., 2000). In studies with artificial baits (fruit fly pupae or tuna fish), ants have been found to rapidly remove them, suggesting that they could potentially remove living herbivores (Armbrecht and Perfecto, in press; Philpott, unpubl. data).

With the reduction of ant diversity along the coffee intensification gradient (Nestel and Dickschen, 1990; Perfecto and Vandermeer, 1994; Perfecto and Snelling, 1995; Perfecto et al., 1997), a diverse community in shaded coffee plantations changes to one dominated by only a few species, mainly *Solenopsis geminata* and a few *Pheidole* species (Nestel and Dickschen, 1990; Perfecto and Vandermeer, 1994; Perfecto and Snelling, 1995). These few species are generalists, of approximately the same size, that nest in the ground and forage on the ground and lower vegetation. Both ant species have been reported to be effective predators and to cause reduction in herbivores in other agroecosystems (Risch and Carroll, 1982; Perfecto, 1990; Perfecto and Sediles, 1992). However, their effectiveness in controlling a variety of herbivores and potential pests in coffee is limited. For example, the coffee berry borer may be protected from *Solenopsis* germination once it is inside the seed, (*S. geminata* is too big to enter the holes, although this species could still have some impact on the adult coffee berry borer when they are outside the coffee berry.

It is also important to distinguish between two major groups of ants that forage in coffee bushes: the nondominant twig nesting ants, which have relatively small colonies, such as those in the genera *Pseudomyrmex*, some *Solenopsis*, and *Lepthotorax;* and the numerically dominant ants that make carton nests or nest in tree trunks and form large colonies, such as *Azteca*, *Camponotus,* and *Crematogaster* (Philpott, unpubl. data). These swarming ants usually are mutually exclusive, forming a mosaic of dominant ants with associated nondominant species in the canopy of shade trees (Leston, 1973; Majer, 1972, 1976). Recent studies in Mexico suggest that these swarming ants nest primarily in the shade trees and their foraging area includes several adjacent coffee bushes (Philpott, unpubl. data). A particularly interesting genus is *Azteca*, members of which form very large colonies and frequently dominate individual coffee bushes. A recent study suggests that even if these ants do not have a direct density-mediated effect on herbivores, they may have an indirect trait-mediated effect by harassing herbivores to the point that they have to move to another plant, reducing the feeding time of individual herbivores (Vandermeer et al., in press). Although the impact of ants on other arthropods has been well documented for other systems (Way and Khoo, 1992; Perfecto and Castiñeiras, 1998), their effect on potential insect pests in coffee is still unknown. Furthermore, some ant species are known to tend scales and other homopterans in coffee, and therefore their effect on coffee plants could be both negative and positive. More controlled experimental studies should be undertaken to evaluate these contradictory effects of ants on coffee herbivores.

Impact of Spiders on Other Coffee Arthropods

Ibarra-Núñez's (1990) baseline study reported 65 species of spiders belonging to 26 families, representing 14% of all the individual arthropods sampled in a coffee plantation in Chiapas, Mexico. In a more recent study of four spider families on three coffee farms in the same region, Ibarra-Núñez and Garcia-Ballinas (1998) reported 87 species belonging to 36 genera. In censuses of web-building spiders in coffee plantations of New Guinea, Robinson and Robinson (1974) estimated that there were 58,050 web-building spiders per hectare in coffee plantations and that these were responsible for the consumption of almost 40 million insects per year per hectare! Although these estimates are derived from extrapolations (from Robinson and Robinson, 1970) and not direct measurements of insect consumption by spiders, they give a general idea of the potential impact of spiders in controlling insect populations. In a recent study of prey analysis of seven common web-building spiders and two ant species, Ibarra-Núñez et al. (2001) found that the bulk of the prey of these nine predator species belonged to the insect orders Hymenoptera (primarily ants), Diptera, Homoptera, and Coleoptera. In general, the frequency of relative predation of any type of prey was proportional to their relative abundance. Herbivores and detritivorous and polyphagous arthropods (mainly ants) constituted the major part (84.7%) of the identified prey items (Ibarra-Núñez et al., 2001). This study also suggested that even though there was some overlap in prey species between different species of web-building spiders, their predation activity appears to be complementary. Essentially, different spider species occupy different microhabitats within an individual coffee bush. Recently,

Riechert et al. (1999) experimentally demonstrated that prey abundance is reduced to lower levels by spider assemblages than by single populations. They suggested that species assemblages are better at sustaining levels of pest suppression because of temporal synchronies.

Relevant Trophic Interactions

Thus far we have discussed the separate effects of birds, ants, and spiders as natural enemies and potential regulators of herbivores in coffee plantations. However, it is important to note that these groups are for the most part generalist predators and will prey on each other. For example, in the bird exclosure study in Guatemala (Greenberg et al., 2000), birds caused a significant reduction of ants and spiders. Likewise, Ibarra-Núñez et al. (2001) reported ants as significant prey items of web spiders. Ants have been reported to eat spiders (Ibarra-Núñez et al., 2001) and each other (Hölldobler and Wilson, 1990), and some species, such as army ants and fire ants, can even kill bird nestlings (pers. obs.). Preliminary results from our study of the trophic structure in coffee plantations in Chiapas suggest that birds may be reducing spiders and spiders may be reducing parasitic wasps. In the first year of a large enclosure experiment, we found significantly higher numbers of spiders and significantly lower numbers of parasitic wasps inside bird exclosures than outside, where they were exposed to bird predation. The fact that we also found a significantly higher herbivory inside the exclosures suggests that other interactions are at play. Controlled exclosure experiments combining ants, spiders, wasps, and birds would be necessary to sort out the trophic web structure in coffee. Figure 6.6 shows a diagrammatic representation of the suspected main trophic interaction in the coffee agroecosystem, based on preliminary results from studies in Chiapas, Mexico. It is still too early to come to definite conclusions about the effects of diversity of the trophic structure of coffee and to accept the hypothesis that a high associated biodiversity functions as a buffer mechanism against pest outbreaks in coffee plantations. However, in spite of the high complexity inherent in this system, preliminary results point in that direction.

COFFEE AGROFORESTS AS A HIGH-QUALITY AGRICULTURAL MATRIX

Up to this point we have focused attention on the biodiversity contained within coffee plantations themselves and the potential of coffee agroforests to serve as a refuge. However, coffee agroforests may be important for the conservation of biodiversity within forest fragments. Conservation biologists are increasingly aware that the matrix within which forest fragments exist may be as important for conservation as the forest fragments themselves (Laurance, 1991; Bierregaard et al., 1992; Franklin, 1993; Weins et al., 1993; Gustafson and Gardner, 1996; Jules, 1997; Vandermeer and Perfecto, 1997; Vandermeer and Carvajal, 2001). Theoretically, the matrix may affect the rate of migration of organisms among forest patches and thus influence extinction rates on a regional level, or the matrix may create conditions

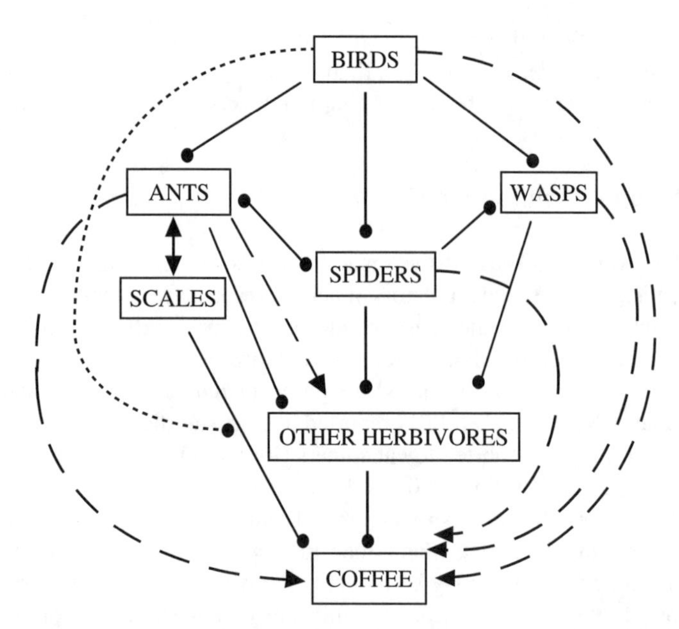

Figure 6.6 Diagram of the proposed main trophic interactions in a coffee plantation of the Soconusco region of Chiapas, Mexico. Arrowheads indicate positive effects and small closed circles indicate negative effects. Solid lines indicate direct effects, dashed lines indicate indirect effects acting through density or biomass modification, and the dotted line indicates the indirect effect acting through the modification of a direct effect.

that alter extinction rates within the forest fragments themselves (Gustafson and Gardner, 1996; Cantrell, Cosner, and Fagan, 1998). For example, the majority of frogs in the Manaus area of Brazil are maintained in forest fragments because they appear to use the surrounding matrix (Tocher et al., 1997).

Many of the mid-elevation regions in Latin America have lost large extensions of the original forest cover. However, many forest fragments remain scattered among the agricultural matrix. These small patches may not be suitable for megafauna, but they could be critical for the conservation of other organisms such as insects and other invertebrates, small- and medium-sized mammals, birds, and plants. Although it is desirable to develop programs to encourage the preservation of these many fragments, an additional problem is the management of the matrix in which they occur. In many tropical montane situations in northern Latin America, coffee plantations occupy most of the matrix. In the context of the matrix within which forest fragments are located, the coffee system thus represents what could be a variety of matrix qualities, from the rustic system to the sun system. It has been recently proposed that the different levels of coffee intensification (rustic coffee being the least intense, sun coffee being the most intense) represent different matrix qualities with respect to a selected bioindicator group (Perfecto and Vandermeer, 2002).

In this context, the quality of the matrix takes on a particular meaning. Focusing on the eventual need for interfragment migration, a high-quality matrix is one in which the barriers to migration are small. The matrix may not provide a source habitat for a particular species (i.e., the species may not be able to persist indefinitely there), but it may not be a perfect sink either (i.e., a propagule landing there may not perish immediately). The process of interfragment migration can be separated into two categories: direct and indirect migration (Perfecto and Vandermeer, 2002). For example, for ants, the principal migratory event is the time of nuptial flight. Flying ants may be carried long distances by the wind, but they mostly fly short distances to locate a nest site and establish a new colony (Hölldobler and Wilson, 1990). Direct migration occurs when a queen originating from a forest fragment is fertilized, flies, and lands on another forest fragment. The likelihood of this occurring depends on the size and spatial separation of forest fragments. Indirect migration occurs when a fertilized queen establishes a colony in the matrix and the colony survives in the matrix at least until it reaches maturity and produces new queens that will mate and disperse to find new nesting sites. Some of the fertilized queens will establish new colonies in the matrix to repeat the process. Eventually, one of the future generation offspring of the original queen establishes a colony in a new forest fragment. It is evident that the quality of the matrix is especially important for indirect migration, at least in this particular case. Even if the matrix is not a sufficiently high-quality habitat to maintain the population of a particular species in perpetuity (a source habitat), it may be sufficiently benign such that populations may be temporarily established, enabling indirect migration to occur.

Recent studies support the idea that shaded coffee represents a high-quality matrix for ants that live in forest fragments. Examining two farms, one an organic farm with considerable shade, the other a conventional farm with only spotty shade, Perfecto and Vandermeer (2002) found that species richness of ground-foraging ants attracted to tuna baits decreased with distance from the forest fragment in both matrix types. However, the rate of decrease in species richness was greater in the conventional farm (low-quality matrix) than in the organic farm (high-quality matrix). A similar study with leaf litter ants had even more dramatic results: a significant decline in species richness concomitant with distance from forest fragment in the technified coffee plantation but no significant distance relationship in the polycultural shaded plantation (Armbrecht and Perfecto, in press). On the other hand, a study with moths focusing on an agricultural matrix represented by a mosaic of systems that included coffee monocultures and shade coffee with one species of shade tree concluded that diversity was lower in all agricultural habitats, regardless of their composition, and that distance from the forest was the most important variable for species diversity (Ricketts et al., 2001). It is important to point out, however, that in this study the forest sampled was not a small fragment but actually a large reserve. Furthermore, the shaded coffee consisted of plantations with a single species of shade tree.

Based on these results, Perfecto and Vandermeer (2002) propose placing the focus on the quality of the agricultural matrix as an alternative strategy for dealing with habitat fragmentation. Rather than attempting to promote corridors of high-quality habitat (usually tacitly assumed to be necessarily the same as the fragment habitat itself), the question should be framed in terms of matrix quality. Although a matrix may be formally a sink for most populations of concern, if it provides for sufficient survivorship to ensure travel from fragment to fragment, that is all that is required for some organisms. Although a high-quality matrix is frequently difficult to construct, in the case of coffee, many of the plantations are already what might be considered a high-quality matrix, and the challenge would be to preserve them in the face of significant pressures pushing for their technification.

SHADE, BIODIVERSITY, YIELD, AND CERTIFICATION PROGRAMS

In the past 5 years, several biodiversity-friendly coffee certification programs have emerged as market-based strategies to promote either the conservation of diverse agroforests or the restoration of sun plantations to shade plantations. The appearance in the international market of organic and fair trade coffee led the way to the marketing of shade coffee as a green product. Several coffee labels, such as Bird Friendly and Eco-OK (Figure 6.7) began to appear in the market in the late 1990s and generated interest among consumers and producers alike (McLean, 1997; Koelle, 1998; Kotchen et al., in press). The following section reviews the earlier certification programs that set the stage for shade coffee certification and then discusses the shade certification programs themselves.

Organic Coffee

Organic coffee is coffee that has been grown using organic, chemical-free methods. Organic coffee occupies a sector of the gourmet or specialty coffee market that now represents 30% of the total coffee market and continues to grow (McLean, 1997). While standards for organic production have been set by larger governmental bodies, such as the European Economic Community and the U.S. Organic Food Production Act, it has been third-party certifiers like the Organic Crop Improvement Association, Quality Assurance International, and Naturland whose seals have become synonymous with chemical-free production. As a result, coffee grown on certified organic farms throughout Latin America and the world can earn premium prices, as much as 50% higher than conventional production, although it is more common for the premium to be in the range of 10 to 15% (Rice and Ward, 1996).

Figure 6.7 Certification coffee labels. (From (A) Smithsonian's Bird Friendly, (B) Rainforest Alliance's Eco-OK, and (C) the brand logo for Equal Exchange. With permission.)

The primary goal of organic certification programs has been to protect the health of the consumers, not the environment or the producers' well-being. Although this is changing somewhat, the emphasis is still on consumer health. More recently, however, the International Federation of Organic Agricultural Movements (IFOAM), which accredits affiliated organizations that certify farms, in addition to their basic standards (IFOAM, 1996) has developed special criteria for coffee. These criteria include the planting of shade trees and the composting of the coffee pulp, as well as guidelines on social rights and fair trade (McLean, 1997). The way in which these will be implemented remains to be seen.

Fair Trade

Fair trade certification is newer and less established than organic certification, particularly in the U.S. It is based on the principle that the international trading system is unfair and that structures should be set up to provide small producers in the Third World with fair prices for their produce. Organizations like Max Havelaar and the Fair Trade Foundation in Europe have created reputable programs that certify that producers receive a fair price for their produce by cutting out the middle person and by guaranteeing prices in advance, thus benefiting cooperatives. The accompanying focus on development assistance for building schools and health-care centers has created enough added value to the product that consumers are willing to pay more, though not quite the premiums earned by organic coffees. While there are at present only a few fledgling fair trade certifiers in North America, at least one company, Equal Exchange (Figure 6.7), has been working for over 10 years to develop this market in the U.S. and make the model work. In Europe, fair trade certified coffee represents only about 3% of the total coffee market (McLean, 1997); in the U.S., this percentage is much lower.

Shade Coffee

The new developments in IFOAM aside, most organic certification programs do not guarantee environmental conservation or elements of social justice. A farmer can have a completely shadeless plantation that employs hundreds of migrant workers housed under inhumane conditions, paying miserable wages, and still be certified organic. Likewise, fair trade certification has no guidelines to ensure environmental conservation, although most small producers in Latin America have shaded coffee farms. Furthermore, notably missing from both of the certification systems is the attention to other pressing ecological issues, like the loss of biodiversity that accompanies the removal of shade trees. To address this issue, and to spread information on the ecological value of shaded coffee farms, the First Sustainable Coffee Congress was organized by the Smithsonian Migratory Bird Center in 1996. This meeting inspired the formation of two new certification programs for sustainable or shade coffee.

Smithsonian's Bird Friendly Label

In its efforts to conserve bird habitats for neotropical migratory birds in Latin America, the Smithsonian Migratory Bird Center (SMBC) developed an innovative program to certify coffee as bird-friendly. Initially, the SMBC developed a series of guidelines mostly related to the shade, and teamed up with Cafe Audubon, an ecoorganic coffee company. The rationale was that by adding the SMBC guidelines to already certified organic farms, the ecological integrity of the farms would be ensured. The main problem with this model is that social justice issues are omitted and left to the fair trade labels, resulting in a certification program that is more for birds than for people. The logo and name of Smithsonian bird-friendly coffee make the focus on bird conservation clear (Figure 6.7). Although the association with the Audubon Society was terminated after a few years, the focus is still on bird conservation, and the certification criteria are designed to maintain habitats for birds (http://www.natzoo.si.edu/smbc/Research/Coffee/Thoughtpaper/thoughtpaper.htm).

Rainforest Alliance's Eco-OK Label

The certification program developed by the Rainforest Alliance has broader goals that stretch beyond just coffee. As stated on their web page (http://www.rainforest-alliance.org/index.html), their idea is to integrate all of the criteria mentioned so far — regulating chemical use, shade canopy, overall ecological integrity of the farm and its surroundings, and social justice. The Rainforest Alliance, a nonprofit organization dedicated to the conservation of tropical forests, is not new to the certification business. It developed the first international program for inspecting and certifying tropical woods that come from ecologically managed forests, the Smart Wood® label. It is also responsible for the Eco-OK coffee, bananas, cacao, citrus, and other export crops. Its Better Banana Program best illustrates the Alliance's philosophy about green labels. In this program the Alliance has worked very closely with Chiquita Brands, a large multinational corporation with operations in Latin America. The argument is that in order to save habitat on a large scale, it is necessary to form alliances with large, influential producers as well as small farmers. This philosophy has guided the Eco-OK program in coffee, and the Alliance begun to certify large producers in Central America. The Eco-OK program has certified 3500 ha of coffee in Mesoamerica (27% in El Salvador) and 4355 ha more are waiting to be certified (Belloso, 2001). A major criticism of the Eco-OK label is that in its efforts to expand and work with large producers, the Alliance's standards have become too lenient and only echo already existing practices and laws (McLean, 1997).

There is another major problem with the Rain Forest Alliance's approach that has received little attention. By certifying large coffee producers, who, because of their size, produce large quantities of coffee, they may be saturating an already small niche market for shade coffee. This could have a negative impact on the initiatives that focus on small producers and that have a higher potential for addressing issues of social justice. It also illustrates the danger of banking on the environmental concerns of consumers in the industrialized nations to address social justice issues for small farmers in developing countries. Furthermore, large producers are more prone to respond to economic incentives (i.e., profits) than small producers, who may have many reasons for producing coffee in diverse agroforestry systems, including risk avoidance, tradition, and reliance on other noncoffee products from their plantations. By working with large coffee producers whose primary motivation is profit maximization, the Rainforest Alliance is taking the risk that under conditions of high coffee prices in the international market, the producers will abandon the Eco-OK certification to increase production as fast a possible and take advantage of the favorable prices. It is important to point out that the premium price for Eco-OK certified coffee is not much, and, as with many other certification programs, the percentage of the premium declines with higher prices of the noncertified commodity. The higher the coffee prices in the international market, the less advantageous it will be for a producer to have an Eco-OK or any other green certification.

Striking a Balance between Conservation and Economic Goals

Data currently available do not allow us to say with confidence what levels of shade or what qualitative vegetative structure are the best for maintaining biodiversity in coffee plantations. As shown in Figure 6.4, different organisms may exhibit different patterns of diversity loss along an intensification gradient. Therefore, it is very difficult to propose all-encompassing criteria that would enhance the conservation of all biodiversity. A possible solution to this problem is to certify only the so-called rustic plantations, which have been shown to maintain a high diversity of most taxa studied so far. While this approach may preserve the most biodiversity, plantations with very dense shade canopies may also have very low coffee yield.

There are very few studies that examine the relationship between shade management and coffee yield (Escalante, 1995; Hernández et al., 1997; Muschler, 1997a,b; Soto-Pinto et al., 2000), yet producers' perception is that dense shade significantly reduces yield. Therefore, many producers will not be inclined to seek shade coffee certification unless the price premium is sufficiently high to overcome yield losses, which could discourage consumption in consumer countries. A recent economic analysis of the financial feasibility of investing in the certification criteria for a biodiversity-friendly coffee in El Salvador indicates that investment was financially viable for all types of plantations investigated (including sun or unshaded plantations) (Gobbi, 2000). However, this study also highlights the importance of yield for the financial viability of the investment. Of all types of farms, only the traditional shaded plantation was risk-free, primarily due to no change in yield associated with the certification criteria. The higher risk was obtained for the sun coffee, because the investment for complying with biodiversity-friendly criteria was higher and yields were assumed to be reduced due to an increase in shade cover (Gobbi, 2000), even though the shade cover that is required under the current certification programs is below 50%.

Shade coffee certification programs have emerged primarily from conservation concerns, and this bias is reflected in the certification criteria that have been developed for the different programs (Mas, 1999). However, for these programs to be widely accepted by farmers, they have to incorporate the economic goals of the producers in addition to the broader environmental goals. The success of shade coffee certification programs will depend on the adoption of these programs by coffee producers in Latin America and the willingness of consumers in the North to pay premium prices for environmental services. Certification criteria should thus be based on scientific knowledge regarding the response of biodiversity to vegetative structure as well as realistic assessments of the willingness of farmers to satisfy those criteria. Therefore, it is important for those establishing the criteria to have some information about the interactions between shade, biodiversity, and yield. In Figure 6.8 we illustrate an example of the relationship between percent yield and percent of species richness based on the percent canopy openness and its relationship to yield and species richness. This example is based on the relationship between percentage of shade and species richness illustrated in Figure 6.4 for ants and butterflies in the Soconusco region of Chiapas, and on data from Soto-Pinto et al. (2000) of the relationship between percent shade and yield for coffee farms in the Chilon region of Chiapas. Based on this example, maintaining 80% of the yield (based on the highest yield that can be achieved within a range of canopy openness) results in the maintenance of 33% and 82% of the species richness of butterflies and ants, respectively.

With this approach it would be possible to examine how yield and species richness are related in a particular region. This type of information can guide farmers' management decisions in terms of how much shade to have in their plantations. It could also help certification organizations in setting more realistic criteria for shaded coffee certification.

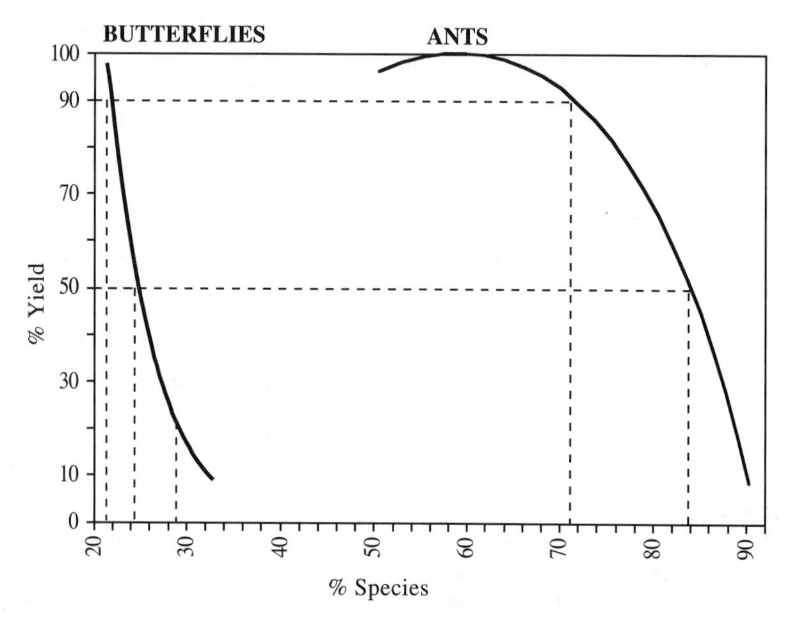

Figure 6.8 Relationship between percent of species of ants and butterflies and percent coffee yield. (Based on data from Figure 6.4 and yield data from Soto-Pinto et al., 2000.)

CONCLUSION

This chapter discussed the economic and ecological importance of the coffee agroecosystem in northern Latin America. As one of the most traded commodities in the world, coffee affects the economic life of hundreds of thousands of farmers and farm workers in northern Latin America. Coffee is grown in countries that have been identified as megadiverse and in regions with high levels of endemism. Furthermore, in countries that have been devastated by deforestation, shade coffee represents one of the few forested habitats left for wildlife. As was shown above, the extensive literature on biodiversity in coffee plantations provides ample evidence that the coffee agroforests are important habitats for biodiversity and that they are particularly critical in midelevation areas where natural forests are highly fragmented. By providing a high-quality matrix, coffee agroforests serve as either permanent or temporary habitat to forest species, or at least as safe passage for species that require closed-canopy forests. Furthermore, some evidence suggests that the high diversity of the coffee agroforests has important ecological functions, especially with regard to pest regulation. The trend toward coffee technification has encouraged a deforestation process, which has resulted in a dramatic loss of biodiversity in the montane landscapes of northern Latin America. Ironically, it has also contributed to an overproduction of coffee, which has depressed prices internationally and inflicted much misery on small producers and workers alike. In 2001, the International Coffee Council (2001) predicted that the coffee crisis will continue and that in the medium term there could be a dismanteling or weakening of the coffee sector in countries that depend heavily on this commodity. Among the many possible devastating consequences of the continuation of this crisis are social and political instability, increased external debt, and increased violence. Ecologically, the consequences of the crisis could be even more devastating than the technification process as producers leave coffee production altogether and establish cattle ranches or annual crops. At this crossroad, a sensible long-term solution to the coffee crisis may be to promote shade-grown coffee with the goal of reducing worldwide overproduction of coffee while simultaneously promoting biodiversity conservation in sensitive tropical and subtropical areas where coffee is produced. Furthermore, linking shade-grown coffee certification programs with the already-established organic and fair trade systems may offer the best solution for sustainable development in the coffee sector. Producers would not only receive the benefits of higher prices for their coffee, but would contribute to lowering global coffee production to better match consumer demand while conserving biodiversity and promoting a truly sustainable export crop in tropical nations.

REFERENCES

Adis, J.Y., Lubin, D., and Montgomery, G.G., Arthropods from the canopy of inundated and terra firme forest near Manaus, Brazil, with critical considerations on the pyrethrum-fogging technique, *Stud. Neotrop. Fauna Environ.*, 19:223–236, 1984.

Aguilar-Ortíz, F., Estudio ecológico de las aves del cafetal, in *Estudios Ecológicos en el Agroecosistema Cafetalero,* Jiménez-Avila, E. and Gómez-Pompa, A., Eds., INIREB, Xalapa, Veracruz, México, 1982, p. 103–127.

Altieri, M.A., Ed., *Crop Protection Strategies for Subsistence Farmers,* Westiew Press, Boulder, CO, 1993.

Armbrecht, I., personal observations.

Armbrecht, I. and Perfecto, I., Ant biodiversity and function in Colombian coffee plantations, paper presented at the Research summaries, of the "Coffee and Biodiversity" evening session, 87th Annual Meeting of the Ecological Society of America, Tucson, AZ, 2002.

Armbrecht, I. and Perfecto, I., Litter-twig dwelling ant species richness and pedation potential within a forest fragment and neighboring coffee plantations of contrasting habitat quality in Mexico, Agriculture, Ecosystems and Environment, in press.

Askins, R.A., Lynch, J.R., and Greenberg, R., Population declines in migratory birds in eastern North America, *Curr. Ornithol.,* 7:1–57, 1990.

Atlegrim, O., Exclusion of birds from bilberry stands: impact on insect larval density and damage to the bilberry, *Oecologia,* 79:136–139, 1989.

Belloso, G., La certificación socioambiental de café: El programa ECO-OK en el Salvador, paper presented at Resúmenes del V Congreso de la Sociedad Mesoamericana para la Biología y la Conservación, Simposio Café y Biodiversidad, San Salvador, El Salvador, 2001.

Benítez, J. and Perfecto, I., Efecto de diferentes tipos de manejo de café sabre las camunidades de hormigas. Agroecología Neotropical, 1:11–15, 1990.

Bierregaard, R.O., Jr. et al., The biological dynamics of tropical rain forest fragments, *Bioscience,* 42:859–866, 1992.

Bock, C.E., Bock, J.H., and. Grant, M.C., Effects of bird predation on grasshopper densities in Arizona grasslands, *Ecology,* 73:1706–1717, 1992.

Borrero, H., La substitución de cafetales de sombrío por caturrales y su efecto negativo sobre la fauna de vertebrados, *Caldasia,* 15:725–732, 1986.

Botero, J.E. and Baker, P.S., Coffee and biodiversity: a producer-country perspective, in *Coffee Futures,* Baker, P.S., Ed., Cabi-Federacafe, USDA-ICO, Chinchiná, Colombia, 2001.

Cantrell, R.S., Cosner, C., and Fagan, W.F., Competitive reversals inside ecological reserves: the role of external habitat degradation, *J. Math. Biol.,* 37:491–533, 1998.

Celedonio-Hurtado, H., Aluja, M., and Liedo, P., Adult population fluctuations of *Anastrepha* species (*Diptera: Tephritidae*) in tropical orchard habitats of Chiapas, Mexico, *Environ. Entomol.*, 24:861–869, 1995.

Consejo Salvadoreño del Café, Ambiente Cafetalero Nacional, Reporte trimestral estado caficultura nacional, El Salvador, 1997.

Corredor, G., Estudio Comparativo entre la Avifauna de un Bosque Natural y un Cafetal Tradicional en el Quindío, tesis de licenciatura, Universidad del Valle, Cali, Colombia, 1989.

DeJong, B.H. et al., Community forest management and carbon sequestration: a feasibility study from Chiapas, Mexico, *Interciencia,* 20:409–416, 1995.

DeJong, B.H. et al., Forestry and agroforestry land-use systems for carbon mitigation: an analysis in Chiapas, Mexico, in *Climate-Change Mitigation and European Land-Use Policies,* Adger, W.N., Whitby, M., and Pettenella, D.M., Eds., CAB International, New York, 1997, p. 269–276.

De Palma, A., For coffee traders, disaster comes in pairs, *The New York Times*, October 28, 2001.

Dietsch, T.V., Assessing the conservation value of shade-grown coffee: a biological perspective using neotropical birds, Endangered Species Update, 17:50–72, 2000; www.umich.edu/~esupdate/

Dietsch, T. and Mas, A., Assessing conservation success for neotropical birds across a coffee management intensity gradient in Chiapas, Mexico, paper presented at Resúmenes del V Congreso de la Sociedad Mesoamericana para la Biología y la Conservación, Simposio Café y Biodiversidad, San Salvador, El Salvador, 2001.

Erwin, T.L. and Scott, J.C., Seasonal and size patterns, trophic structure, and richness of Coleoptera in the tropical arboreal ecosystem: the fauna of the tree *Luechea seemannii* Triana and Planch in the Canal Zone of Panama, *Coleopterist Bull.*, 34:305–322, 1980.

Escalante, E.E. Coffee and agroforestry in Venezuela, *Agrofor. Today,* 7:5–7, 1995.

Estrada, A., Non-flying mammals and landscape changes in the tropical rain forest region of Los Tuxtlas, *Ecography,* 17:229–241, 1994.

Estrada, A., Coates-Estrada, R., and Merrit, D., Bat species richness and abundance in tropical rain forest fragments and in agricultural habitats in Los Tuxtlas, Mexico, *Ecography,* 16:309–318, 1993.

Estrada, A. et al., Dung and carrion beetles in tropical rain forest fragments and agricultural habitats at Los Tuxlas, Mexico, *J. Trop. Ecol.,* 14:577–593, 1998.

FAO, *Food and Agriculture Production Yearbook,* Food and Agriculture Organization of the United Nations, Rome, 1997.

FAO, *State of the World's Forests 2001,* Food and Agriculture Organization of the United Nations, Rome, 2001.

Fournier, L.A., Fluctuación de carbono y diversidad biológica en el agroecosistema cafetalero, paper presented at Resumenes, XVII Simposio Sobre Caficultura Latinoamericana, San Salvador, El Salvador, Octubre, 1995.

Franklin, J.F., Preserving biodiversity: species, ecosystems or landscapes?, *Ecol. Appl.,* 3:202–205, 1993.

Gallina, S.E., Mandujano, S., and González-Romero, A., Importancia de los cafetales mixtos para la conservación de la biodiversidad de mamíferos, *Bol. Soc. Ver. Zool.,* 2:11–17, 1992.

Gallina, S.E., Mandujano, S., and González-Romero, A., Conservation of mammalian biodiversity in coffee plantations of Central Veracruz, Mexico, *Agrofor. Syst.,* 33:13–27, 1996.

Gobbi, J.A., Is biodiversity-friendly coffee financially viable? An analysis of five different coffee production systems in western El Salvador, *Ecol. Econ.,* 33:267–281, 2000.

González, D., A coffee crisis' devastating domino effect in Nicaragua, *The New York Times,* August 29, 2001.

Gradwohl, J. and Greenberg, R., The effect of a single species of avian predator on the arthropods of aerial leaf litter, *Ecology,* 63:581–583, 1982.

Greenberg, R., Insectivorous migratory birds in tropical ecosystems: the breeding currency hypothesis, *J. Avian Biol.,* 26:260–264, 1995.

Greenberg, R., Bichier, P., and Cruz Angon, A., The impact of avian insectivory on arthropods and leaf damage in some Guatemalan coffee plantations, *Ecology,* 81:1750–1755, 2000.

Greenberg, R., Bichier, P., and Sterling, J., Bird populations in rustic and planted shade coffee plantations of Eastern Chiapas, México, *Biotropica,* 29:501–514, 1997b.

Greenberg, R. and Rice, R., *Shade-Grown Coffee and Biodiversity in Peru,* Smithsonian Migratory Bird Center, Washington, D.C., 2000.

Greenberg, R. et al., Bird populations in shade and sun coffee plantations in Central Guatemala, *Conserv. Biol.,* 11:448–459, 1997a.

Gustafson, R.J. and Gardner, R.H., The effect of landscape heterogeneity on the probability of patch colonization, *Ecology,* 77:94–107, 1996.

Hernández, G.O., Beer, J., and von Platen, H., Rendimiento de café (*Coffea arabica* cv Caturra), producción de madera (*Cordia alliodora*) y análisis financiero de plantaciones con diferentes densidades de sombra en Costa Rica, *Agrofor. Am.*, 4:8–13, 1997.

Hölldobler, B. and Wilson, E.O., *The Ants*, Cambridge University Press, Cambridge, 1990.

Holmes, R.T., Ecological and evolutionary impacts of bird predation on forest insects: an overview, *Stud. Avian Biol.*, 13:6–13, 1990.

Holmes, R.T., Schultz, J.C., and Nothnagle, P., Bird predation on forest insects: an exclosure experiment, *Science,* 206:462–463, 1979.

Horváth, A., March, I.J., and Wolf, J.H.D., Rodent diversity and land use in Montebello, Chiapas, Mexico, Department of Lanscape Ecology and Planning, El Colegio de la Frontera Sur, San Cristobal de las Casas, Chiapas, Mexico, unpublished manuscript.

Human Rights Watch, http://www.hrw.org, 2001.

Huston, M.A., Hidden treatments in ecological experiments: re-evaluating the ecosystem function of biodiversity, *Oecologia,* 110:449–460, 1997.

Huston, M.A. et al., No consistent effect of plant diversity on productivity, *Science,* 289:1255a, 2000.

Ibarra-Núñez, G., Los artrópodos asociados a cafetos en un cafetal mixto del Soconusco, Chiapas, México, Variedad y abundancia, *Folia Entomol. Mex.,* 79:207–231, 1990.

Ibarra-Núñez, G. and Garcia-Ballinas, J.A., Diversidad de tres familias de arañas tejedoras (*Araneae: Araneidae*, Tetragnathidae, Theridiiae) en cafetales del soconusco, Chiapas, Mexico, *Folia Entomol. Mex.*, 102:11–20, 1998.

Ibarra-Núñez, G., García-Ballinas, J.A., and Moreno, M.A., Diferencias entre un cafetal orgánico y uno convencional en cuanto a diversidad y abundancia de dos grupos de insectos, paper presented in Memorias Primera Conferencia Internacional IFOAM sobre Café Orgánico, Universidad Autónoma Chapingo, Mexico, 1995, p. 115–129.

Ibarra-Núñez, G. et al., Prey analysis in the diet of some ponerine ants (*Hymenoptera: Formicidae*) and web-building spiders (Araneae) in coffee plantations in Chiapas, Mexico, *Sociobiology,* 37:723–755, 2001.

International Coffee Council (ICC), Examen de la situación del mercado cafetero, Organización Internacional del café, 84° periodo de sesiones, ICC 84-3, London, 2001.

International Federation of Organic Agriculture Movements (IFOAM), IFOAM Basic Standards for Organic Agriculture and Food Processing and Guidelines for Social Rights and Fair Trade: Coffee, Cocoa and Tea, Decided by the IFOAM General Assembly at Christchurch, New Zealand, 1996.

Johnson, M.D., Effects of shade-tree species and crop structure on the winter arthropod and bird communities in a Jamaican shade coffee plantation, *Biotropica,* 32:133–145, 2000.

Jordan, M., Coffee glut and drought hit Nicaragua, *Washington Post Foreign Service*, 2001.

Jules, E., History and Biological Consequences of Forest Fragmentation: A Study of *Trillium ovatus* in Southwestern Oregon, Ph.D. dissertation, University of Michigan, Ann Arbor, MI, 1997.

Koelle, A.V., Coffee, Consumerism, and Conservation: An Environmental Discourse Analysis of the Sustainable Coffee Movement, M.S. thesis, University of Montana, 1998.

Komar, O. and Domínguez, J.P., Efectos del estrato de sombra sobre poblaciones de anfibios, reptiles y aves en plantaciones de café de El Salvador: implicaciones para programas de certificación, paper presented at Resúmenes del V Congreso de la Sociedad Mesoamericana para la Biología y la Conservación, Simposio Café y Biodiversidad, San Salvador, El Salvador, 2001.

Koptur, S., Floral and extrafloral nectars of Costa Rican Inga trees: a comparison of their constituents and composition, *Biotropica,* 26:276–284, 1994.

Kotchen, M.J., Moore, M.R., and Messer, K.D., Green products as impure public goods: shade-grown coffee and tropical biodiversity, *J. Environ. Econ. Manage.,* in press.

Laurance, W.F., Ecological correlates of extinction proneness in Australian tropical rainforest mammals, *Conserv. Biol.,* 5:79–89, 1991.

Lazcano-Barrero, M.E., Góngora-Arones, E., and Vogt, R., Anfibios y reptiles de la Selva Lacandona, in *Reservade la Biósfera Montes Azules, Selva Lacandona: Investigación para su Conservación. Special Publication Ecosfera I,* Vázquez-Sánchez, M.A. and Ramos, M.A., Eds., Universidad Nacional Autónoma de México, México City, 1992, p. 145–171.

Lenart, L.A. et al., Anoline diversity in three differently altered habitats in the Sierra de Barouco, Republica Dominicana, *Hispaniola Biotropica,* 29:117–123, 1997.

Le-Pelley, R.H. Coffee insects, *Annu. Rev. Entomol.*, 18:121–142, 1973.

Leston, D., The ant mosaic-tropical tree crops and the limiting of pest and diseases, *PANS,* 19:311–341, 1973.

López Méndez, J.A., La Actividad y la Depredación de Artrópodos Asociados al Cultivo del Café-Cacao por dos Hormigas Ponerinas (*Hymenoptera: Formicidae*) en el Municipio de Tuxtla Chico, Chiapas, B.S. thesis, Universidad Autónoma de Chiapas, Mexico, 1989.

Loreau, M. and Hector, A., Partitioning selection and complementarity in biodiversity experiments, *Nature*, 412:72–76, 2001.

Majer, J.D., The ant mosaic in Ghana cocoa farms, *Bull. Entomol. Res.*, 62:151–160, 1972.

Majer, J.D., The maintenance of the ant mosaic in Ghana cocoa farms, *J. Appl. Ecol.*, 13:123–144, 1976.

Márquez-Barrientos, L.I., Validación de Campo de los Métodos del Instituto Winrock para el Establecimiento de Parcelas Permanentes de Muestreos para Cuantificar Carbono en Sistemas Agroforestales, B.S. thesis, Universidad del Valle de Guatemala, Ciudad de Guatemala, 1997.

Marquis, R.J. and Whelan, J.C., Insectivorous birds increase growth of white oak through consumption of leaf-chewing insects, *Ecology*, 75:2007–2014, 1994.

Mas, A., Butterfly Diversity and the Certification of Shade Coffee in Chiapas, Mexico, M.S. thesis, University of Michigan, Ann Arbor, MI, 1999.

McLean, J., Merging Ecological and Social Criteria for Agriculture: The Case of Coffee, M.S. thesis, University of Maryland, 1997.

Mittermeier, R. et al., Biodiversity hotspots and major tropical wilderness areas: approaches to setting conservation priorities, *Conserv. Biol.,* 12:516–520, 1998.

Moguel, P. and Toledo, V.M., Biodiversity conservation in traditional coffee systems of Mexico, *Conserv. Biol.,* 13:11–21, 1999.

Molina, J., Diversidad de escarabajos coprofagos (*Scarabaeidae: Scarabaeinae*) en matrices de la zona cafetera (Quindio-Colombia), in *Memorias Foro Internacional Café y Biodiversidad*, Federación Nacional de Cafeteros de Colombia, Ed., Chinchiná, Colombia, 2000, p. 29.

Monro, A. et al., Arboles de los cafetales de El Salvador, The Natural History Museum of London, Tecnoimpresos, S.A., El Salvador. 2001.

Moore, F.R. and Yong, W., Evidence of food-based competition among passarine migrants during stopover, *Behav. Ecol. Sociobiol.,* 28:85–90, 1991.

Morón, M.A. and Lopéz-Méndez, J.A., Análisis de la entomofauna necrófila de un cafetal en el Soconusco, Chiapas, México, *Folia Entomol. Mex.,* 63:47–59, 1985.

Muschler, R.G., Efectos de sombra de *Erytrhina poeppigiana* sobre *Coffea arabica* vars. Caturra y Catimor, paper presented in Memorias del XVIII Simposio Latinoamericano de Cafeticultura, San José, Costa Rica, 1997*a,* p. 157–162.

Muschler, R.G., Sombra o sol para un cafetal sostenible: un nuevo enfoque de una vieja discusión, paper presented in Memorias del XVIII Simposio Latinoamericano de Cafeticultura, San José, Costa Rica, 1997*b,* p. 471–476.

Nestel, D., Coffee in Mexico: international market, agricultural landscape and ecology, *Ecol. Econ.,* 15:165–178, 1995.

Nestel, D. and Dickschen, F., Foraging kinetics of ground ant communities in different Mexican coffee agroecosystems, *Oecologia,* 84:58–63, 1990.

Nestel, D., Dickschen, F., and Altieri, M.A., Diversity patterns of soil macro-coleopterans in Mexican shaded and unshaded coffee agroecosystems: an indication of habitat perturbation, *Biodiv. Conserv.,* 2:70–78, 1993.

Nestel, D., Dickschen, F., and Altieri, M.A., Seasonal and spatial population loads of a tropical insect: the case of the coffee leaf-miner in Mexico, *Ecol. Entomol.,* 19:159–167, 1994.

Nolasco, M., *Café y Sociedad en México,* Centro de Ecodesarrollo, Mexico City, 1985.

Oxfam, *Bitter Coffee: How the Poor are Paying for the Slump in Coffee Prices,* 2002.

Panayotou, T., Faris, R., and Restrepo, C., *El Desafío Salvadoreño: de la Paz al Desarrollo Sostenible*, San Salvador, 1997.

Pérez-Higereda, G.C., Vogt, R., and Flores, O.V., Lista Anotada de los Anfibios y Reptiles de Los Tuxtlas, Veracruz, Instituto de Biología, Universidad Nacional Autónoma de México, México City, 1987.

Perfecto, I., personal observations.

Perfecto, I., Indirect and direct effects in a tropical agroecosystem: the maize-pest-ant system in Nicaragua, *Ecology,* 71:2125–2134, 1990.

Perfecto, I., Foraging behavior as a determinant of asymmetric competitive interaction between two ant species in a tropical agroecosysem, *Oecologia*, 98:184–192, 1994.

Perfecto, I. and Castiñeiras, A., Deployment of the predaceous ants and their conservation in agroecosystems, in *Conservation Biological Control,* Barbosa, P., Ed., Academic Press, Washington, D.C., 1998, p. 269–289.

Perfecto, I. and Sediles, A., Vegetational diversity, the ant community and herbivorous pests in a tropical agroecosystem in Nicaragua, *Environ. Entomol.*, 21:61–67, 1992.

Perfecto, I. and Snelling, R., Biodiversity and tropical ecosystem transformation: ant diversity in the coffee agroecosystem in Costa Rica, *Ecol. Appl.,* 5:1084–1097, 1995.

Perfecto, I. and Vandermeer, J., Understanding biodiversity loss in agroecosystems: reduction of ant diversity resulting from transformation of the coffee ecosystem in Costa Rica, *Entomology (Trends Agric. Sci.),* 2:7–13, 1994.

Perfecto, I. and Vandermeer, J., Microclimatic changes and the indirect loss of ant diversity in a tropical agroecosystem, *Oecologia,* 108:577–582, 1996.

Perfecto, I. and Vandermeer, J., The quality of the agricultural matrix in a tropical montane landscape: ants in coffee plantations in southern Mexico, *Conserv. Biol.,* 16:174–182, 2002.

Perfecto, I. et al., unpublished data.

Perfecto, I. et al., Shade coffee as refuge of biodiversity, *BioScience,* 46:698–608, 1996.

Perfecto, I. et al., Arthropod diversity loss and the technification of a tropical agroecosystem, *Conserv. Biodiv.,* 6:935–945, 1997.

Perfecto, I. et al., Species richness along an agricultural intensification gradient: a tri-taxa comparison in shade coffee in southern Mexico, *Biodiv. Conserv.*, in press.

Perfecto, I. et al., Diverse agricultural systems offer insurance against future pest outbreaks: the role of birds in shade coffee plantations, unpublished manuscript.

Philpott, S., unpublished data.

Pinkus-Rendón, M., Spider Diversity: Is There a Decline with Increasing Shade in Coffee Plantations? M.S. tesis, El Colegio de la Frontera Sur, Tapachula, Mexico, 2000.

Ramos, M.P., Diversidad de hormigas en bosque mesófilo y cafetales en la reserva La Sepultura, Chiapas, paper presented in Resúmenes del V Congreso de la Sociedad Mesoamericana para la Biología y la Conservación, Simposio Café y Biodiversidad, San Salvador, El Salvador, 2001.

Rendón-Rojas, M.G., Estudio de la Herpetofauna de la Zona Cafetalera de Santiago Jalahui, Oaxaca, B.S. thesis, Escuela Nacional de Ciencias Biológicas, Instituto Politécnico, Nacional, México City, 1994.

Rice, R., Transforming Agriculture: The Case of Coffee Leaf Rust and Coffee Renovation in Southern Nicaragua, Ph.D. dissertation, University of California, Berkeley, 1990.

Rice, R., A place unbecoming: the coffee farm of northern Latin America, *Geogr. Rev.,* 89:554–579, 1999.

Rice, R. and Ward, J., Coffee, Conservation and Commerce in the Western Hemisphere: How Individuals and Institutions Can Promote Ecologically Sound Farming and Forest Management in Northern Latin America, Smithsonian Migratory Bird Center and Natural Resources Defense Council, Washington, D.C., 1996.

Ricketts, T.H. et al., Countryside biogeography of moths in a fragmented landscape: biodiversity in native and agricultural habitats, *Conserv. Biol.,* 15:378–388, 2001.

Riechert, S., Provencher, E.L., and Lawrence, K., The potential of spiders to exhibit stable equilibrium point control of prey: test of two criteria, *Ecol. Appl.,* 9:365–377, 1999.

Risch, S.J. and Carroll, C.R., Effect of a keystone predaceous ant, *Solenopsis geminata,* in arthropods in a tropical agroecosystem, *Ecology,* 63:1979–1982, 1982.

Robbins, C.S. et al., Comparison of Neotropical migrant landbird populations wintering in tropical forests, isolated forest fragments, and agricultural habitats, in *Ecology and Conservation of Neotropical Migrant Landbirds,* Hagan, J.M. and Johnson, D.W., Eds., Smithsonian Institution Press, Washington, D.C., 1992, p. 207–220.

Roberts, D.L., Cooper, R.J., and Petit, L.J., Use of premontane moist forest and shade coffee agroecosystems by army ants in western Panama, *Conserv. Biol.,* 14:192–199, 2000.

Robinson, M.H. and Robinson, B., Prey caught by a sample population of the spider *Argiope argentata* (*Araneae: Araneidae*) in Panama: a year's census data, *Zool. J. Limn. Soc.,* 49:345–358, 1970.

Robinson, M.H. and Robinson, B., A census of web-building spiders in a coffee plantation at WAU, New Guinea, and an assessment of their insecticidal effect, *Trop. Ecol.,* 15:95–107, 1974.

Rojas, L. et al., Diversity of hoppers (*Homoptera: Auchenorrhyncha*) in coffee plantations with different types of shade in Turrialba, Costa Rica, in *Multistrata Agroforestry Systems with Perennial Crops,* Jiménez, F. and Beer, J., Com., CATIE, Turrialba, Costa Rica, 1999, p. 216–219.

Sadeghian, S., Diversidad de mesoorganismos del suelo en la zona cafetera, paper presented in Memorias Foro Internacional Café y Biodiversidad, Federación Nacional de Cafeteros de Colombia, Ed., Chinchiná, Colombia, 2000, p. 20.

Sanint, L.R., Crop biotechnology and sustainability: a case study of Colombia, http://www1.oecd.org/dev/PUBLICATION/tp/Tp104.pdf, 1994.

Seib, R., Feeding Ecology and Organization of Neotropical Snake Faunas, Ph.D. dissertation, University of California, Berkeley, 1986.

Sossa, J. and Fernandez, F., Himenopteros de la franja cafetera del departamento del Quindio, in *Biodiversidad y Sistemas de Producción Cafetera en el Departamento del Quindio,* Numa, C. and Romero, L.P., Eds., Instituto Alexander von Humboldt, Dic, 2000. Bogota, Colombia, 2000, p. 168–180.

Soto-Pinto, L. et al., Shade effects on coffee production at the Northern Tzeltal zone of the state of Chiapas, Mexico, *Agric. Ecosyst. Environ.,* 80:61–69, 2000.

Tocher, M.D., Gascon, C., and Zimmerman, B.L., Fragmentation effects on a central Amazonian frog community: a ten year study, in *Tropical Forest Remnants: Ecology, Management, and Conservation of Tropical Forest Fragments*, Laurance, W.F. and Bierregaard, R.O., Eds., The University of Chicago Press, Chicago, 1997, p. 124–137.

Torres, J.A., Diversity and distribution of ant communities in Puerto Rico, *Biotropica,* 16:296–303, 1984.

Vandermeer, J. H. and Carvajal, R., Metapopulation dynamics and the quality of the matrix, *Am. Nat.*, 158:211–220, 2001.

Vandermeer, J. and Perfecto, I., The agroecosystem: a need for the conservation biologist's lens, *Conserv. Biol.,* 11:1–3, 1997.

Vandermeer, J. and Perfecto, I., Biodiversity and pest control in agroforestry systems, *Agrofor. Forum,* 9:2–6, 1998.

Vandermeer, J. et al., Ants (*Azteca* sp.) as potential biological control agents in organic shade coffee production in southern Chiapas, Mexico: complications of indirect effects, *Agrofor. Syst.,* in review.

Vélez, M., Bustillo, A.E., and Posada, F.J., Predacion sobre *Hypothenemus hampei*, (Ferrari) de las hormigas *Solenopsis* spp., *Pheidole* spp. y *Dorymyrmex* spp. durante el secado del café, paper presented in Resumenes XXVII Congreso, Soc. Col. Entomol. Medellin, Colombia, 2000, p. 17.

Vennini, J.P., Neartic avian migrants in coffee plantations and forest fragments of southwestern Guatemala, *Bird Conserv. Int.*, 4:209–232, 1994.

Way, M.J. and Khoo, K.C., Role of ants in pest management, *Annu. Rev. Entomol.*, 37:479–503, 1992.

Weins, J. et al., Ecological mechanisms and landscape ecology, *Oikos,* 66:369–380, 1993.

Wilson, E.O., The arboreal ant fauna of Peruvian Amazon forests: a first assessment, *Biotropica*, 19:245–251, 1987.

Wilson, J., Colombian coffee growers start sowing poppies, *Financial Times*, October 25, 2001.

Witt, E., Seed Dispersal by Small Terrestrial Mammals in Shaded Coffee Farms in Chiapas, Mexico, M.S. thesis, University of Michigan, Ann Arbor, 2001.

Wunderle, J. and Latta, S., Overwinter turnover of Neartic migrants wintering in small coffee plantations in Dominican Republic, *J. Ornith.*, 135:477, 1994.

Wunderle, J. and Latta, S., Avian abundance in sun and shade coffee plantations and remnant pine forest in the Cordillera Central, Dominican Republic, *Ornith. Neotrop.*, 7:19–34, 1996.

Wunderle, J. and Latta, S., Avian resource use in Dominican shade coffee plantations, *Wilson Bull.*, 110:271–281, 1998.

Wunderle, J. and Wide, R.B. Distribution of overwintering Neartic migrants in the Bahamas and Greater Antilles, *Condor*, 95:904–933, 1993.

CHAPTER 7

Tropical Agricultural Landscapes

Robert A. Rice

CONTENTS

INTRODUCTION

Landscapes hold within them traces of the constellation of forces to which they have been exposed. Whether natural or managed, the physical landscape is subjected to the dynamism of human agency and natural processes. Agricultural landscapes by their very nature emerge from forces both natural and human. Climate and geography obviously affect the degree and distribution of agricultural imprints upon

0-8493-1581-6/03/$0.00+$1.50

the Earth's surface (Rice and Vandermeer, 1990). Yet, in the quest to evaluate or understand tropical agricultural landscapes, recognizing the interplay of human agency and natural forces is only part of the equation. There is a hidden landscape as well. It is the socioeconomic landscape lying behind or alongside the physical landscape features. It is what Don Mitchell has deemed the "lie of the land" (Mitchell, 1996). In considering tropical agricultural landscapes, it is worthwhile — imperative, even — that we seek to understand both the physical and the social.

Agriculture, that most direct and intimate complex involving human agency and the Earth, reflects the results of a globalized world in the physical and social landscapes it both creates and absorbs. An array of trajectories linked to food and fiber production contends with realities grounded in specific coordinates upon the Earth's surface, and through such actions simultaneously recontours agroecosystems and social relations. In short, a restructuring of economies, local and global, works to rearrange the physical and social landscapes. The driving force behind this process, which some call globalizing food, is the "many headed beast known as global capitalism" (Watts and Goodman, 1999). The world's garden patch continues to expand into ever more remote and heretofore unexploited regions. Those tending it often find themselves drawn into closer contact with forces and interests far removed from their specific locale. Tropical agriculture provides a veritable buffet of examples depicting changes in the physical and social landscapes of this process.

As managed lands supplying food, fiber, oils, and seed, tropical agricultural landscapes obviously offer a glimpse into our closest relations with the land. Moreover, given the history of much of the tropical agricultural regions, these landscapes carry with them the marks of social relations extending back into colonial times and before. The physical surface reveals both tradition and change — and the quest to interpret them has a rich history in the literature (Parsons, 1949; Sauer, 1963; Denevan, 1992). Some students of these changes either implicitly or explicitly call upon a web of causality approach, in which attempts to see behind the landscape changes are made (Williams, 1986; Blaikie and Brookfield, 1987; Wright, 1990; Vandermeer and Perfecto, 1995). And a technique recently borrowed from sociology is the actor-network theory (ANT), which intriguingly dissolves the distinction between society and nature by giving equal weight to all actors, animate or otherwise, involved in a given system undergoing changes (Woods, 1997; Murdoch, 1997; Sousa and Busch, 1998; Goodman, 1999). One truism prevails: the land and the people who work it confront and conform to the consequences, for better or for worse, of the expansion of the global garden.

This chapter aims to lay out recent research trends that have addressed tropical agricultural landscapes and directions for future research. It is organized around the general theme of market forces and, specifically, the degree to which cropping systems have been penetrated by capital-intensive measures such as chemical inputs, mechanized harvest, or modern labor techniques. The sections presented focus on subsistence farming, traditional exports, and nontraditional export cropping systems. In a general sense, there is an argument that the hidden landscape changes — the socioeconomic and/or cultural changes that accompany the physical landscape transformations — are positively related to the degree of market orientation. That is, the more a landscape's orientation tilts toward the market or the higher degree of capital penetration involved, the greater the potential for uncovering changes in the physical and social landscape.

In presenting recent works that focus upon tropical agricultural landscapes, we first examine traditional systems such as shifting cultivation, agroforestry, and home gardens. A second grouping of traditional systems features a number of agroexport cropping systems, including the spread of permanent pasturelands for beef production. Next, we see how the spread of nontraditional export cropping systems affects both the physical and socioeconomic landscapes in various tropical regions. A final section presents a conceptualization of how the social and physical landscapes can be affected by ever-increasing exposure to global market forces.

WIDESPREAD CHANGE IN CROPPING AREA

The United Nations Food and Agriculture Organization production statistics (FAO, 1961a, 2000a) show an increase in crop area for the 23 primary crops of the world — mainly grains and tubers, many of which are tropical (Table 7.1). On average, the area increase for these crops is 28%. Examining those 25 selected crops that are principally tropical in origin, many of which are cash crops (Table 7.2), we see an even greater average percent increase in area, namely 127% growth across all crops.

Since the 1960s, arable land in the world increased 9%. For developing countries, the increase was 21%, with the developing nations of Africa and Asia showing 27% and 12% increases, respectively (FAO, 1961b, 1999b). Latin America as a region during this same period increased its arable land by a whopping 54%. Concomitantly, global food production increased dramatically during the last half of the 20th century. While Africa's production trend mirrored that of world production by more than doubling between 1960 and 1996, Asia and Latin America increased production nearly threefold (Thrupp, 1998). Production can increase via either more land being put into use or more intensive methods being applied to that already under production (or both). A look at some of the major food, fiber, and oil crop groupings from the Food and Agriculture Organization's database reveals that where areal expansion remained steady or fell since 1960, production increased due to yield improvement (Table 7.3).

But what is behind these changes? On the ledger sheets, such changes satisfy the productionist notion so prevalent to Western ideology. Yet, the underpinnings of these changes involve tremendous alteration of the production process, usually featuring the displacement of age-old practices by labor-saving or yield-increasing activities and inputs. More and more, the components of production — animal traction, animal waste, manual weeding — have been replaced by capital-intensive inputs such as tractors and combines, synthetic fertilizer, and herbicides. It is what Goodman, Sorj, and Wilkinson (1987) refer to as appropriationism.

The global information in Tables 7.1 and 7.2 shows crop group trends of tropical agricultural landscapes by region — notably Africa, Asia, and Latin America. For the crop groups citrus, oil crops, and vegetables/melons, we find that in every case

Table 7.1 Area Devoted to FAO-Defined Primary Crops, 1961–2000

Crop	Area Harvested (ha) in 1961	Area Harvested (ha) in 2000	Percentage Change 1961–2000
Wheat	204,209,850	215,180,486	5
Rice, paddy	115,501,150	153,457,686	33
Barley	54,518,640	55,697,658	2
Maize	105,584,151	137,548,910	30
Rye	30,254,816	9,896,288	–67
Oats	38,260,751	14,416,329	–62
Millet	43,394,559	36,161,260	–17
Sorghum	46,009,146	42,805,487	–7
Buckwheat	4,640,230	2,790,409	–40
Quinoa	52,555	68,779	31
Fonio	307,957	363,671	18
Canary seed	91,713	209,050	128
Mixed grain	3,059,440	1,720,448	–44
Cereals NES[a]	2,261,550	2,582,858	14
Potatoes	22,147,776	18,777,209	–15
Sweet potatoes	13,387,283	9,577,018	–28
Cassava	9,631,856	16,611,913	72
Yautia (coco yam)	23,111	29,969	30
Taro (coco yam)	669,300	1,458,352	118
Yams	1,149,364	3,916,248	241
Roots and tubers NES[a]	603,400	1,118,521	85
Sugar cane	8,911,879	19,083,690	114
Sugar beets	6,926,098	6,446,987	–7
Total (average for % change)	711,596,575	749,919,226	28

[a] NES = not elsewhere specified.

Source: Data taken from FAO, *Agricultural Production Statistics*, http://apps.fao.org/, 1961a and 2000a.

except for Africa's oil crop category, the increase in either area or in yields is at least 100% (Table 7.4). Obviously, where area expansion is relatively high, the changes to the agricultural landscape would be evident. Where yield increases dominate, cultivation changes most likely involve chemical inputs. Readily discernable physical changes to the landscape may prove elusive unless and until soil analyses, diversity assessments of local biota, and examination of forest removal rates are made.

One major factor contributing to visible change in many tropical landscapes is the expansion of cattle lands — often via forest conversion. Globally, permanent pasture land totaled 3.4 billion hectares in 1999 (FAO, 1999b), a 10% increase since 1961. For developing countries overall, the increase was 14%. While pastureland area in the developing countries of Africa remained steady during this period, that in Asia increased by 30%. Latin America and the Caribbean as a region show a 19% increase in permanent pasture area for this period (FAO, 1961b, 1999b). More pasture area, of course, means more cattle. The global herd grew by 42% in the last four decades of the 20th century. Latin America and the Caribbean showed nearly a 100% increase in stocks, while Africa's developing countries increased their herd

Table 7.2 Area Devoted to Selected Tropical Crops, 1961 and 2000

Crop	Area Harvested (ha) in 1961	Area Harvested (ha) in 2000	Percentage Change 1961–2000
Abaca (Manila hemp)	187,036	126,220	–33
Agave fibers NES[a]	32,100	48,661	52
Areca nuts (betel)	300,557	468,316	56
Avocados	76,297	323,135	324
Bananas	2,030,193	3,844,524	89
Brazil nuts	1,800	1,000	–44
Carobs	223,622	128,380	–43
Cashew nuts	516,550	2,602,401	404
Cinnamon (canella)	37,100	132,970	258
Cloves, whole and stems	80,800	492,984	510
Cacao beans	4,403,334	7,053,169	60
Coconuts	5,234,813	10,778,417	106
Coffee, green	9,755,805	11,505,503	18
Fruit tropical fresh NES[a]	711,773	1,829,691	157
Kolanuts	155,000	367,800	137
Mangoes	1,275,081	2,759,119	116
Natural rubber	3,879,860	7,308,292	88
Nutmeg, mace, cardamom	73,450	233,396	218
Papayas	111,042	318,409	187
Pimento, allspice	1,136,335	1,822,811	60
Plantains	2,403,073	4,966,230	107
Sisal	887,426	378,794	–57
Tea	1,366,126	2,405,551	76
Tung nuts	59,700	172,850	190
Vanilla	16,483	40,380	145
Total (average for % change)	34,955,356	60,109,003	127

[a] NES = not elsewhere specified.

Source: Data taken from FAO, *Agricultural Production Statistics*, http://apps.fao.org/, 1961a and 2000a.

Table 7.3 Percent Change in Area Harvested and Yields Obtained in Food, Fiber, and Oil Crop Groups, 1961–1999

Crop Group[a]	Percentage Change, Area	Percentage Change, Yield
Cereals	4.90	124.34
Citrus fruit	227.26	20.14
Coarse grain	–5.76	113.65
Fiber crops	–4.85	69.21
Oil crops	96.32	112.95
Pulses	10.89	31.48
Roots and tubers	5.17	35.71
Vegetables and melons	72.13	65.08

[a] FAO groupings.

Source: Data taken from FAO, *Agricultural Production Statistics*, http://apps.fao.org/, 1961a and 1999a.

Table 7.4 Percent Change (1961–1999) in Area and Yield for Selected Crop Groups in Countries of Different Regions and Industrial Status

Region/ Category	Area/ Yield	Cereals	Citrus	Coarse Grain	Fiber	Oil Crops	Pulses	Roots and Tubers	Vegetable and Melons
Africa	Area	71	177	70	6	56	140	133	116
developing	Yield	52	14	37	41	13	–3	40	50
Asia	Area	15	295	–17	11	70	–2	–4	125
developing	Yield	176	99	214	85	217	23	124	77
Latin America	Area	26	517	28	–51	257	37	23	60
and Caribbean	Yield	124	6	129	175	106	20	27	102
Oceania	Area	–9	0	96	N/A	21	113	36	120
developing	Yield	48	–12	168	N/A	102	40	9	7
Developing	Area	25	312	11	0	87	21	39	118
countries	Yield	148	47	135	85	150	14	66	76
Industrialized	Area	–6	60	–15	–5	151	33	–60	–5
countries	Yield	123	29	141	56	64	139	94	76

Source: Data taken from FAO, *Agricultural Production Statistics*, http://apps.fao.org/, 1961a and 1999a.

collectively by 91% (obviously intensifying production, since total pasture area remained steady) and those of Asia saw the herd grow by 39% (FAO, 1961a, 1999a).

It is these general changes to tropical agricultural landscapes to which we now turn. We will see that the factors behind these statistical changes linked to land use involve transformations in both the physical and social landscapes. Yet not all agricultural landscapes or the stewards involved are equal with respect to the forces of global markets. Subsistence farmers, a group that still abounds in the worldwide agricultural scheme, face distinct sets of challenges and opportunities when compared to their contract-farming brethren more closely allied with global market forces. We turn now to a number of agricultural categories to examine the current status of tropical agricultural landscapes and the research attention they have received.

TRADITIONAL AGRICULTURAL LANDSCAPES

Traditional agriculture, as Altieri (1990) points out, has captured the attention of anthropologists, geographers, and other social scientists for decades. Some researchers have opined that the accumulated knowledge, technology, and talents associated with traditional practices might better inform developers and decision makers who plan and carry out agricultural policies in the tropics. More recently, agroecologists see a dual benefit in studying traditional agroecosystems. These benefits are that (1) investigation can provide an understanding of the traditional management practices and cropping patterns, which are being lost as a result of landscape changes linked to inevitable agricultural modernization, and can generate important information that may "be useful for developing appropriate agricultural strategies more sensitive to the complexities of agroecological and socioeconomic processes and tailored to the needs of specific peasant groups and regional agroecosystems"; and (2) ecological principles derived from these studies can inform development of sustainable agroecosystems in industrial nations, helping to counteract "the many deficiencies affecting modern agriculture" (Altieri, 1990: 551–552).

Subsistence Cropping Systems

Subsistence cropping systems are those oriented toward and maintained for survival of the farming family. Based *de facto* upon small growers' strategies, subsistence farming systems include swidden (also known as shifting or slash-and-burn agriculture), polycultural systems, paddy production, and agroforestry systems. As market forces expand into remote areas of the globe, many subsistence systems have become modified or eliminated entirely. New plant varieties, agrochemicals, and increased pressure upon forest, soil, and water resource bases work to transform traditional, often indigenous, agricultural practices. More often than not, the landscapes change accordingly.

Shifting agriculture is a case in point. Slash-and-burn agriculture is recognized for its universality (Nye and Greenland, 1965). Researchers in the 1970s and 1980s, estimated that around half the land area in the tropics was modified by slash-and-burn agriculture, with 250 to 300 million farmers involved (Dove, 1983). In Southeast Asia in the 1960s, some 12 million families practiced slash-and-burn agriculture (Spencer, 1966). Today, shifting agriculture is thought to embrace 2.9 billion hectares, and one estimate has a probable total of 1 billion people — more than one fifth of the population of the developing world in tropical and subtropical nations — relying directly or indirectly on shifting cultivation in some fashion (Thrupp et al., 1997).

The urgent need for research focused on shifting agriculture has been noted for several decades (Ruddle, 1974; Brookfield and Padoch, 1994). Much work on shifting cultivation has aimed to find alternatives to it, seeing it as a threat to biodiversity and global climate change, as well as a poor candidate for sustained production (ASB, 1999). This view rests upon a neo-Malthusian premise, a view rife with assumptions about the operative forces involved in deforestation and its consequences, which need to be examined before blaming those directly involved for particular agricultural practices (Jarosz, 1993). In fact, some works point to significant differences in forest management that can be attributed to folk ecology and social relationships with the environment — even when groups are faced with similar exogenous pressures (Atran et al., 1999).

A popular attitude about shifting agriculture, reinforced by international efforts like those of the ASB program headquartered in Nairobi, Kenya, is that exploding population levels push people into the agricultural frontier and beyond, where forests are sacrificed for a few years of subsistence production. In Latin America, 25 to 30% of the forest cover was lost between 1850 and 1985, with one half of that occurring after 1960. Shifting cultivation accounted for 10% of the forest reduction but ranked well behind the expansion of pastures, croplands, and degraded lands (Houghton et al., 1991). Moreover, these same researchers relate that the greatest uncertainties in assessing the causes of forest reduction came in quantifying the historical rates of degradation and shifting agriculture.

Undoubtedly, there are instances in which shifting agriculture and an accompanying shortened fallow period may be driven by population pressure, but an examination of government policies must also be part of any attempt to understand the forces behind such practices (Hecht, 1985; Jarosz, 1993). Moreover, researchers must spend time in areas to understand the dynamics involved. Shifting cultivation presents an especially dynamic set of practices within tropical landscapes (Dufour, 1990; Thrupp et al., 1997). Padoch et al. (1998) report that a reinterpretation of present practices could be in order for what researchers might report as destructive activity upon the land. They cite their own experience in Southeast Asia, where careful attention to the dynamics in place found that a shortening of the fallow period was actually part of a deliberate plan toward a more productive *sawah* system. Additionally, we should note that shifting cultivation makes use not only of the cultivation portion of the cycle. The fallow is a resource from which a number of food, fiber, and other products can be obtained (Brookfield and Padoch, 1994). The use of the fallow needs more investigation.

The prevailing and dismissive attitude that shifting agriculture is destructive needs to be rethought. Ruddle (1974) noted that swidden could serve as a teacher of land management. Indeed, though much has been claimed about the impact of shifting agriculture upon nutrient stocks in the soil, work on nutrient dynamics shows not only that swidden plots tend to have high nutrient levels throughout the cycle, but that even at abandonment the nutrient stocks are relatively high (Jordon, 1989). Other myths abound about shifting cultivation's negative aspects, yet most of the preconceived notions are not borne out by the research that has been done (Thrupp et al., 1997).

More detailed work on the mechanisms that make shifting cultivation the preferred way of life for so much of humanity is certainly in order. The environmental impact of swidden and the ways in which resources are used in it are two obvious themes that need concerted attention from researchers, as is the investigation of various factors concerning productivity (nutrient cycles, disease and pest management, ecological interactions within the system, etc.).

The Question of the Small Farm

Whether a swidden system or a farm permanently situated, a common feature of tropical agricultural landscapes is that of the small land manager. The importance and tenacity of agriculture in general and small producers in particular — especially in the face of ever more industrialized farming practices — has been recognized for more than a century (Kautsky, 1988). Small producers dominate major internationally traded commodities based on traditional cash crops such as coffee and cacao (Rice and Ward, 1996; Rice and Greenberg, 2000). And even outside the confines of the tropics, in countries such as the U.S., we find that small landowners are the acknowledged managers of significant numbers of holdings, responsible for crop and cultural diversity, thoughtful land stewardship, and economic vitality (U.S. Department of Agriculture, 1998).

Perhaps one of the most important factors associated with small farms, regardless of whether temperate or tropical, is what agricultural economists refer to as "the inverse relationship between farm size and output" (Rosset, 1999a,b). Differentiating between the conventional measurement of yield (which focuses on a single crop) and total output (which takes into account all products derived from a given unit of land), we see that for many countries it is the smaller farm that prevails. And in terms of efficiency, we also see that large farms do not necessarily outcompete the smallholdings (Rosset, 1999a).

Small farms in the tropics can be, and often are, based on simple subsistence. Many regions, however, display vast numbers of small producers involved in some way with market-oriented crops, be they traditional cash crops or the more recent nontraditional export crops such as melons, broccoli, snow peas, cut flowers, etc. (to be addressed below). A common feature of tropical agricultural landscapes is the species and structural diversity of agroforestry systems.

Agroforestry Systems

Production from and management of agroforestry systems run the gamut in terms of scale, function, and structure — as well as their socioeconomic *raison d'etre*. The National Research Council (1993), basing its categorization on Nair (1999), identifies three major categories based on structure and function: agrisilvicultural systems, those combining food crops and trees; silvopastoral systems, those mixing trees with pasturelands; and agrisilvopastoral systems, those combining food crops, pastures, and trees. Home gardens, one example of agrisilviculture, have enjoyed moderate attention within the agroforestry student community (Mercer and Miller, 1998). They represent an intimate example of agroforestry, in which the managed or artificial forest lies in close proximity to the farmer's house. These are time-tested systems that have provided food, fiber, and general sustenance to millions of people in far-flung regions for centuries, yet have escaped focused scientific scrutiny (Nair, 1999).

Even though all these categories represent age-old management strategies involving trees, there are many details about their functions that science has simply not addressed. Even within the biophysical and agronomic studies, which prevail in terms of focus, we find descriptive, qualitative reports.* And once we step away from the biophysical and agronomic sketchiness of such systems (Nair, 1999), the economic and sociocultural details, not to mention the ecological value, show an even greater lack of attention. For all agroforestry systems, research priorities should include a better understanding of the interrelationship among tree species used, as well as attention paid to specific interactions such as the competition for light, water, and nutrients (as discussed in Chapter 2, García-Barrios, this volume).

Biophysical research dominates the field, with nutrient cycling accounting for a great portion of the work. There seems to be a certain aura of mysticism involved with most discussions, however (Nair et al., 1999). While a substantial body of work exists on nutrient cycling in tropical agroforesty systems, there nonetheless remains a lack of appropriate research methodologies. Evaluation of four types of agroforestry systems, alley cropping hedgerow intercropping systems, tree/cropland (parkland) systems, improved fallows, and shaded perennial systems, concluded that we

* For instance, at excellent research centers such as CATIE in Costa Rica, CIFOR in Indonesia, and ICRAF in Kenya.

know trees can help provide nitrogen for crop production. Concomitantly, however, adequate levels of phosphorus are not provided by the agroforestry system. While "major tree-mediated processes of the [nutrient cycling] mechanisms" are known, there is much in terms of the dynamics that remains shrouded — a situation that warrants much more concerted and rigorous research (Nair et al., 1999: 25).

The socioeconomic aspects of agroforestry systems, by contrast, continue to receive less attention than they deserve. Echoing this sentiment, Diane Rocheleau (1999) calls for a social science mandate for researching agroforestry collaboratively. Part of this mandate calls for examining agroforestry technology within its social context — a view based on the belief that all science is local. The social context of agroforestry technologies, issues of gender, class, age, religion, race, and ethnicity, must be part of any endeavor to understand the benefits, challenges, and pitfalls of agroforestry. Another tenet of Rocheleau's mandate is local participation in confronting the complexity of agroforestry systems. An analysis of the first 14 years of research presented in the journal *Agroforestry Systems* (Mercer and Miller, 1998) reveals that 22% of all articles relate to socioeconomics, although recent years show a gradual improvement in the scope and quality of socioeconomic research focused on agroforetry. To understand these agroforestry landscapes that cut across so much of the tropical agricultural terrain, more rigorous economic analysis based on larger sample sizes is needed.

Here it is worth commenting upon the concept of natural forestlands as they relate to historic human agency. Even those pristine tropical forest areas previously romanticized in the popular (and academic) conservation literature have in fact been subjected over human history to significant intervention by culture groups exploiting the resource base (Denevan, 1992; Dufour, 1990). While some might see this as a sullied landscape in some way, the presence of people in forest for millennia translates into a potential treasure trove of local information relating to these ecosystems. That local people today still command huge lexicons of their faunal and floral resources and can employ agricultural strategies that defy assumptions has received little attention (Zimmerer and Young, 1998). Relatively unaffected by the market forces of globalized agricultural production, such populations may be the richest reservoir of knowledge left to us.

TRADITIONAL AGROEXPORT CROPPING SYSTEMS

A number of tropical crops rank as important export-earning commodities. Some cover significant area; others are relative patches upon the Earth. However, there is a *quality* issue involved alongside the *quantity* issue. While total area of coffee, cacao, bananas, and sugarcane combined makes up less than 2% of that covered by pasture (Table 7.5), it is worth noting that coffee expansion often occurs in mid-elevational forests — one of the more biodiverse ecosystems of the world. Moreover, the lowland confines of cacao means that producers target lowland humid forest for expansion (Ruf, 1995), another ecosystem harboring significant portions of the world's biodiversity. Depending upon how these agroforestry systems are managed, they can either result in total removal of tropical habitat or act, to limited degrees, as refuges for biodiversity (Perfecto et al., 1996; Rice and Greenberg, 2000).

Table 7.5 Hectares Devoted to Major Tropical Crops and Land Uses, 1961 and 1998

Crop or Category	World Area 1961	World Area 1998	Hectareage Change	Percentage Change
Bananas	2,030,193	3,898,364	1,868,171	92
Cocoa beans	4,403,334	6,864,172	2,460,838	56
Coffee, green	9,755,805	10,762,980	1,007,175	10
Sugarcane	8,911,879	19,460,812	10,548,933	118
Tea	1,366,126	2,326,871	960,745	70
Pasture[a]	1,934,433,000	2,207,065,000	272,632,000	14
Forest/woodland[a,b]	2,381,910,000	2,254,462,000	–127,448,000	–5.4
Agricultural land[a]	2,610,612,000	3,062,556,000	451,944,000	17

[a] Developing countries only.
[b] "1998" data are from 1994.

Source: Data taken from FAO, *Agricultural Production Statistics* and *Land Use Statistics,* http:// apps.fao.org/.

Tropical agricultural landscapes are dotted with an array of traditional export crop production systems. Many of these attest to humans' power of landscape transformation, in which organisms have been moved around the globe, usually for reasons of commerce. Ecologically, removing a specific crop from its native area to an analogous environment in a distant region releases it from the relational confines posed by its natural pests and diseases. Thus we see the explosion of coffee into the Americas after its transatlantic voyage from Africa and cacao's spread into west Africa and Asia once it slipped beyond its native range of the neotropics.

Regardless of whether native or introduced, however, traditional agroexport crops often result in nearly identical cropping patterns and/or structures in far-flung regions of the world. Coffee presents an excellent example of a crop place diffusing into different areas (Rice, 1999). Table 7.5 shows the areal change of a number of traditional agroexports in the tropics over the past four decades. Except for the category of forest/woodland, which is included to provide additional information about what is happening to the landscape, all these crop categories show an increase during the last half of the 20th century. (Industrialized countries, it is worth noting, account for about 5% of the sugarcane area globally.)

All of the traditional tropical export crops show an increase in area over the past four decades, mirroring the general increase in agricultural lands in developing countries. At the same time, forest and woodland area has decreased. While the forest and woodland category shows what might be construed as a relatively small decrease during this time (5.4%), a more telling figure is the ratio of forest/woodland area to agricultural land in the developing countries. In 1961, the two were close to parity, with forest/woodlands covering 91% of the area covered by agricultural lands. By the 1990s, this figure had decreased to 74%.

Coffee

Tropical landscapes reveal significant changes over recent years in some of the traditional export crops such as coffee, cacao, and sugarcane. The coffee agroecosystem (and to a lesser extent, cacao) has received attention from an unlikely source in recent years, namely from ecologists and conservation biologists interested in the conservation value of managed lands. Central to this interest is that shade coffee systems, those incorporating an artificial forest or agroforest as the canopy for the coffee, can provide a refuge for biodiversity (Perfecto et al., 1996; see also Chapter 6, Perfecto and Armbrecht, this volume). Other work focusing on specific groups of organisms has recently uncovered some of the details and nuances of biodiversity and coffee — especially that related to avian communities and resource use (Wunderle and Latta, 1996, 1998, 2000; Greenberg et al., 1997b, 2000; Greenberg et al., 1997a, Wunderle, 1999; Roberts et al., 2000b; Sherry, 2000). The importance of shade coffee to other taxa supports the notion that managed systems seem to have a functional role in conservation (Ibarra-Nuñez, 1990; Nestel, Dickschen, and Altieri, 1993; Vandermeer and Perfecto, 1997; Roberts et al., 2000a). Coffee systems occupy nearly 11 million hectares worldwide (Table 7.5). Although we have some notion of the area managed with shade in parts of Latin America (Rice, 1999), global figures for such details are not yet available.

Cacao

Cacao systems offer similar conditions, albeit usually at a lower altitude, and consequently with a shade component quite distinct from that of coffee. Worldwide cacao systems cover close to 7 million hectares (Table 7.5). Produced mainly by small growers, different regions can be characterized as to general management strategies (Rice and Greenberg, 2000). Cacao is grown under agronomic and socio-economic circumstances that Ruf (1995) identifies as recurrent themes in a cacao cycle theory. The critical environmental outcome is the incessant advance of cacao farmers into naturally forested areas in search of untapped forest rent — the benefits that befall anyone taking advantage of the stored (yet elusive) nutrients bound up in lowland tropical forest soils. The result, of course, has been deforestation in these tropical areas in exchange for cacao systems, which after a couple of decades are abandoned in search of soils with unspent forest rent.

A trend within some cacao-producing countries has been to modernize to increase yields, which translates into monocultural plantation production in the open sun. Work in Indonesia, a showcase example of such changes, suggests that open-sun systems "may adversely affect biodiversity, long-term agricultural productivity and sustainability, and local livelihood security" (Belsky and Siebert, n.d.). Others, working in Central America, where the economic importance of cacao has diminished greatly compared to its historical levels, point to the potential of cacao as an agricultural system with tremendous ecological and biodiversity value (Parrish et al., 1999).

As a cash crop covering only about 7 million hectares worldwide, cacao's potential as a conservation tool, as well as that of coffee, which accounts for around 11 million hectares, may not be obvious immediately. However, it is not a simple matter of *quantity* of area involved. Rather, the *quality* of the ecological zones associated with these crops matters. Oftentimes, tropical crops coincide spatially with what many conservation groups refer to as biological or biodiversity hot spots. Moreover, the shaded crops' proximity to intact natural forest seems to play a role in the degree to which such systems harbor diverse communities (Parrish et al., 1999).

Cane Sugar

Sugarcane blankets nearly 20 million hectares globally, with 95% being in developing, nonindustrial countries (FAO, 1999). Since the early 1960s, the area devoted to its production has jumped 118% (Table 7.5). In contrast to coffee or cacao, sugarcane has seen relatively greater penetration of capital into the production process, especially in relation to irrigation and mechanization of the harvest. Moreover, certain cases reveal a truly industrial, factory-in-the-field approach, in which the labor force has received unprecedented attention.

Taking a page from the gold mines of South Africa and the sugarcane industry of Colombia's Cauca Valley, the sugar processing mills of Guatemala's south coast recently redefined labor contracting via computerized migrancy (Oglesby, 2000). Electronic monitoring of individual workers now tracks the daily productivity of cane cutters, rewarding those who meet or exceed minimum productivity standards and dismissing those who do not. A Tayloristic, factory-oriented approach to field work defines details of a cane cutter's job that include how to hold the machete, how to sharpen the machete, how to sit when sharpening the machete, how to hold the file when sharpening the machete, and how to grab hold, cut, and arrange sugarcane stalks in the cutting process (Oglesby, 2001). On the positive side, good workers are treated like marathon athletes, receiving 4000 calories per day in a balance of protein and carbohydrates, as well as oral rehydration drinks. To ensure that workers remain in top physical condition, workers are bussed to the work site, provided with water tanks in the field, and are weighed periodically during the season to make sure they are not losing weight (Oglesby, 2000: 21). Such facets of the production process certainly qualify as the hidden (social) landscape underlying the visible agricultural landscape — what Mitchell (1996) calls the "lie of the land".

Bananas

The world today consumes many more bananas than it did in the early 1960s, even though in absolute area bananas exceed only that of tea when compared to other major tropical exports. Still, the 92% increase in banana area over the past 40 years is surpassed only by that of sugarcane (Table 7.5). Between 1961 and 2000, world production of bananas increased 202%, much of which can be attributed to the entrance of Asian countries onto the scene. In 1961, Asia did not even warrant mention of FAO record keepers as a region of banana production. By the year 2000, Asia accounted for 42% of the harvested area. During this same period, Latin America and the Caribbean maintained about one third of the harvested global banana area.

Yet, these figures mask much of what has occurred in banana production. One reason for the 202% increase in production derives from the increased yields. Worldwide, yields grew by more than 53%, mirroring the modernization trends presented in tropical export agriculture around the globe. Much of the impetus for increased yields relates to trade. Competition in trade between producing countries — a situation which is always advantageous for consuming counries and next to disastrous for producing countries, regardless of the commodity — only serves to promote the dependence upon the inputs and genetic material needed to survive in free market conditions. Trade agreements which have traditionally protected some producers are easily overturned with today's policymakers' enchantment with free trade (Oxfam, 1998).

While it might be tempting to attribute the bulk of social and environmental ills to export agriculture, the situation can be more nuanced in certain cases. In the St. Vincent banana sector, for instance, Grossman (1993) found that declining food crop area and increased food imports could not be laid directly at the feet of increased banana area. Rather, a host of social and political forces, as well as farmer choice, worked to create these symptoms. Still, we should realize that such cases may be the exception to the rule that export agriculure often displaces food production and discrupts social tradition.

Considering that managed ecosystems and human settlements combine to account for some 95% of the Earth's land surface (Pimentel et al., 1992), examination of such systems for their potential contribution to conservation is not only intriguing, but may spell hope for conservation in many areas. The destruction of natural forest areas via commercial logging, swidden systems, and other operations associated with the extractive frontier makes conservation research findings in managed lands all the more important. We need not pretend that managed lands such as coffee with a shade canopy can *replace* natural forest — they certainly cannot. But if economically active and productive land-use practices can also be shown to have an ecological or conservation value, then more work on such systems is clearly warranted. Research obviously needs to address issues of how biodiversity and production interrelate and, as some have pointed out (Zimmerer and Young, 1998), how development with conservation might proceed.

Moreover, as globalization moves to draw increasing numbers of producers into the global supermarket, understanding the workings behind the obvious physical setting can greatly inform us as to what is taking place throughout the tropical agricultural landscapes. Future work on traditional export cropping systems must take into account the social backdrop and explore the causal web which, while often invisible when studying the biophysical aspects of agriculture, forms an integral part of the landscape mosaic.

IRRIGATION, NONTRADITIONAL, AND TEMPERATE CROP LANDSCAPES IN THE TROPICS

FAO statistics show that irrigated agricultural land in developing countries jumped more than 100% between the 1960s and the 1990s, adding more than 103 million hectares to the area cultivated (Table 7.6). All regions except Africa developing, Africa south of the Sahara, and East and Southeast Asia show at least twofold increases.

Irrigation can certainly increase production by opening up previously unused or temporally underused lands, as witnessed by the 2.4-fold increase in global grainland productivity between 1950 and 1995 (Postel, 1999:165). Depending upon the site characteristics, however, impacts can range from health and disease problems for

Table 7.6 Irrigated Lands

Region	1961 (1000s ha)	1998 (1000s ha)	Percentage Change
Developing countries	101,953	205,358	101.4
Africa developing	6,602	11,170	69.2
Africa south of Sahara	2,709	5,169	90.8
Asia developing	87,090	175,874	101.9
East and Southeast Asia	9,704	19,209	97.9
Latin America and Caribbean	8,260	18,311	121.7
Central America and Caribbean	3,599	8,268	129.7
Caribbean	441	1,282	190.7
Least developed countries	5,853	14,435	146.6
Low-income countries	77,481	159,785	106.2
Industrialized countries	25,897	41,436	60.0

Source: Data taken from FAO, *Agricultural Production Statistics* and *Land Use Statistics,* http://apps.fao.org/.

local residents, salinization of large areas, and, often undiscussed, the use and waste of irrigation water itself. The aforementioned rise in grain production was coupled to a 2.2-fold jump in irrigation water use (Postel, 1999:165).

Salinization involves the deposition of naturally occurring salts found in soils within the upper portion of an irrigated soil. Studies show that 10 to 15% of the world's irrigated land suffers some degradation in the form of salinization or waterlogging (Alexandratos, 1998, cited in World Resources Institute, 1998). Researchers now surmise that the fall of the Sumerian civilization was closely linked to salt deposition caused by irrigation in the Mesopotamia region (Postel, 1999:18*ff*). India has some 7 million hectares affected by salt poisoning, and Pakistan spends more than $1 billion annually to combat the salinization problem (Postel, 1999:96*ff*).

Health problems related to irrigation projects in the tropics are legion, with more than 30 diseases having links to such endeavors (World Resources Institute, 1998; also see Chapter 9 in this volume). Malaria, schistosomiasis, Japanese encephalitis, cholera, and a host of gastrointestinal maladies have been associated with irrigation canals or wastewater, expanding the natural ranges of such diseases into areas heretofore unaffected by such problems (World Resources Institute, 1998:47*ff*).

A number of irrigated crops arguably qualify as nontraditional — at least in the ecosystem in which they grown. Granted, the mere practice of irrigation does not elicit images of nontraditional crops. Yet irrigation has changed the agricultural landscape throughout the tropics. From 1961 to 1998, irrigated agricultural land in all developing countries grew 101%, as opposed to 60% in industrialized countries (FAO, various years). While those countries the UN lists as least developed saw a 146% increase in irrigated lands during this period, Latin America, the Caribbean, and the developing countries of Asia and Africa realized 122%, 102%, and 69% increases, respectively (Table 7.6). While irrigation systems certainly have their place in traditional agricultural systems, the water control regimes generally depict historical precedents and practices steeped in local knowledge. Capital-intensive irrigation, by contrast, having exploded across the tropical belt since the mid-1900s, depend upon imported technology and knowledge oriented toward the production of feed crops as opposed to food crops. In countries such as Mexico, irrigated land area increased 117% between 1961 and 1998 (FAO, various years).

While not all feed grains are irrigated, their production has expanded in arid regions due to capital intensive irrigation projects. Barkin et al. (1990) provide an excellent overview of how expansion of feed crops like sorghum and maize, linked to the green revolution of industrial animal feed, has hampered the nutritional status of many developing world residents, led to food insecurity, and shifted grain production in many countries toward upper and middle income markets. Brazil, for example, saw a tremendous jump in industrial and nontraditional export grain production over the last half of the 20th century. Owing principally to the expansion of soybeans, the transformation has been grounded in financing and export links with Japan (Barkin et al., 1990:44). Between 1961 and 2000, the area, yields, and production of soybean in Brazil increased 56-fold, 2-fold, and 120-fold, respectively (FAO, various years).

An interesting assessment of Brazil's production and consumption of soybeans was done by Sousa and Busch (1998), in which they employed the actor-network theory (ANT) to unravel the complexities of how soybean production evolved such high status within Brazil's agricultural sector. Their work provides an excellent view of how this approach might inform other questions or cases of concern to agroecologists. ANT allowed them to consider humans and nonhumans as actors within the network, arguing that nature and technology, while not acting in any purposeful way, act (or fail to act) in ways that mediate human interactions. Indeed, the ANT approach allows us to formulate questions and problematize the case of soybean production in Brazil in ways that go beyond diffusion theory or political economy in providing a rich texture to the changes that occurred there. Many working on agroecological questions might do well to consider ANT as an approach in some of their own work.

FRUITS, VEGETABLES, FLOWERS, SEEDS: QUINTESSENTIAL NONTRADITIONAL AGRICULTURAL EXPORTS (NTAEs)

The last several decades abound with examples of tropical countries developing economies upon the foundation of crops either heretofore not present upon the physical landscape or formerly not part of the economic export package. Countries of South America and Africa have devoted substantial resources and areas to the export of flowers, fruits, and vegetables with measurable success (Barham et al., 1993; Jaffee, 1993; Thrupp, 1995; Conroy et al., 1996). Some even point to such crops as playing influential roles in national peace efforts (Maggs, 1999).

There is no doubt that these nontraditional export crops, often characterized by high-value fruits, vegetables, or flowers, have evolved to cover relatively large areas of the tropical agricultural landscape during the past few decades. Data in Table 7.2 show that, while industrialized countries' area in fruits and vegetables decreased 5% between 1961 and the end of the century, Africa, Asia, and Latin America/Caribbean rang up an average increase of 100% in the fruit and vegetable area. In Central America, the Caribbean, and Mexico, this same period shows an increase of 172% for the category vegetables and melons (FAO, various years). Lest we conclude that our tropical neighbors developed a late-20th century craving for such crops, the data reveal that the increase accompanies an export explosion. Cantaloupes and melons exit the Central America/Mexico region in trade to the tune of over 614,000 metric tons per year, a 1200% increase over the 1961 figures. (In terms of value, the increase has been even greater, showing a 27-fold increase.) Fresh vegetable exports from this region show a 700% increase during this same time frame (FAO, various years).

Such changes, however, do not benefit all concerned. While the physical landscape shows impressive changes that might suggest all is well in the garden patch, several students of shifts in tropical agricultural report less-than-rosy pictures of the socioeconomic and human sides of these changes (Murray, 1994; Green, 1995; Conroy et al., 1996; Carletto et al., 1999; Rosset et al., 1999). Production is often mediated via contractual agreements, which are part of a global restructuring of agricultural production, as well as a motor force in restructuring local livelihoods (Raynolds, 1999; Rosset et al., 1999; Watts and Goodman, 1999). Benefits realized from the nontraditional export market accrue disproportionately to larger, wealthier growers. In fact, some of the very policies designed ostensibly to aid the rural poor by becoming involved in NTAE can be detrimental to them. Food security dissolves, economic risk and economies of scale work against them, and, if they do manage to surmount some formidable barriers to entry, they face personal risk from the agrochemicals involved.

In spite of these difficulties, the strategy to implement NTAEs in Central America found powerful institutional support from the United States Agency for International Development (USAID), the International Monetary Fund (IMF), and the World Bank. The thrust of these development and lending interests was to privatize state-owned enterprises, liberalize capital markets, and form business-led entities capable of assuming traditional government agency tasks (Conroy et al., 1996). Many of these efforts were successfully realized, enriching larger producers and the export companies, while simultaneously working against smaller growers. Moreover, satellite production (contacted outgrower arrangements), in which exporters provide agrochemical inputs, seeds (or seedlings), credits, and technical assistance to producers, transfers the risks of production and market fluctuations to the farmer (Murray, 1994).

Studies that examine the social and cultural landscape transformations involved in the promotion and production of these types of crops are sorely needed to allow for a better understanding of *total* landscape changes occurring. These crops may be no more international in their distribution than traditional crops such as coffee, tea, or cacao. Yet, due to their novelty in certain areas of production, as well as their perishability and consumer demands when produced for far-flung fresh markets, producers confront an array of physical/agronomic, financial, and commercial challenges. The varied infrastructural components that pose these challenges, be they local or global, must be addressed by researchers if we are to understand the forces and players affecting tropical agricultural landscapes.

CONCEPTUALIZING THE PROCESS

Figure 7.1 attempts to show graphically the various positive and negative effects that being drawn into the marketplace, especially where global forces are concerned,

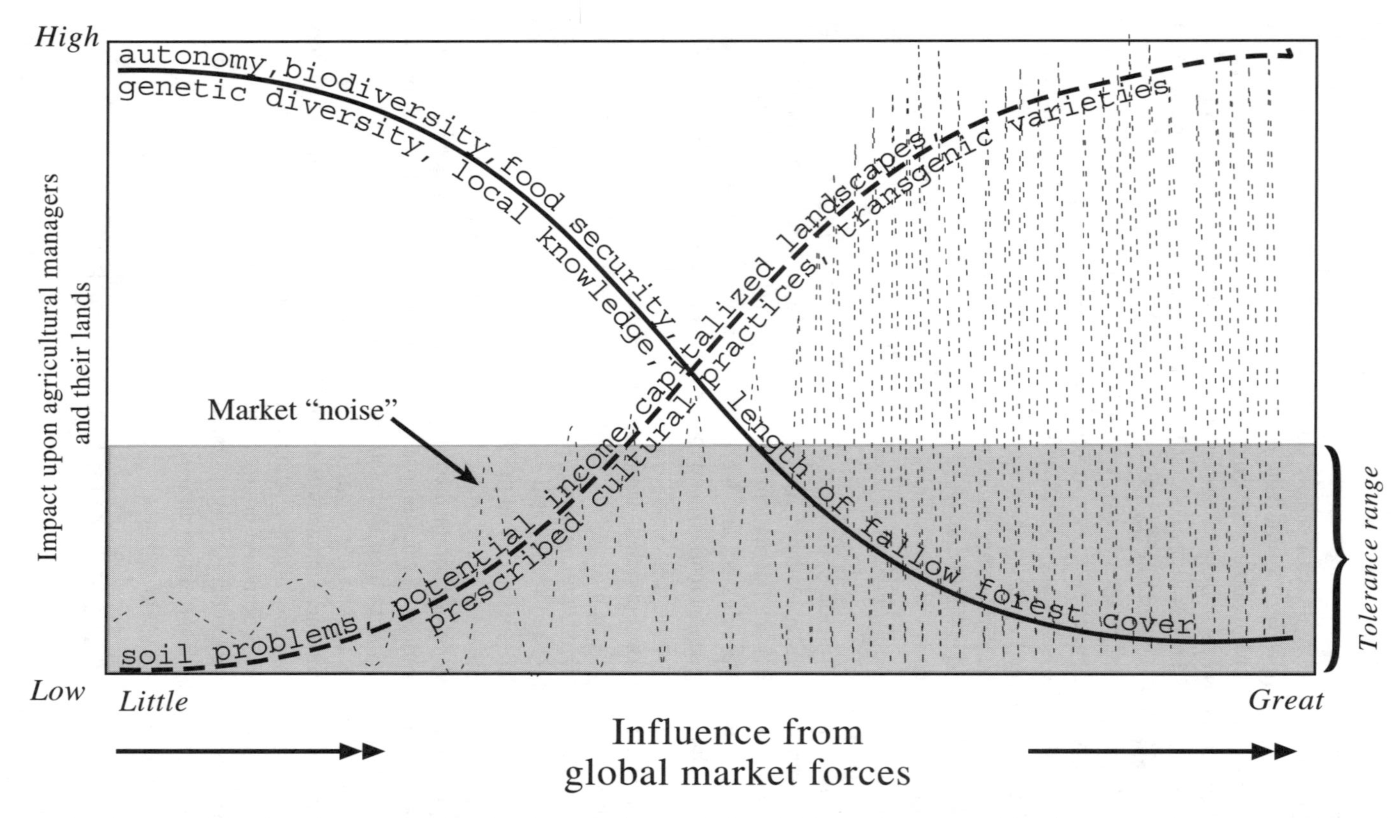

Figure 7.1 Conceptualizing the effects of global market forces and the sociofinancial demands accompanying them on tropical agriculturalists and their managed lands.

might have on farmers and the tropical agricultural landscapes they manage. At issue is the tolerance farmers have to the forces involved in the ever-widening reaches of the global market. Any market generates demands and concomitant sociopolitical constraints to which those involved must respond.

A farmer might decide to grow some commodity like snow peas, say, only to discover once he has committed himself, his land, and his family's labor that the contract is written (and honored) in ways that only serve the company. He might be faced with increasing demands on the cosmetic appearance of his produce by the quality control agents of the company, all the while trying to produce in an area where certain insect pests constantly cause cosmetic damage to the crop. If he uses chemical pesticides to protect the snow peas from thrips or other insect damage, his produce might be rejected when tested for pesticide residues.

There is, in essence, a certain amount of "market noise" with which producers must contend. It might come in the form of pressure from government institutions trying to establish and maintain production levels of specific crops through imposed policies or international food-web firms looking to expand or maintain production via contract farming might constitute the major force in some areas. Local elites with global connections electing to control production through economic, social, and political cronyism are often involved as well. Regardless of the source, farmers face and must learn to tolerate a battery of challenges. We can also imagine that there is a "range of tolerance" within which most farmers can withstand whatever the market throws their way.

Figure 7.1 depicts these challenges in graphic form. On the horizontal axis is the influence from global market forces. These forces are multitudinous and might manifest themselves in terms of production demands, restrictive social relations, use of one's children (who might otherwise be in school) in the field in order to harvest a crop on time, etc. But collectively, we can refer to them as market influences. The vertical axis represents a measure of the relative levels of the variables listed upon the two curves (there are undoubtedly other variables that we can place on these curves). As the influence from market forces increases — that is, the "market noise" — these aspects associated with the physical landscape, socioeconomic well-being, community relations and environmental health can change accordingly.

For instance, a farmer in Chiapas, Mexico, who has traditionally grown corn, beans, and other food crops for his own family and perhaps for the local marketplace, might elect to grow cabbage (or flowers or hybrid vegetable seeds) for distant markets in Mexico City or beyond. Assuming that he chooses to do this with a contract agreement, his potential income might well increase as he puts more time, effort, and (borrowed) money into the landscape. If he finds that continuous cropping with irrigation allows him to sell cabbages year round, he might find that soil erosion also increases along with his income. Moreover, the contractor might insist on specific varieties of cabbage and predetermined cultivation practices (see the ascending curve in Figure 7.1). At the same time, as the descending curve shows, several cultural and/or landscape variables might decrease in a relative sense. Whatever autonomy enjoyed while producing food crops might disappear under the contract arrangement–not necessarily a bad thing, considering the potential economic gains. But unless land is also available to continue the food production or unless enough money is earned to purchase the necessary food stuffs previously produced on his land, a farmer might face food insecurity issues. If he has to use previously untilled or little-used land to honor the contractual agreement, forest cover might be eliminated or fallow periods shortened in order to comply. If these market forces act for a number of years or cross into the next generation, we might also find that genetic diversity in indigenous crops falls, along with local knowledge about how to cultivate such food stuffs.

There exists, we can argue, a "range of tolerance" within which the market forces acting upon the social and physical landscapes will not lead to any long term or irreversible transformations. As long as the outside influences from the market do not reach the point of pushing a land manager out of this tolerance zone, participation in global markets may well proceed without incident. As the forces become sronger, more frequent, or more directive, however, effects upon a host of variables (biodiversity levels, soil erosion, conrol over decision making, length of fallow, etc.) can present themselves. Any given site will have inherent characteristics — social, economic, and physical — which will determine at what point the market forces will push a given system out of the range of tolerance. Depending upon the frequency and amplitude of the "market noise" — as well as what point in the process we attempt to assess the influence of the market, it is most likely that some intermediate form of social and landscape transformation will be found.

Such "noise," of course, can be enhanced by a number of factors, such as falling world prices in a particular crop, consecutive years of bad luck from weather or disease, out-migration of labor forces from a region, etc. Decibel levels might understandably increase to unacceptable ranges. The noise-enhancing factors could propel farmers to feel more pressure from global market forces, resulting in the noise exceeding the range of tolerance and thus changing various tropical agriculture landscape features — be they physical or social. Not only can the noise increase in amplitude, but in frequency as well, resulting in a farmer being hit from all sides with information, choices, challenges, and demands from the market.

The degree of participation in the global marketplace, of course, plays a central role in how the physical and social landscapes associated with tropical agriculture look and operate. As Figure 7.1 suggests, these changes associated with the increasing influence of global market forces can also affect the long-term sustainability of these systems. A challenge for planners, policymakers, and decision makers, of course, is how to allow land managers and the communities in which they operate to benefit from linkages to the global market while keeping the market noise decibel level within the range of tolerance.

REFERENCES

Alexandratos, N., *World Agriculture: Towards 2010,* FAO and John Wiley & Sons, Chichester, U.K., 1998, p. 359.

Altieri, M., Why Study Traditional Agriculture? in *Agroecology*, Carroll, R., Vandermeer, J., and Rosset, P., Eds., McGraw-Hill, New York, 1990, p. 551–564.

ASB (Alternatives to Slash and Burn Agriculture Program), wwwscas.cit.cornell.edu/ecf3/Web/AF/ASBMain.html, 1999.

Atran, S. et al., Folk ecology and commons management in the Maya lowlands, *Proc. Natl. Acad. Sci. U.S.A.,* 96:7598–7603, 1999.

Barham, B. et al., Non-traditional agricultural exports in Latin America: toward an appraisal, *Latin Am. Res. Rev.*, 27:43–82, 1993.

Barkin, D., Batt, R.L., and DeWalt, B.R., *Food Crops vs. Feed Crops: Global Substitution of Grains in Production*, Rienner, Boulder, CO, 1990.

Blaikie, P. and Brookfield, H., *Land Degradation and Society*, Methuen, New York, NY, 1987.

Brookfield, H. and Padoch, C., Appreciating agrodiversity: a look at the dynamism and diversity of indigenous farming practices, *Environment,* 36:6–11, 37–45, 1994.

Carletto, C., de Janvry, A., and Sadoulet, E., Sustainability in the diffusion of innovations: smallholder nontraditional agro-exports in Guatemala, *Econ. Dev. Cult. Change,* 47:345, 1999.

Conroy, M.E., Murray, D.L., and Rosset, P.M., *A Cautionary Tale: Failed U.S. Development Policy in Central America,* Food First Books, Oakland, CA, 1996.

Denevan, W.M., The pristine myth: the landscape of the Americas in 1492, *Ann. Assoc. Am. Geogr.,* 82:369–385, 1992.

Dove, M.R., Theories of swidden agriculture, and the political economy of ignorance, *Agrofor. Syst.,* 1:85–99, 1983.

Dufour, D.L., Use of tropical rainforests by native Amazonians, *Bioscience,* 40:652–659, 1990.

FAO (Food and Agriculture Organization of the United Nations), *Agricultural Production Statistics*, http://apps.fao.org/, 1961a, 1999a, 2000a.

FAO (Food and Agriculture Organization of the United Nations), *Land Use Statistics,* http://apps.fao.org/, 1961b, 1999b, 2000b.

Goodman, D., Agro-food studies in the 'age of ecology': nature, corporeality, bio-politics, *Sociol. Ruralis,* 39:17–38, 1999.

Goodman, D., Sorj, B., and Wilkinson, J., *From Farming to Biotechnology: A Theory of Agro-Industrial Development,* Basil Blackwell, New York, 1987.

Green, D. Trapped by trade (Latin American food exports), *Geogr. Mag.*, 67:12–16, 1995.

Greenberg, R., Bichier, P., and Sterling, J., Bird populations in rustic and planted shade coffee plantations of eastern Chiapas, Mexico, *Biotropica,* 29:501–514, 1997a.

Greenberg, R. et al., Bird populations in shade and sun coffee plantations in central Guatemala, *Conserv. Biol.*, 11:448–459, 1997b.

Greenberg, R. et al., The impact of avian insectivory on arthropods and leaf damage in some Guatemalan coffee plantations, *Ecology*, 81:1750–1755, 2000.

Grossman, L.S., The political ecology of banana exports and local food production in St. Vincent, Eastern Caribbean, *Annals of the Association of American Geographers,* 83(2):347–468, 1993.

Hecht, S.B., Environment, development and politics: capital accumulation and the livestock sector, *World Dev.*, 13:663–684, 1985.

Houghton, R.A., Lefkowitz, D.S., and Skole, D.L., Changes in the landscape of Latin America between 1850 and 1985. I. Progressive loss of forests, *For. Ecol. Manage.*, 38:143–172, 1991.

Ibarra-Nuñez, G., Arthropods associated with coffee plants in mixed coffee plantations at Soconusco, Chiapas, Mexico, *Folia Entomol. Mex.*, 97:207–233, 1990.

Jaffee, S., Exporting High-Value Food Commodities: Success Stories from Developing Countries, World Bank Discussion Paper 189, World Bank, Washington, D.C., 1993.

Jarosz, L., Defining and explaining tropical deforestation: shifting cultivation and population growth in colonial Madagascar (1896–1940), *Econ. Geogr.,* 69:366–379, 1993.

Jordan, C.F., Ed., An Amazonian rain forest: the structure and function of a nutrient stressed ecosystem and the impact of slash-and-burn agriculture, Parthenon Publishing Group, Park Ridge, NJ, 1989, Chapter 5.

Kautsky, K., *The Agrarian Question,* Zwan Publications, London, 1988.

Maggs, J. The power of flowers to bring peace (Colombia's cut flower industry), *Natl. J.,* August 21:2440, 1999.

Mercer, D.E. and Miller, R.P. Socioeconomic research in agroforestry: progress, prospects, priorities, *Agrofor. Syst.,* 38:177–193, 1998.

Mitchell, D., *The Lie of the Land: Migrant Workers and the California Landscape,* University of Minnesota Press, Minneapolis, 1996.

Murdoch, J., Inhuman/nonhuman/human: actor-network theory and the prospects for a non-dualistic and symmetrical perspective on nature and society, *Environ. Plann. D: Soc. Space,* 15:731–756, 1997.

Murray, D., *Cultivating Crisis: The Human Cost of Pesticides in Latin America*, University of Texas Press, Austin, TX, 1994.

Nair, P.K.R., Do tropical home gardens elude science, or is it the other way round?, paper presented at the International Symposium on Multi-strata Agroforestry Systems with Perennial Crops, Turrialba, Costa Rica, 1999.

Nair, P.K.R. et al., Nutrient cycling in tropical agroforestry systems: myths and science, in *Agroforestry in Sustainable Agricultural Systems,* Buck, L.E., Lassoie, J.P., and Fernandes, E.C.M., Eds., CRC Press/Lewis Publishers, Boca Raton, FL, 1999, p. 1–32.

National Research Council, Agroforestry systems, in *Sustainable Agriculture and the Environment in the Humid Tropics,* National Academy Press, Washington, D.C., 1993, pp. 92–99.

Nestel, D., Dickschen, F., and Altieri, M.A., Diversity patterns of soil macro-Coleoptera in Mexican shaded and unshaded coffee agroecosystems: an indication of habitat perturbation, *Biodiversity Conserv.,* 2:70–78, 1993.

Nye, P.H., and Greenland, D.J., *The Soil under Shifting Cultivation,* Commonwealth Agricultural Bureau, Farnham Royal, U.K., 1965.

Oglesby, E., Managers, Migrants and Work-place Control: the Politics of Production on Guatemalan Sugar Plantations, 1980 to the present, paper presented at the Latin American Labor History Conference, Duke University, Durham, NC, 2000.

Oglesby, E., Machos and machetes in Guatemala's cane fields, *NACLA: Rep. Am.,* 34:16–17, 2001.

Oxfam, A Future for Caribbean Bananas: The Importance of Europe's Banana Market to the Caribbean, Oxfam GB Policy Paper–Mar. 98; http://www.oxfam.org.uk/policy/papers/bananas/bananas.htm

Padoch, C., Harwell, E., and Susanto, A., Swidden, sawah, and in-between: agricultural transformation in Borneo, *Hum. Ecol. Interdisciplinary J.,* 26:3–21, 1998.

Parrish, J.D. et al., *Cacao as Crop and Conservation Tool in Latin America: Addressing the Needs of Farmers and Forest Biodiversity,* Working Paper 3, Latin America and Caribbean Region, International Conservation Program, The Nature Conservancy: Arlington, VA, 1999.

Parsons, J.J., Antioqueño colonization in western Colombia, in *Ibero-americana:32,* University of California Press, Berkeley, 1949.

Perfecto, I. et al., Shade coffee: a disappearing refuge for biodiversity, *Bioscience,* 46(8):598–609, 1996.

Pimentel, D. et al., Conserving biological diversity in agricultural/forestry systems, *Bioscience,* 42:354–362, 1992.

Postel, S., *Pillar of Sand: Can the Irrigation Miracle Last?*, WW. Norton, New York, 1999.

Raynolds, L., Restructuring national agriculture, agro-food trade, and agrarian livelihoods in the Caribbean, in *Globalizing Food: Agrarian Questions and Global Restructuring,* Goodman, D. and Watts, M.J., Eds., Routledge, New York, 1999, p. 119–132.

Rice, R., A place unbecoming: the coffee farm of northern Latin America, *Geogr. Rev.,* 89:554–579, 1999.

Rice, R. and Greenberg, R., Cacao cultivation and the conservation of biological diversity, *Ambio,* 29:167–173, 2000.

Rice, R. and Vandermeer, J., Climate and the geography of agriculture, in *Agroecology,* Carroll, C.R., Vandermeer, J.H., and Rosset, P.M., Eds., McGraw-Hill, New York, 1990, p. 21–64.

Rice, R. and Ward, J., *Coffee, Conservation and Commerce in the Western Hemisphere* (Policy White Paper 2), Smithsonian Migratory Bird Center/National Zoological Park, Washington, D.C., 1996.

Roberts, D., Cooper, R.J., and Petit, L.J., Use of premontane moist forest and shade coffee agroecosystems by army ants in western Panama, *Conserv. Biol.,* 14:192–199, 2000*a*.

Roberts, D., Cooper, R.J., and Petit, L.J., Flock characteristics of ant-following birds in premontane moist forest and coffee agroecosystems, *Ecol. Appl.,* 10:1414–1425, 2000*b*.

Rocheleau, D., Confronting complexity, dealing with difference: social context, content, and practice in agroforestry, in *Agroforestry in Sustainable Agricultural Systems,* Buck, L.E., Lassoie, J.P., and Fernandes, E.C.M., Eds., CRC Press/Lewis Publishers, Boca Raton, FL, 1999, p. 191–235.

Rosset, P., The multiple functions and benefits of small farm agriculture in the context of global trade negotiations, paper presented at Cultivating Our Futures, the FAO/Netherlands conference on the Multifunctional character of Agriculture and Land, Maastricht, The Netherlands, 1999a.

Rosset, P., Small if bountiful, *Ecologist,* 29:452–459, 1999b.

Rosset, P., Rice, R., and Watts, M., Thailand and the world tomato: globalization, new agricultural countries (NACs) and the agrarian question, *Int. J. Soc. Agric. Food,* 8:71–94, 1999.

Ruddle, K., *The Yukpa Cultivation System: A Study of Shifting Cultivation in Colombia and Venezuela,* University of California Press, Berkeley, 1974.

Ruf, F., From forest rent to tree-capital: basic "laws" of cocoa supply, in *Cocoa Cycles: The Economics of Cocoa Supply,* Ruf, F. and Siswoputranto, P.S., Eds., Woodhead Publishing, Ltd., Cambridge, England, 1995, p. 1–53.

Sauer, C.O., The morphology of landscape, in *Land and Life: A Selection from the Writings of Carl Ortwin Sauer,* Leighly, J., Ed., University of California Press, Berkeley, 1963, p. 315–350.

Sherry, T.W., Shade coffee: a good brew even in small doses, *The Auk,* 117:563–568, 2000.

Sousa, I.S.F. de and Busch, L., Networks and agricultural development: the case of soybean production and consumption in Brazil, *Rural Soc.,* 63:349–371, 1998.

Spencer, J.E., *Shifting Cultivation in Southeastern Asia,* University of California Press, Berkeley, 1966.

Thrupp, L.A., *Bittersweet Harvests for Global Supermarkets: Challenges in Latin America's Agricultural Export Boom,* World Resources Institute, Washington, D.C., 1995.

Thrupp, L.A., *Cultivating Diversity: Agrobiodiversity and Food Security,* World Resources Institute, Washington, D.C., 1998.

Thrupp, L.A., Hecht, S., and Browder, J., *The Diversity and Dynamics of Shifting Cultivation: Myths, Realities, and Policy Implications*, World Resources Institute, Washington, D.C., 1997.

U.S. Department of Agriculture, *A Time to Act: A Report of the USDA National Commission on Small Farms,* USDA Miscellaneous Publication 1545, USDA, Washington, D.C., 1998.

Vandermeer, J. and Perfecto, I., *Breakfast of Biodiversity: The Truth About Rain Forest Destruction,* Food First, Oakland, CA, 1995.

Vandermeer, J. and Perfecto, I., The agroecosystem: the need for a conservation biologist's lens, *Conserv. Biol.,* 11:591–592, 1997.

Watts, M.J. and Goodman, D., Agrarian questions: global appetite, local metabolism: nature, culture, and industry in *fin-de-siècle* agro-food systems, in *Globalizing Food: Agrarian Questions and Global Restructuring,* Goodman, D. and Watts, M.J., Eds., Routledge, New York, 1999, p. 1–34.

Williams, R.G., *Export Agriculture and the Crisis in Central America,* University of North Carolina Press, Chapel Hill, 1986.

Woods, M., Researching rural conflicts: hunting local politics and actor-networks, *J. Rural Stud.,* 14:321–340, 1997.

World Resources Institute (WRI), *World Resources: 1998–99, Environmental Change and Human Health,* Oxford University Press, New York, 1998, p. 47–49; 157.

Wright, A., *The Death of Ramón González,* University of Texas Press, Austin, 1990.

Wunderle, J.M., Jr., Avian distribution in Dominican shade coffee plantations: area and habitat relationships, *J. Field Ornith.,* 70:58–70, 1999.

Wunderle, J.M., Jr. and Latta, S.C., Avian abundance in sun and shade coffee plantations and remnant pine forest in the Cordillera Central, Dominican Republic, *Ornith. Neotrop.,* 7:19–34, 1996.

Wunderle, J.M., Jr. and Latta, S.C., Avian resource use in Dominican shade coffee plantations, *Wilson Bull.,* 110:271–281, 1998.

Wunderle, J.M., Jr. and Latta, S.C., Winter site fidelity of Nearctic migrants in shade coffee plantations of different sizes in the Dominican Republic, *The Auk,* 117:596–614, 2000.

Zimmerer, K. and Young, K., *Nature's Geography: New Lessons for Conservation in Developing Countries,* University of Wisconsin Press, Madison, 1998.

CHAPTER 8

Interactions between Wildlife and Domestic Livestock in the Tropics

Johannes Foufopoulos, Sonia Altizer, and Andrew Dobson

CONTENTS

INTRODUCTION

The Ngorongoro Crater and the greater Serengeti ecosystem of East Africa harbor some of the most diverse and captivating wildlife assemblages in the tropics. In this

0-8493-1581-6/03/$0.00+$1.50

area as in other parts of Africa, wildlife ventures outside protected parks during part of each year and commonly grazes the same lands as livestock. The apparent coexistence of wildlife and domestic animals, although superficially peaceful, poses complicated challenges for local people using the land. In the Ngorongoro area, Masai tribe members who depend on domestic animals for their livelihoods face many problems associated with lion predation and disease, and they themselves have caused dramatic losses for wildlife (Rodgers and Homewood, 1986; Homewood et al., 1987). One historical avenue through which wildlife and farmers interact is pathogens such as rinderpest, an exotic virus introduced into the area around the turn of the 20th century. This disease has had wide-ranging effects on both the Masai cattle and the local antelope populations, triggering secondary changes on predators and native plant communities. This complex relationship between the local tribe, their animals, and the native wildlife is typical of sub-Saharan Africa and other tropical regions.

Worldwide, livestock production is one of the primary uses of terrestrial ecosystems — almost one quarter of the total land area, or 60% of the world's agricultural land, is used for grazing cattle, sheep, and goats (e.g., Vitousek et al., 1997; Lutz et al., 1998; Voeten, 1999; Tilman et al., 2001). In addition, up to one fifth of all crops are currently grown to feed livestock, and a major episode of agricultural expansion is predicted to ensue during the next 50 years (Tilman et al., 2001). Historically in arid tropical regions, livestock and wildlife have coexisted for many thousands of years. Although not always stable (several taxa have gone extinct in the more arid habitats of the Sahara and the Sahel), this coexistence has been facilitated by relatively low human densities in most tropical regions. This coexistence between wildlife and livestock produces multiple benefits for local societies: in addition to direct nutritional and economic advantages, humans reap many indirect benefits from big-game hunting and ecotourism, as well as the various ecosystem services provided by a stable natural environment (e.g., Daily, 1997). Conversely, natural ecosystems benefit when humans and domestic animals sustainably coexist with wild vertebrate populations, and ultimately this coexistence is critical in maintaining biodiversity over long periods of time, especially in tropical regions.

As a result of improved health services, peace, and easier access to technology, human populations burgeoned throughout the second half of the 20th century, generally with catastrophic impacts on the local biodiversity (Armesto et al., 1998; Balmford et al., 2001). Furthermore, livestock production and meat consumption are rising faster than the increase in human population size, and this is especially true for goats, pigs, and poultry in developing regions of Asia and Latin America. Livestock is becoming increasingly common and important to sustaining farmers in the tropics, as it provides manual power, manure, and capital reserve in addition to food. Moreover, there is increasing evidence that areas of great conservation importance (rich in endemic wildlife and species diversity), particularly in Africa and Latin America, coincide with high human densities and intense land use in the form of farming and raising livestock (Armesto et al., 1998; Balmford et al., 2001). Consequently, interactions and conflicts between wildlife and livestock are likely to become more intense as wild animals become sectioned between urban areas and managed farmlands.

The complete domination of the landscape by humans has contributed to the collapse of wildlife populations in many parts of the tropics. Currently, many herbivores are either threatened or close to extinction (e.g., tapirs in South America; blackbuck (*Antilope cervicapra*) in India; gaur cattle (*Bos frontalis*), kouprey cattle (*Bos sauveli*), and wild water buffalo (*Bubalus bubalis*) in Southeast Asia; and several medium-sized marsupials in Australia), contributing to secondary declines in many carnivore species. Asian lions and cheetahs have been reduced to critically low numbers, and other predators like leopards, wolves, dhole, and tigers now have been placed on the endangered species lists (IUCN Red List, 2000; Gittleman et al., 2001). Environmental problems arising from livestock production are particularly severe in developing nations. Although deforestation for rearing livestock is primarily a concern in Latin America (Armesto et al., 1998), overgrazing and land degradation occur in most areas where humans manage domestic stock. Particularly in Africa, livestock and wildlife graze the same lands and compete for similar resources (Voeten, 1999). Weaker and fewer links between humans and their land contribute to land flight, favoring short-term resource exploitation over long-term, sustainable land use. Fences and other human-made barriers interfere with wildlife migrations or natural movements and impede tracking of ephemeral resources. A suite of parasites and infectious diseases are shared between wild and domestic animals, and elevated densities of domestic cattle and dogs have triggered and sustained major epidemics in wild ungulates and carnivores (Dobson and Hudson, 1986; Packer et al., 1999; Funk et al., 2001). Finally, hunting and removal of both herbivores and predators to eliminate competition and predation of livestock have taken a heavy toll on wild animal populations.

History of Livestock–Wildlife Interactions

Livestock and other domesticated animals have interacted with wildlife since domestication began. The earliest domestication of animals and plants most likely occurred in the Near East when hunting and gathering tribes began to domesticate dogs, goats, and sheep at least as early as 12,000 years ago (e.g., Ucko and Dimbleby, 1969; Diamond, 1997). The process of domestication in the New World occurred independently and much later than in the Old World (Sauer, 1952). In fact, archaeological evidence indicates that plant and animal domestication arose independently in at least five separate locations, including the Near East, Southeast Asia, eastern North America, highland Mexico, and the Peruvian coast and highlands (Diamond, 1997).

Evidence suggests that the first domesticated species were used for meat, bones, and fur, much in the same way that hunter-gatherers used animals (Clutton-Brock, 1981). Sheep and goats were used for food in the initial stages of domestication and only later became valued for milk and wool. The principal aim of cattle breeding in ancient times was to obtain meat, skin, and work animals, which greatly assisted agricultural development. In contrast, the first domesticated fowl were probably used for sport and as a religious symbol; high egg yield and improved meat quality developed later (Mason, 1984). Selection of domestic species focused on several common features, including a docile or tame demeanor, products or services provided to humans, and breeding and care that can be almost totally regulated by humans (Mason, 1984). The range of characteristics produced by artificial selection on domesticated species can be quite stunning, although some domesticated animals (e.g., Bali cattle, water buffalo) remain close to their wild phenotypes (Clutton-Brock, 1981).

Of the more than 45,000 vertebrate species that exist (Smith et al., 1993), approximately 40 have been domesticated by different human cultures, with as few as 14 species dominating 90% of current livestock production (Anderson, 2001). Of all current species of domestic animals, five terrestrial herbivores are the most widespread and have the greatest economic and historical importance: sheep, goats, cattle, pigs, and horses. Nine other terrestrial mammals, including camels, llamas, donkeys, reindeer, and buffalo, have more limited geographic distributions or are less common relative to the dominant mammals (Diamond, 1997). The ancestors of many of these species had ranges that coincided with tropical (or subtropical) regions, including the aurochs and wild boars in North Africa, wild asses and camels in North Africa and Southwest Asia, and water buffalo and banteng in Southeast Asia. Other domestic species with historical prominence in tropical regions include chickens (wild jungle fowl of Southeast Asia and Indonesia), turkeys (wild turkeys of Central America), goats, and sheep (both of the latter occurring in Southwest Asia; Isaac, 1970; Mason, 1984).

TYPES OF INTERACTIONS AND IMPACTS

Predation and disease are the major conflicts between wildlife and livestock, although competition for space and resources plays an increasingly important role. Extinctions of wild ancestral species historically and repeatedly followed the development and expansion of new animal breeds (MacPhee, 1999). For example, the extinction of wild aurochs (*Bos primigenius*) followed the worldwide spread of domestic cattle (Epstein and Mason, 1984), and wild horses also vanished after domestication of modern horses (*Equus caballus*). Though ultimately exterminated via hunting, competition for space and resources during the last three centuries likely played a role in their demise (Day, 1981). In South America, wild camelids (vicunas and guanacos) declined rapidly following the Spanish conquest due to hunting and competition with sheep, and remaining wild populations are either endangered or extremely threatened (Wheeler, 1995). Finally, the spread of European settlers and their domestic animals throughout northern Europe and North America during the past four centuries was followed by the deliberate extermination of large predators, including seven subspecies of wolf (*Canis lupus*) (Day, 1981).

Although interactions between livestock and wildlife can take on many forms, the two groups most commonly interact through one of the following four modes: direct competition for food, predation (generally from wildlife on livestock), pathogen exchange, or hybridization. Most interactions involve direct conflict, but there are regions where livestock and wildlife have coexisted for hundreds of years with relatively few tensions (Boyd et al., 1999). These regions, including much of Africa, have also supported some of the most abundant wildlife populations during the past few centuries. Historically, human populations were small and widely dispersed, but competition for grazing and water resources has risen in recent decades. Expanding cultivation and human establishments in parts of Africa have recently pushed agriculture and ranching into the edges of protected areas and natural habitats. In other tropical regions, livestock and agriculture have recently and rapidly invaded, causing dramatic losses for both wildlife and their natural habitats.

Genetic Interactions between Wildlife and Domestic Animals

Although most domesticated species differ phenotypically from their wild relatives, even the most distinct of breeds owe their origins to natural variation among wild ancestors. During the 12,000 years that followed initial domestication, many breeds underwent changes so extreme that differences between them often exceed those that separate wild species. Genetic changes associated with breed diversification originated from the expression of recessive alleles often masked in wild populations, in combination with directional selection on traits valued by humans. Characteristics selected most strongly by humans include increased docility (or reduced aggressiveness), reduced time between birth and reproduction, reduced sexually related displays, and increased productivity of meat, milk, eggs, fur, and feathers. Another key result of animal domestication is evidenced by dramatic changes in seasonal breeding behavior and molting (Mason, 1984), and modifications continue to the present time with new advances in animal cloning and genetic engineering.

Domestic species are not always reproductively isolated from their wild relatives. For example, most of the world's important food crops can cross with related wild plant species, with such gene flow having potentially disastrous consequences. These include the extinction of rare species and the evolution of aggressive or invasive hybrids (Ellstrand et al., 1999). This problem is not isolated to plants, and hybridization between feral or domestic animals and wildlife has caused undesirable gene flow that threatens the existence of rare species in both recent and ancient times (Rhymer and Simberloff, 1996). For example, stallions of the Tarpan (*Equus ferus*), ancestor to modern horses, were reported to herd off large groups of domestic mares, thus leading to substantial gene introgression before their extinction in Poland in 1879 (Day, 1981; Mason, 1984). Indigenous wildcats and domestic cats have been sympatric and interbreeding in Great Britain for over 2000 years, confusing characteristics between the two species (Daniels et al., 1998).

Captive environments of domestic species are often quite different from those in the wild, and behavioral and morphological traits that perform best in captivity are unlikely to be favored in nature. Some characteristics such as reduced seasonality in reproduction, high growth rates, and early maturation may be deleterious in resource-limited or seasonally fluctuating environments. Even semidomestic animal populations (e.g., reindeer, red deer, and ferrets) can experience selective environments different enough from those of wild populations that the risk of nonadaptive alleles spreading into wild populations via hybridization remains a concern (e.g., Knut, 1998). Moreover, particular combinations of alleles form co-adapted gene complexes that can be broken down in hybrid crosses between wild and domestic stock (Lynch, 1996).

Evolutionary differences among domestic animals that mix with their wild relatives can also exacerbate ecological problems. For example, animals reared in high densities are more prone to disease epidemics than those in low-density wild populations. If genetically resistant or tolerant animals escape into low-density populations, they may carry pathogens to naturally unexposed animals. Such a scenario has happened more than once, with native Atlantic salmon threatened by resistant fisheries stock from the Baltic (Johnsen and Jensen, 1986) and endangered Ethiopian wolves exposed to diseases from more resistant domestic dogs (Gotelli et al., 1994; Wayne, 1996).

The problem of hybridization between domestic species and wildlife has intensified in recent decades as humans continue to expand into wild areas and splinter natural habitats. Many wild relatives of livestock in Nepal, including the arnee (*Bubalus arnee*), gaur (*Bibos gaurus*), wild boar (*Sus scrofa*), jungle fowl (*Gallus gallus*), and rock dove (*Columba livia*), have been hybridizing increasingly with domestic species (Wilson, 1997). This hybridization has been implicated in the genetic endangerment or dramatic losses of several tropical or semitropical species, including the Simian jackal (Ethiopian wolf), jungle fowl, and dingo (Table 8.1). The most convincing evidence of hybridization comes from domestic dogs, wild dogs, and wolves. Hybridization between dingoes (*Canis familiaris dingo*) and domestic dogs in Australia exists wherever human settlements are close to wild populations (Newsome and Corbett, 1985). Seasonal breeding among dingoes persists in parts of Australia, although hybridization has led to earlier age at sexual maturity, odd coat color patterns, and changes in skull morphology (Jones and Stevens, 1988; Jones, 1990). The Ethiopian wolf (*Canis simensis*), a close relative of gray wolves and coyotes, is currently the world's most endangered canid. Human growth and agriculture are accelerating its decline, and domestic dogs are sympatric with these wolves in parts of their remaining habitat (Gotelli et al., 1994). The presence of odd coat coloration in up to 17% of wolves in conjunction with domestic dog microsatellite markers indicates that a number of female *C. simensis* have mated with male domestic dogs (Gotelli et al., 1994; Wayne, 1996). Genetic dilution

Table 8.1 Recognized Cases for which Hybridization between Wild and Domestic Species Poses Serious Conservation Concerns

Wild Taxa	Domestic Species	Location	Evidence of Hybridization	Status of Wild Species
Dingo (*Canis familiaris dingo*)	Domestic dog (*Canis familiaris*)	Inland and southeastern Australia	Coat coloration, skull morphology	
Ethiopian wolf (*Canis simensis*)	Domestic dog (*Canis familiaris*)	Ethiopian highlands	Microsatellite markers	Highly endangered
Wildcat (*Felis sylvestris*)	Domestic cat (*Felis catus*)	Scotland	Length of limb bones and intestines, molecular evidence	
Red junglefowl (*Gallus gallus*)	Domestic chicken	Southeastern Asia	Reduced eclipse plumage	Genetic endangerment
Wild yak (*Bos grunniens*)	Domestic yak	Tibetan plateau	Size, color patterns	Threatened

Note: References for each example are cited in the text.

between Ethiopian wolves and domestic dogs threatens the genetic integrity of this species and has prompted calls for the control of domestic dogs in and around national parks.

Hybridization also threatens ungulates and avian species. In Tibet, wild yaks persist in only a few small populations in the alpine steppe and desert, with livestock encroachment and hybridization between domestic and wild yaks threatening the remaining populations (Schaller and Wulin, 1996). Modern chickens were originally domesticated from red junglefowl (*Gallus gallus*), which still can be found throughout parts of southern and southeastern Asia. These wild birds have plumage and calls distinct from domestic fowl, including male eclipse plumage and a lack of prominent combs. However, extensive interbreeding between domestic stocks and wild junglefowl has caused genetic contamination of wild populations, resulting in loss of eclipse plumage from birds in the Philippines and extreme Southeast Asia during the past century (Peterson and Brisbin, 1999).

Although hybridization between wild and domestic animals poses problems for the agricultural industry, the abundance of livestock on human-dominated landscapes and controlled breeding of domestic species render this a minor concern (see also Table 8.2). More likely, wild species can be increasingly viewed as genetic resources for domestic lineages, countering the loss of genetic diversity and inbreeding depression in specialized breeds (e.g., Weigund et al., 1995). In fact, the current biodiversity crisis has been extended to domesticated species, with over 30% of livestock breeds becoming threatened, endangered, or extinct in recent decades (Scherf, 2000). Genetic erosion in livestock (caused by the loss of local breeds or dilution of distinct lineages) may not be reversed easily because most wild relatives are rare or extinct. For a few domesticated species, however, wild relatives allow humans to isolate and transfer new alleles to crops and livestock that enhance disease resistance or promote vigor in stressful environments. Advances in genetic engineering take this application to the extreme, and future bioprospecting efforts are likely to isolate novel traits in wild species that can be transferred and expressed in crops or captive-bred animals.

Finally, domestic species may be useful in rescuing wildlife from the brink of extinction. Recent advances in endocrinology and reproductive biology originally developed for domestic animals have been considered as potential tools for restoring

Table 8.2 Types of Wildlife and Domesticated Animals in the Tropics

Human Dependence	Type of Animal: Native	Type of Animal: Exotic
Free-ranging	Regular wildlife taxa	Introduced or exotic wildlife species (red deer, pheasants, foxes), feral taxa
In human care	Mainly semi-domesticated, or tamed species[a] (green iguanas, ocellated turkeys, Asian elephants, reindeer)	Traditional domesticated taxa (cattle, cats, dogs, pigeons, llamas)

[a] spp. in this category are used only within their native range.

endangered or extinct wild birds and mammals. For example, techniques for artificial insemination, *in vitro* gamete storage, and nuclear and embryo transfer have been proposed to rescue the crested ibis (*Nipponia nippon*), giant panda (*Ailuropoda melanoleuca*), and wild felids in captive breeding programs and zoos (Fujihara and Xi, 2000; Goodrowe et al., 2000).

Competition

In many areas of the Paleotropics and Neotropics, domestic herbivores share open land with a diverse group of wild mammals. Although pastoralists assert that wildlife species belonging to equid, bovid, and camelid families compete with domestic animals for forage, very little research addressed this issue until the second half of the 20th century. Most published work on competition between wild and domesticated ungulates has been conducted in temperate ecosystems (e.g., Schwartz and Ellis, 1981; Osborne, 1984; Loft, Menke, and Kie, 1991; Yeo et al., 1993), but more recent studies have been initiated in eastern and southern Africa and in tropical and subtropical Australia.

Both wild and domesticated herbivores do not feed indiscriminately but have distinct dietary preferences related to food quality, quantity, and location. Food preferences and dietary niche are determined both by gastrointestinal tract architecture (e.g., hindgut fermentation versus rumination) and by muzzle morphology (Skinner, Monro, and Zimmermann, 1984). Whereas cattle (with their broad muzzle) are relatively nonselective roughage grazers (e.g., Hofmann, 1989; Van Soest, 1994; Voeten and Prins, 1999), narrow-snouted antelope selectively forage on higher-quality vegetation. In general, allometric constraints on gut size dictate that smaller herbivores must consume higher-quality vegetation like buds, shoots, and young leaves. Dietary preferences and niche dimensions of each species are flexible, however, and depend significantly on season, habitat, food availability, and the presence of other herbivores. Although ecologists have shown that interspecific food competition among sympatric herbivores is a central factor structuring ungulate communities (at least in African savannas), other factors such as weather, predators, and overall food availability also play a key role (Fritz and Duncan, 1994; Fritz, De Garine, and Letessier, 1996).

In East and South African savanna ecosystems, cattle are the main domestic herbivores; they overlap in diet with several wild ungulates, including impala, plains zebra, and wildebeest. This overlap is most prominent during periods of severe food limitation. Although common resource use does not necessarily imply interspecific competition, all studies examining this issue suggest that competition does occur. In the Ngorongoro Crater Conservation Area, for example, resource use by Masai cattle closely resembles that of the resident wildlife (Homewood, Rodgers, and Arhem, 1987). The strongly seasonal conditions dictate a nomadic or migratory strategy, and both cattle and wild herbivores range widely across the landscape tracking ephemeral vegetation. Direct competition may be ameliorated by disease (malignant catarrhal fever — MCF) that keeps certain regions seasonally off-limits for cattle, as well as additional government-imposed constraints on grazing. In the western Kalahari desert in Botswana, where such legal protections do not exist, wild ungulates are absent from a radius of 10 km from human settlements (Parris and Child, 1973; Bergström and Skarpe, 1999). This is primarily attributed to lack of suitable food, competition with cattle, and, to a lesser degree, human disturbance.

Fritz, De Garine, and Letessier (1996) demonstrated that in Zimbabwe, cattle, kudu, and impala overlap in habitat and resource use. Despite different dietary preferences, impala were forced to change feeding habits in the presence of cattle — lowering their food selectivity, decreasing their group size, reducing overall density, and moving to refuge habitat to avoid competition. This is an example of the general trend of habitat and resource loss among ungulate wildlife following displacement by livestock and pastoralist actions. When accompanied by human encroachment into increasingly marginal habitats, displacement eventually leads to irreversible declines among wild herbivores, as occurred with Bactrian camels and Prezwalski's horses in Asia and the nailtail wallaby (Ellis, Tierney, and Dawson, 1992; Dawson et al., 1992) and two species of stick-nest rats in Australia (Copley, 1999). Exotic herbivores such as cattle or goats do not invariably translate to competitive displacement, and situations exist in which native and domestic herbivores co-exist without problems (Payne and Jarman, 1999).

Fortunately, distinct dietary preferences often allow domestic livestock and wild herbivores to coexist given a variety of available resources. In fact, mixed herding strategies are often part of traditional societies and capitalize on different vegetation strata (Skinner et al., 1984). Mixed ranching practices not only increase income (especially if a market for wildlife products is available) but may also be ecologically beneficial because wildlife grazing has been shown to promote the diversity of semi-arid grassland plant communities. Such management requires careful planning and monitoring, especially in strongly seasonal or arid environments where interactive grazing of different species must be carefully weighed against a fluctuating resource base or varying environmental conditions.

Predation

Historically, predation is probably the most important venue through which wildlife and domestic animals interact (Reynolds and Tapper, 1996). One of the first activities European settlers instigated after colonizing new areas was the relentless removal of native predator populations. This attitude still persists in most areas of the world where modest predator populations exist. A literature review reveals that predator size roughly corresponds to the domestic prey size, so that not all predators pose equal risks to livestock. Typically, adult domesticated bovids are hunted only by lions and tigers (Singh and Kamboj, 1996; Srivastava et al., 1996; Veeramani et al., 1996); whereas smaller livestock such as calves, sheep, and goats can be captured by smaller predators such as leopards (Veeramani et al., 1996), wolves (Kumar and Rahmani, 1997), coyotes (Nass et al., 1984), dingoes (Corbett and Newsome, 1987), jackals (Roberts, 1986), and even wedge-tailed eagles (*Aquila audax*; Brooker and Ridpath, 1980). Nevertheless, this review also suggests that predators prefer native prey species over domesticated animals, in part because they are more abundant, familiar, and of optimal size (Mizutani, 1999). A study in Asian lions also suggests that individual predators imprint on different prey species (domestic or otherwise), which they prefer to the point of starvation (Singh and Kamboj, 1996). As a result, the mere presence of predators does not automatically cause domestic animal losses, especially if native prey is available (Mizutani, 1999).

One common conclusion among many studies that evaluate predator impacts on domesticated herbivores is the surprisingly large effect of feral or exotic mammals (such as pigs, cats, foxes, and dingoes or wild dogs) on livestock. Feral animals, defined as non-native domesticated species that reverted to a free-ranging lifestyle, are often generalists that can attack livestock whenever the opportunity arises. Careful evaluation of bite marks on sheep carcasses in South Africa demonstrated that dogs rather than jackals or caracals were responsible for the overwhelming majority of kills (Roberts, 1986). Feral pigs in arid regions of Australia are important predators of newborn lambs, and their presence can have a significant negative impact on sheep-ranching profits (Choquenot et al., 1997). Feral cats, dogs, and pigs, as well as exotic predators like foxes and mongoose, also have a similarly negative influence on the native wildlife populations and are largely responsible for the endangerment or the extinction of endemic species such as rock wallabies (Dovey et al., 1997), stick-nest rats (Copley, 1999), and various island birds (Rodriguez et al., 1996). Although situations exist where native predators have significant impacts on livestock numbers, they are frequently held responsible for losses inflicted by feral predators (Roberts, 1986) or even cattle rustlers (Rasmussen, 1999).

Exchange of Pathogens and Parasites between Wildlife and Livestock

A stunning variety of pathogens can be transferred between domesticated animals and wildlife (Table 8.3), posing great concern for pastoralists and ranchers and generating complicated problems for conservation biologists. Historically, transfer of pathogens from wildlife reservoirs may have limited (at least transiently) human colonization and use of new regions for grazing cattle and other domesticated livestock. As an example, the vast grasslands of South America and eastern and southern Africa were a huge temptation both to the estranged younger sons of European farmers and to those escaping political persecution in their home countries. Land prices were cheap and often subsidized by governments enthusiastic to establish an imperial presence on relatively underexploited continents (Simon, 1962). Unfortunately, they failed to consider the potential impact of the large diversity of infectious pathogens that infected Africa and South America's native wildlife on domestic crops and livestock (Thomson, 1999). From a pathogen's perspective, livestock simply represented a novel, sedentary, and often conveniently aggregated resource. Thus, ranchers were repeatedly locked into combat with diverse pathogens that had suddenly been supplied with an abundant new population of hosts with little natural resistance to their depredations (Grootenhuis, 1991). A typical example was trypanosomiasis, an African pathogen circulating in native ungulate populations that has ravaged the populations of introduced cattle. In fact, it appears that in many areas of Africa, trypanosomiasis (together with the tsetse fly, its vector) has been a critical factor limiting human activities and therefore determining overall use of the landscape (Wilson et al., 1997; Reid et al., 2000).

Pathogen transfer also occurs from being introduced to native species. The most dramatic example was when imported domestic cattle introduced rinderpest into sub-Saharan Africa (Plowright, 1968). This led to the widespread devastation of Africa's ungulates, particularly the artiodactyls. Rinderpest is a morbillivirus, closely related to canine distemper and human measles, and all three have been responsible for devastating epidemics in unexposed populations that lack immunity (Anderson, 1995). For example, ungulates infected with rinderpest develop symptoms after 4 to 5 days, grow sick, dehydrate, and either die or, if rehydrated, survive and are then immune for the rest of their lives.

Until the advent of relatively rapid transportation by steamship, sub-Saharan Africa had been spared exposure to rinderpest because the population density of artiodactyls in the Sahara was too low to sustain the spread of the disease. In contrast, outbreaks of cattle plague were common in Europe and India. Rinderpest was introduced to the Horn of Africa in 1888 by Europeans. It took 10 years to spread to the Cape, but this pandemic was arguably the largest ever recorded (Plowright, 1982). Many artiodactyl species declined in abundance by as much as 80%, disrupting the social system of many pastoralist tribes (Simon, 1962). These extreme circumstances created opportunities for the European colonists to expand and establish a variety of agricultural practices that reflected their origins despite the prevailing soil and climatic conditions.

Although rinderpest failed to spread through South America and Australia, European settlers engaged in the same process of colonization and agricultural modification. In analogy to Africa, development of agriculture and ranching was followed by the introduction of exotic pathogens (Grainger and Jenkins, 1996; Almeida et al., 2001). However, because both of these continents supported smaller large-mammal populations than Africa, these exchanges were less dramatic (see, e.g., Karesh et al., 1998; Courtenay et al., 2001).

Pathogen Life History Characteristics and Mechanisms of Transmission

Pathogens can be classified in a number of ways — indeed, much of the history of tropical medicine has focused on the business of taxonomic classification of parasite species (and their vectors) and the painstaking elucidation of their life cycles. The simplest ecological classification differentiates between microparasites and macroparasites (Anderson and May, 1979; May and Anderson, 1979; Altizer et al., 2001). While microparasites (which include viruses, bacteria, and protozoa) usually have simple life cycles, macroparasites (which include helminths, flukes, and various ectoparasites) have life cycles characterized by distinct and sometimes dramatically different stages.

What life history characteristics are associated with pathogens that underlie many wildlife–domestic animal conflicts? Although existing data are limited and biased toward large and charismatic wildlife, it is clear that such pathogens are a nonrandom sample of all parasitic organisms (see Table 8.3). The majority of these pathogens are opportunistic microparasites (Dobson and Foufopoulos, 2001) with the ability to infect an unusually large number of hosts species (crossing different host genera and, at times, families). They are also characterized by a high basic reproductive

Table 8.3 Pathogens Important in Livestock–Wildlife Interactions

Pathogen Name	Wildlife Host	Domestic Host	Geographic Area	Comments	Selected References
Anthrax (*Bacillus anthracis*)	Many (e.g., various ungulates)	Cattle, horses, goats, sheep	Worldwide	Transmission through soil contamination but also at wells or through insects	WHO, 1994
Bovine tuberculosis (*Mycobacterium bovis*)	Rare in ungulates and carnivores	Mostly cattle, swine	Worldwide	Spreading into wildlife populations	O'Reilly and Daborn, 1995
	Brushtail opossum (*Trichosurus vulpecula*)	Cattle	New Zealand	Opossum depopulation used to control epidemic	Barlow et al., 1997; Kean et al., 1999;
Brucellosis (several species; mostly *Brucella abortus*)	Many taxa, rodents, carnivores, ungulates	Many, mostly ruminants, swine, dogs, and cats	Worldwide	Various biotypes exist	Acha and Szyfres, 1987
	Cape buffalo, wildebeest	Cattle	Kenya, Tanzania		Waghela and Karstad, 1986;
Canine distemper (morbillivirus)	Several rare African carnivores, (lions, hunting dogs)	Dogs	Worldwide	Broad host range	Anderson, 1995
Foot and mouth disease	Cape buffalo, various antelopes	Cattle	Several African countries	Transmission likely only during acute stages of an infection	Bengis et al., 1986; Anderson et al., 1993; Gainaru et al., 1986
Heartwater (*Cowdria ruminantium*)	Many spp.	Cattle	Zimbabwe	Tick-borne disease, control through acaricide applications	Norval et al., 1994; Peter et al., 1998
Nipah virus	*Pteropus* bats	Pigs, humans	Malaysia	Climate change, habitat fragmentation associated with virus emergence	Chua et al., 2000
Rabies	Vampire bats, other wildlife	Cattle, humans	Argentina	Transmission dependent on bat and cattle densities	Delpietro and Russo, 1996

	Various carnivores	Cattle, dogs, humans	Africa	Dogs identified as primary reservoir hosts in some areas	Cumming, 1982
Rinderpest	Several ungulate spp., Capebuffalo	Goats, sheep, cattle	Africa, Tanzania, Kenya	Cattle important reservoir hosts	Anderson, 1995
Theileriosis (*Theileria annulata*)	Several wild ungulate species	Cattle	Africa, Uganda	Tick-borne disease	Ocaido et al., 1996
Trypanosomes	Several wild ungulate species	Cattle	Sub-Saharan Africa	Transmitted by tsetse flies	Murray and Njogu, 1989

rate (R_o), often by virtue of their spread via insect vectors. It is this rapid transmission coupled with high virulence toward novel hosts that can cause both rapid and devastating losses in wildlife and domestic animals.

Pathogens circulating between domesticated animals and wildlife capitalize on several mechanisms for transmission. A suite of classic problem pathogens is transmitted by ectoparasitic vectors (e.g., the tsetse fly *Glossina* for trypanosomiasis, *Amblyomma* ticks for heartwater, and vampire bats [*Desmodus rotundus*] for rabies). However, a majority of these pathogens are transmitted through either direct contact (e.g., brucellosis, foot and mouth disease, and bovine tuberculosis) or fomites (either contaminated fodder or water sources). Contact between wildlife and domesticated animals is facilitated in areas where humans have expanded into native wildlife habitat. Another crucial factor promoting epizootics is the presence of feral or exotic (e.g., nonnative) animals in an area. Feral organisms typically harbor a diverse community of serious pathogens (Mckenzie and Davidson, 1989). Because their activity patterns put them in contact with both wildlife and domestic animals, feral animals can efficiently shuttle pathogens between these two groups and promote the spread of epidemics.

Finally, evidence from the temperate zone indicates that transmission of pathogens between domestic animals and wildlife is facilitated by habitat fragmentation and degradation (Dobson and Foufopoulos, 2001). These ongoing processes increase direct contact between wildlife and domestic animal populations and place stress on wildlife populations, further increasing their susceptibility to exotic diseases.

The Dynamics of Pathogens Shared between Wild and Domestic Hosts

Experience has shown that wildlife–domestic epizootics can take wildly different trajectories, depending on a constellation of ecological factors. For example, the development and duration of postinfection immunity (strong and long-lasting for microparasites, weak and short-lived for macroparasites) is of crucial importance in determining the outcome of an epizootic. As a result, it is difficult to provide general models of such multispecies epidemics. Here we illustrate some of the simpler cases, leaving more complex situations for the extensive specialist literature (for a review of the field, see Grenfell and Dobson, 1995; and Hudson et al., 2002).

Important insights into the population dynamics of pathogens of domestic hosts that pose threats to wild species may be gained by a simple modification of the standard *SI* and *SIR* epidemiological models. Here we assume that the pathogen poses no threat to the domestic host and that the domestic host is managed at some constant abundance, *D*. The dynamics of the pathogen in the domestic host may be described by the following pair of coupled differential equations, where *s* denotes the proportion of uninfected (susceptible) hosts and *i* is the proportion of infected hosts.

$$\frac{ds}{dt} = \delta iD - \beta i(1-i)D \qquad (8.1)$$

$$\frac{di}{dt} = \beta i(1-i)D - \delta iD \tag{8.2}$$

Here, δ defines the recovery rate of infected hosts to the susceptible class and β is the transmission parameter. These equations can be easily solved by setting $ds/dt = di/dt = 0$ and solving for i^*, the proportion of domestic hosts that are infected at equilibrium.

$$i^* = 1 - \delta/\beta$$

Note the similarity between this expression and that for the proportion of patches occupied in a simple metapopulation (Levins, 1969).

It is also possible to examine a slightly more complicated case where levels of infection in the domestic host are controlled by vaccination or some other form of veterinary intervention.

We can now develop a simple model for the wild population, which is also divided into susceptible, *S,* and infected, *I,* individuals. Here we assume the wild hosts have constant birth and death rates, *b* and *d,* that are independent of host population density. *N* denotes the total population size of wild hosts. No recovery is possible here, so that infected hosts either die due to infection (α) or other causes (*d*), or remain in the infected class. Transmission can occur via contact with either infected domesticated or wild hosts, with transmission parameters β_{WD} and β_{WW}, respectively.

$$\frac{dS}{dt} = bN - dS - \beta_{WW}SI - \beta_{WD}SDi \tag{8.3}$$

$$\frac{dI}{dt} = \beta_{WW}SI - \beta_{WD}SDi - (\alpha + d)I \tag{8.4}$$

This pair of equations can then be solved at equilibrium, $dS/dt = dI/dt = 0$, to obtain expressions for the proportion of wild hosts infected and the size of the wild host population. (Here, $r = b - d$, or the intrinsic rate of increase for the host.)

$$\frac{I^*}{N^*} = \frac{(b-d)}{\alpha}$$

$$N^* = [(d + \alpha)r/(\alpha - r) - b_{WD}D_i][a/(B_{WW}r)]$$

This implies that the size of the wild population declines with increasing force of infection from the domestic population and with greater transmission within the wild population (Figure 8.1). Note that more virulent pathogens tend to produce lower levels of infection in the wild host population than less virulent pathogens (Figure 8.2). This occurs mainly because virulent pathogens tend to kill their hosts

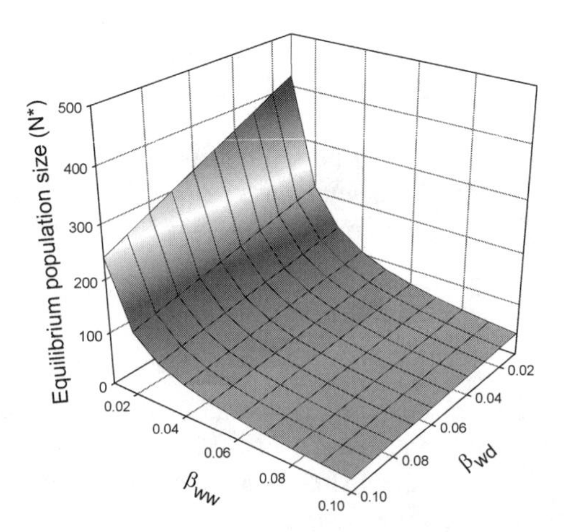

Figure 8.1 Effect of within and among species transmission on the population-level impacts of a shared pathogen. Here, β_{ww} describes pathogen transmission within a wild host population, and β_{wd} is the rate of transmission from domesticated to wild hosts. N* defines the equilibrium population size of wild hosts. Clearly, the wild population declines with increasing transmission rates within and between host species. Other model parameters used were: $b = 1.2$, $d = 0.2$, $\alpha = 2$, $D = 50$, and $i = 0.2$.

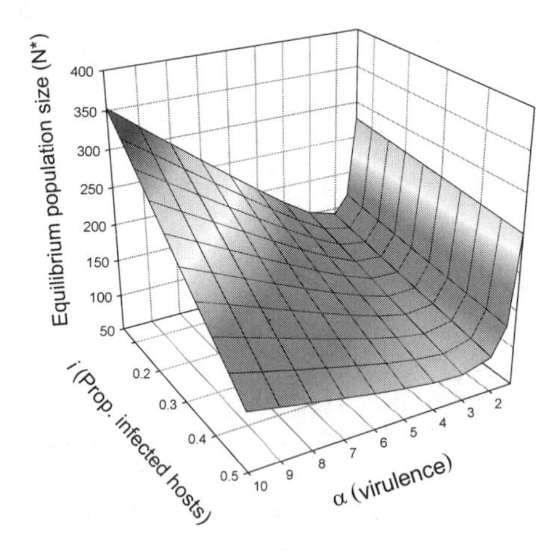

Figure 8.2 Effect of increasing prevalence in the domestic population and pathogenicity toward the wild host on population-level impacts of a shared pathogen. Here, *a* describes pathogen virulence toward the wild host, and *i* defines the proportion of the domestic population that is infectious. *N** defines the equilibrium population size of wild hosts. Other model parameters used were: $\beta_{WD} = 0.3$, $\beta_{WW} = 0.3$, $b = 1.2$, $d = 0.2$, and $D = 50$.

before they are transmitted, while less virulent pathogens may spread more widely in the population. The potential threat to wildlife increases as the size of the domestic livestock population increases and as the duration of time for which hosts are infectious increases. Thus, these models suggest that greater numbers of domestic animals lead to larger negative impacts on the wildlife population.

From an applied management perspective, the reverse situation poses equally pressing challenges: How can one protect susceptible livestock from pathogens circulating in protected wildlife populations? Finding a practical answer to this question cannot only provide the key for the continued coexistence of wildlife and domestic animals, but can also open new avenues for innovative mixed herding regimes. In reality, pathogen transmission is governed by a wide array of ecological factors that defy generalizable solutions. Historically, wildlife to livestock transmission has been reduced or eliminated by indiscriminate use of insecticides (which can have severe repercussions up the trophic chain), elimination of native wildlife through culling or depopulation (with obvious conservation drawbacks), or, in the best cases, the fencing-in of game reserves (Sutmoller et al., 2000) — an approach that can severely disrupt animal migratory patterns. Today, increased quantitative understanding of disease dynamics and improved veterinary methodologies have opened up a variety of disease control alternatives. For example, the spread of epidemics out of game reserves into susceptible livestock populations can be prevented by (1) conducting ring vaccinations of domestic animals, (2) creating buffer zones or lowering stocking densities around reserves below thresholds required for transmission, (3) vaccinating specific wildlife reservoir species, or (4) selectively moving domesticated animals outside specific habitats at certain times of the year. Such alternatives are not always the cheapest, but if applied judiciously they can promote the health of both wildlife and domestic animals.

General Approaches to Improving Wildlife and Livestock Health

Maintaining healthy livestock and wildlife populations requires more than reducing interspecific pathogen transmission. An overall effective and well-organized institutional approach is evident in all countries that have successfully tackled these problems. Below, we discuss methods that can be used to reduce the overall impact of pathogens circulating in wildlife and domestic animal population — many of which derive from basic epidemiological principles.

Research that addresses parasite communities and epidemic potential in domestic, feral, and wild animals can be helpful in predicting disease risk (e.g., see Paling, Jessett, and Heath, 1979; Wells et al., 1981; Paling et al., 1988). Monitoring is extremely important in preventing the spread of an epidemic, as are easy-to-use diagnostic tests. Transferring diagnostic technology from more to less developed nations can also help control important pathogens. Large-scale vaccination and immunization programs (either in domestic animals and/or in wildlife; see e.g., Hedger, Condy, and Gradwell, 1980) can be important in stopping the spread of an epidemic. Control of trade and movement of domestic animals across the landscape (including domestic and international borders) must incorporate a stringent pathogen testing and quarantine system.

Comprehensive education and training programs in veterinary care and animal health must include not only international cooperation in the training of veterinary professionals, but also extensive outreach programs on the community level. The overall efficiency of government agencies can be increased through the establishment of cooperation schemes between the government, the private sector, and various

NGOs. Strengthening international cooperation will increase the effectiveness of transboundary epizootic disease control.

DISCUSSION AND CONCLUSIONS

Almost one third of the Earth's human population relies on livestock for some or all of their daily needs, in the form of meat, milk products, eggs, fiber, or draft power (Anderson, 2001). In developing nations, livestock can provide a majority of income on small farms and offer critical buffers against a background of economic instability. In arid regions of the tropics where soils and rainfall are unsuitable for crops, livestock production has continued to increase despite droughts and other limitations. In fact, demands for livestock and their products in the tropics are predicted to rise most rapidly in eastern Africa, Latin America, and southern Asia (Steinfield, De Haan, and Blackburn, 1997). These demands are likely to increase the negative impacts of livestock on grasslands and other ecosystems by overgrazing, overstocking, decreased herd mobility, and encroachment on natural areas. Although the impacts of livestock on certain habitats are almost always negative, positive impacts on natural ecosystems can result from traditional livestock management or low-to-moderate stocking densities. Moreover, maintaining healthy wildlife populations may have positive effects on local economies and could provide genetic material that may buffer domesticated breeds from further genetic erosion or loss.

Factors Governing Livestock–Wildlife Interactions

Solutions to problems of interspecific disease transmission, genetic hybridization, predation, and competition must be based on a solid understanding of the biological details of species interactions. What may constitute an important conflict at one site or time may be less important in a different region or for a different species assemblage. For example, competition for food between livestock and wild ungulates in Africa is very often a strictly seasonal phenomenon driven by the absence of alternative food sources. Similarly, the spillover of an epidemic from wildlife to cattle may be restricted to seasonal factors, as evidenced by malignant catarrhal fever (MCF) in the Ngorongoro Conservation Area in Tanzania (Homewood, Rodgers, and Arhem, 1987). Transmission of the virus to livestock is most severe during wildebeest calving and postnatal period, so that separation of wild and domestic ungulates is less important during other times of the year (Rossiter, Jessett, and Karstad, 1983). Frequently, conflicts can be addressed by adopting management practices with relatively narrow scope rather than completely eliminating wildlife (as practiced today in many areas of the world). Similarly, hybridization — for seasonally reproducing species at least — can often be avoided if domestic animals and wildlife are kept separate during the breeding period. Nonetheless, it must be stressed that situations *do* exist in which conflict cannot be avoided without radical and sweeping changes in management.

Mechanisms underlying wildlife–livestock conflicts can usually be traced to land productivity, interspecific disease transmission, or animal behavior and habitat requirements. Perhaps the most important factor determining the intensity and outcome of domesticated animal and wildlife interactions is human behavior. Human attitudes determine both stocking rates and herding practices of domesticated animals, in turn affecting the extent of natural resources such as water or forage available to resident wildlife. In regions like the Sahel, where large cattle herds are a status symbol, land overstocking in conjunction with an arid and unproductive environment has caused major degradation of native vegetation and desertification of vast landscapes. Not surprisingly, local wild ungulates such as the addax (*Addax nasomaculatus*) and scimitar-horned oryx (*Oryx dammah*) have greatly suffered under intense competition with livestock. Hunting by humans has dealt a secondary blow, and today both ungulate species are considered highly endangered.

Native wildlife populations are also *directly* affected by a multitude of human factors, ranging from religious beliefs about other vertebrates to utilitarian attitudes toward local fauna. In some areas of Southeast Asia, native species such as the Asian two-horned rhinoceros (*Dicerorhinus sumatrensis*) and wild ungulates have benefited from religious-based protection. Alternatively, other taxa, typically predators, have been ruthlessly eliminated on the basis of superstitious and inaccurate beliefs about the damage they cause (Reynolds and Tapper, 1996). It is therefore difficult to overemphasize the importance of education campaigns as a precursor to establishing new land management or conservation programs.

Potential Resolutions: Adaptive Management and Mixed Herding Regimes

Although ecologically sound land practices can sustain production without sacrificing critical wildlife habitat, subsidies and policies that encourage short-term land tenure will also promote unsustainable land use and accelerate habitat degradation. Increased rates of destruction and degradation will accelerate biodiversity losses and exert severe negative impacts on a handful of wildlife species. Estimates from Central America suggest that ranching in forest frontiers causes 44% of deforestation, and conversion of large forest tracts to pasture has been a major biodiversity threat (Sere, Steinfeld, and Groenwold, 1996). Despite these concerns, opportunities for joint management of both livestock and wildlife population provide feasible income and conservation benefits. In Africa and South America, several examples already demonstrate that cattle and horses may be concurrently managed with antelopes or capybaras in both savanna and wet grasslands characteristic of open tropical habitats.

Several approaches offer potential resolutions to conflicts between wildlife and domestic animals. In most cases, tropical livestock management under traditional practices causes few if any problems. As a matter of fact, if human pressure and stocking densities are not exceedingly high, coexistence between wildlife and livestock appears to be the rule rather than the exception. This absence of conflict, together with the long history of wildlife use in many tropical areas, has led to the development of mixed management strategies that explicitly use both livestock and local wildlife for income and sustenance. Although such diversified management projects may face initially serious obstacles, their benefits outperform exploitative management programs in the long run.

Developing land management systems that use native wildlife in addition to traditionally managed livestock presents many advantages. Pilot projects that rely on mixed herding systems of cattle and wild ungulates in the savannas of east and southern Africa have demonstrated that such systems can produce significantly higher economic returns over traditional cattle ranching operations. Indeed ranchers' attitudes toward mixed game and livestock ranching appear to be frequently positive as long as disease issues are addressed and locals reap some benefits from the local wildlife (see, e.g., Ocaido, Siefert, and Baranga, 1996). Indeed, mixed herding regimes result in a variety of different products and services, including the following:

1. *Wildlife meat.* Although wildlife species are unlikely to replace cattle as the primary source of meat in future decades, wildlife meat procures high prices and can generate significant additional revenue from specialized niche markets such as high-end culinary establishments.
2. *Hunting.* Safari hunting of ranched wildlife, despite the high initial investment in infrastructure (accommodation, transportation, and guide training), is far more lucrative than traditional ranching operations.
3. *Wildlife viewing and ecotourism.* This nonconsumptive alternative to hunting attracts revenue from both photographers and tourists. Although visitor impacts can generate excessive resource demands and effects on wildlife, sustainable programs dedicated to environmental education have been profitable in many regions.
4. *Other wildlife products.* Additional economic benefits may stem from secondary products (such as skins, hides, wool, horn, feathers, medicinal products) and live animals (for zoos or other wildlife parks).

Current evidence suggests that when ranch income is derived from a variety of services and products, the operation is more effectively buffered against market changes in product prices. As a result, diversified ranching operations have a significant economic advantage over traditional ranches that depend on a single (or, at best, few) products. An important caveat to this strategy as a conservation-friendly approach is that once markets become established for wildlife products, strict regulations must be in place to avoid exploitative hunting of game animals. This is particularly important considering the current bushmeat crisis in central Africa, where established urban markets for wildlife meat drive the extermination of all edible wildlife in the forests. Aside from direct economic benefits, mixed herding systems provide two additional advantages over traditional cattle ranching. First, they allow the recovery of native vegetation and, if implemented properly, can promote plant diversity and limit overgrazing and subsequent erosion. Second, the existence of substantial wildlife populations on private land can be of significant conservation importance because these constitute additional population buffers complementing populations in protected areas that further safeguard endangered species against extinction.

REFERENCES

Acha, P.N. and Szyfres, B., *Zoonoses and Communicable Diseases Common to Man and Animals,* 2nd ed., PAHO/WHO, Washington, D.C., 1987.

Almeida, M.F. et al., Neutralizing antirabies antibodies in urban terrestrial wildlife in Brazil, *J. Wildl. Dis.*, 37:394–398, 2001.

Altizer, S., Foufopoulos, J., and Gager, A., Conservation and diseases, in *Encyclopedia of Biodiversity,* Academic Press, San Diego, CA, 2001.

Anderson, E.C., Morbillivirus infections in wildlife in relation to their population biology and disease control in domestic animals, *Vet. Microbiol.,* 44:319–332, 1995.

Anderson, E.C. et al., The role of wild animals, other than buffalo, in the current epidemiology of foot-and-mouth disease in Zimbabwe, *Epidemiol. Infect.,* 111:559–563, 1993.

Anderson, R.M. and May, R.M., Population biology of infectious diseases: Part 1, *Nature,* 280:361–367, 1979.

Anderson, S., Biodiversity, in *Development Brief: Livestock and Biodiversity,* Biodiversity Development Project, World Conservation Union, available at http://europa.eu.int/comm/development/sector/environment, 2001.

Armesto, J.J. et al., Conservation targets in South American temperate forests, *Science,* 282:1271–1272, 1998.

Balmford, A. et al., Conservation conflicts across Africa, *Science,* 291:2616–2619, 2001.

Barlow, N.D. et al., A simulation model for the spread of bovine tuberculosis within New Zealand cattle herds, *Prev. Vet. Med.,* 32:57–75, 1997.

Bengis, R.G. et al., Foot-and-mouth disease and the African buffalo (*Syncerus caffer*). 1. Carriers as a source of infection for cattle, *Onderstepoort J. Vet. Res.,* 53:69–74, 1986.

Bergström, R. and Skarpe, C., The abundance of large wild herbivores in a semi-arid savanna in relation to seasons, pans and livestock, *Afr. J. Ecol.,* 37:12–26, 1999.

Boyd, C. et al., Reconciling interests among wildlife, livestock, and people in Eastern Africa: a sustainable livelihoods approach, *Nat. Res. Perspect.*, 45:1–9, 1999.

Brooker, M.G. and Ridpath, M.G., The diet of the wedge-tailed eagle, *Aquila audax*, in Western Australia, *Aust. Wildl. Res.,* 7:433–452, 1980.

Choquenot, D., Lukins, B., and Curran, G., Assessing lamb predation by feral pigs in Australia's semi-arid rangelands, *J. Appl. Ecol.,* 34:1445–1454, 1997.

Chua, K.B. et al., A recently emergent deadly paramyxovirus, *Science,* 288:1432–1435, 2000.

Clutton-Brock, J., *Domesticated Animals*, University of Texas Press, Austin, TX, 1981.

Copley, P., Natural histories of Australia's stick-nest rats, genus *Leporillus* (*Rodentia: Muridae*), *Wildl. Res.,* 26:513–539, 1999.

Corbett, L.K. and Newsome, A.E., The feeding ecology of the dingo. III. Dietary relationship with widely fluctuating prey populations in arid Australia: a hypothesis of alternation of predation, *Oecologia,* 74:215–227, 1987.

Courtenay, O., Quinnell, R.J., and Chalmers, W.S.K., Contact rates between wild and domestic canids: no evidence of parvovirus or canine distemper virus in crab-eating foxes, *Vet. Microbiol.,* 81:9–19, 2001.

Cumming, D.H.M., A case history of the spread of rabies in an African country, *South Afr. J. Sci.,* 78:443–447, 1982.

Daily, G.C., Ed., *Nature's Services: Societal Dependence on Natural Ecosystems,* Island Press, Washington, D.C., 1997.

Daniels, M.J. et al., Morphological and pelage characteristics of wild living cats in Scotland: implications for defining the 'wildcat', *J. Zool.,* 244:231–247, 1998.

Dawson, T.J., Tierney, P.J., and Ellis, B.A., The diet of the bridled nailtail wallaby (*Onychogalea fraenata*). II. Overlap in dietary niche breadth and plant preferences with the black-striped wallaby *(Macropus dorsalis)* and domestic cattle, *Wildl. Res.,* 19:79–87, 1992.

Day, D., *The Doomsday Book of Animals,* Viking Press, New York, 1981.

Delpietro, H.A. and Russo, R.G., Ecological and epidemiological aspects of attacks by vampire bats in relation to paralytic rabies in Argentina, and an analysis of proposals for control, *Rev. Sci. Tech. Off. Int. Epizooties.,* 15:971–984, 1996.

Diamond, J., *Guns, Germs, and Steel: The Fates of Human Societies,* Norton, New York, 1997.

Dobson, A.P. and Foufopoulos, J., Emerging infectious pathogens of wildlife, *Philos. Trans. R. Soc. London B,* 356:1001–1012, 2001.

Dobson, A.P. and Hudson, P.J., Parasites, disease and the structure of ecological communities, *TREE,* 1:11–15, 1986.

Dovey, L., Wong, V., and Bayne, P., An overview of the status and management of rock-wallabies (*Petrogale*) in New South Wales, *Aust. Mammalogy,* 19:163–168, 1997.

Ellis, B.A., Tierney, P.J., and Dawson, T.J., The diet of the bridled nailtail wallaby (*Onychogalea fraenata*). I. Site and seasonal influences on dietary overlap with the black-striped wallaby (*Macropus dorsalis*) and domestic cattle, *Wildl. Res.,* 19:65–78, 1992.

Ellstrand, N.C., Prentice, H.C., and Hancock, J. F., Gene flow and introgression from domesticated plants into their wild relatives, *Annu. Rev. Ecol. Syst.,* 30:539–563, 1999.

Epstein, H. and Mason, I.L., Cattle, in *Evolution of Domesticated Animals,* Mason, I.L., Ed., Longman, New York, 1984, p. 6–34.

Fritz, H., De Garine, W.M., and Letessier, G., Habitat use by sympatric wild and domestic herbivores in an African savanna woodland: the influence of cattle spatial behaviour, *J. Appl. Ecol.,* 33:589–598, 1996.

Fritz, H. and Duncan, P., On the carrying capacity for large ungulates of African savanna ecosystems, *Proc. R. Soc. London B*, 256:77–82, 1994.

Fujihara, N. and Xi, Y.M., Possible application of animal reproductive research to the restoration of endangered and/or extinct wild animals, *Asian-Aust. J. Anim. Sci.,* 13:1026–1034, 2000.

Funk, S.M. et al., The role of disease in carnivore ecology and conservation, in *Carnivore Conservation,* Gittleman, J. L. et al., Eds., Cambridge University Press, Cambridge, U.K., 2001.

Gainaru, M.D. et al., Foot-and-mouth disease and the African buffalo (*Syncerus caffer*). II. Virus excretion and transmission during acute infection, *Onderstepoort J. Vet. Res.,* 53:75–86, 1986.

Gittleman, J. L. et al., Eds., *Carnivore Conservation,* Cambridge University Press, Cambridge, U.K., 2001.

Goodrowe, K.L. et al., Piecing together the puzzle of carnivore reproduction, *Anim. Reprod. Sci.,* 60–61:389–403, 2000.

Gotelli, D. et al., Molecular genetics of the most endangered canid: the Ethiopian wolf *Canis simensis*, *Mol. Ecol.,* 3:301–312, 1994.

Grainger, H.J. and Jenkins, D.J., Transmission of hydatid disease to sheep from wild dogs in Victoria, Australia, *Int. J. Parasitol.,* 26:1263–1270, 1996.

Grenfell, B.T. and Dobson, A.P., *Ecology of Infectious Diseases in Natural Populations,* Cambridge University Press, Cambridge, U.K., 1995.

Grootenhuis, J.G., Disease research for integration of livestock and wildlife, paper presented in Wildlife Research for Sustainable Development: Proceedings of an International Conference, Grootenhuis, J.G., Njuguna, S.G., and Kat, P.W., Eds., KARI (Kenya Agricultural Research Institute), KWS (Kenya Wildlife Service) and NMK (National Museums of Kenya), Nairobi, Kenya, 1991.

Hedger, R.S., Condy, J. B., and Gradwell, D.V., The response of some African wildlife species to foot-and-mouth disease vaccination, *J. Wildl. Dis.,* 16:431–438, 1980.

Hofmann, R.R., Evolutionary steps of ecophysiological adaptation and diversification of ruminants: a comparative view of their digestive system, *Oecologia,* 78:443–457, 1989.

Homewood, K., Rodgers, W.A., and Arhem, K., Ecology of pastoralism in Ngorongoro Conservation Area, Tanzania, *J. Agric. Sci.,* 108:47–72, 1987.

Hudson, P.J. et al., *The Ecology of Wildlife Diseases,* Oxford University Press, New York, 2002.

Isaac, E., *Geography of Domestication,* Prentice Hall, Englewood Cliffs, NJ, 1970.

IUCN Red List of Threatened Animals (IUCN, Gland, Switzerland, 2000); Available at: http://www.redlist.org/, 2000.

Johnsen, B.O. and Jensen, A.J., Infestations of Atlantic salmon, *Salmo salar*, by *Gyrodactylus salaris* in Norwegian rivers, *J. Fish Biol.,* 29:231–233, 1986.

Jones, E., Physical characteristics and taxonomic status of wild canids, *Canis familiaris*, from the eastern highlands of Victoria (Australia), *Aust. Wildl. Res.,* 17:69–82, 1990.

Jones, E. and Stevens, P.L., Reproduction in wild canids, *Canis familiaris*, from the eastern highlands of Victoria (Australia), *Aust. Wildl. Res.,* 15: 385–394, 1988.

Karesh, W.B. et al., Health evaluation of free-ranging Guanaco (*Lama guanicoe*), *J. Zoo Wildl. Med.,* 29:134–141, 1998.

Kean, J.M., Barlow, N.D., and Hickling, G.J., Evaluating potential sources of bovine tuberculosis infection in a New Zealand cattle herd, *NZ J. Agric. Res.,* 42:101–106, 1999.

Knut, R.H., Influence of selection and management on the genetic structure of reindeer populations, *Acta Theriologica,* 5:179–186, 1998.

Kumar, S. and Rahmani, A.R., Status of Indian grey wolf *Canis lupus pallipes* and its conservation in marginal agricultural areas of Solapur district, Maharashtra, *J. Bombay Nat. Hist. Soc.,* 94:466–472, 1997.

Levins, R., Some demographic and genetic consequences of environmental heterogeneity for biological control, *Bull. Entomol. Soc. Am.,* 15:237–240, 1969.

Loft, E.R., Menke, J.W., and Kie, J.G., Habitat shifts by mule deer: the influence of cattle grazing, *J. Wildl. Manage.,* 55:16–26, 1991.

Lutz, E. et al., *Agriculture and the Environment: Perspectives on Sustainable Rural Development,* World Bank, Washington, D.C., 1998.

Lynch, M., A quantitative-genetic perspective on conservation issues, in *Conservation Genetics: Case Histories from Nature,* Avise, J.C. and Hamrick, J.L., Eds., Chapman & Hall, New York, 1996.

MacPhee, R.D.E., Ed., *Extinctions in Near Time: Causes, Contexts, and Consequences,* Kluwer Academic/Plenum, New York, 1999.

Mason, I.L., *Evolution of Domesticated Animals,* Longman, New York, 1984.

May, R.M. and Anderson, R.M., Population biology of infectious diseases. Part II, *Nature,* 280:455–461, 1979.

Mckenzie, M.E. and Davidson, W.R., Helminth parasites of intermingling axis deer, wild swine and domestic cattle from the island of Molokai, Hawaii (USA), *J. Wildl. Dis.,* 25:252–257, 1989.

Mizutani, F., Impact of leopards on a working ranch in Laikipia, Kenya, *Afr. J. Ecol.,* 37:211–225, 1999.

Murray, M. and Njogu, A.R., African trypanosomiasis in wild and domestic ungulates — the problem and its control, in Symposia of the Zoological Society of London, 61, The biology of large African mammals in their environment, Jewell, P.A. and Maloiy, G.M.O., Eds., Oxford University Press, New York, 1989.

Nass, R.D., Lynch, G., and Theade, J., Circumstances associated with predation rates on sheep and goats, *J. Range Manage.,* 37:423–426, 1984.

Newsome, A.E. and Corbett, L.K., The identity of the dingo (*Canis familiaris dingo*). 3. The incidence of dingoes, dogs, and hybrids and their coat colors in remote and settled regions of Australia, *Aust. J. Zool.,* 33:363–376, 1985.

Norval, R.A.I. et al., Factors affecting the distributions of the ticks *Amblyomma hebraeum* and *A. variegatum* in Zimbabwe: implications of reduced acaricide usage, *Exp. Appl. Acarology,* 18:383–407, 1994.

Ocaido, M., Siefert, L., and Baranga, J., Disease surveillance in mixed livestock and game areas around Lake Mburo National Park in Uganda, *South Afr. J. Wildl. Res.,* 26:133–135, 1996.

O'Reilly, L.M. and Daborn, C.J., The epidemiology of *Mycobacterium bovis* infection in animals and man: a review, *Tubercle Lung Dis.,* 76:1–46, 1995.

Osborne, B.C., Habitat use by red deer (*Cervus elaphus* L.) and hill sheep in the West Highlands, *J. Appl. Ecol.,* 21:497–506, 1984.

Packer, C. et al., Viruses of the Serengeti: patterns of infection and mortality in African lions, *J. Anim. Ecol.,* 68:1161–1178, 1999.

Paling, R.W., Jessett, D.W., and Heath, B.R., The occurrence of infectious diseases in mixed farming of domesticated wild herbivores and domestic herbivores including camels in Kenya. Part 1. Viral diseases. A serologic survey with special reference to foot-and-mouth disease, *J. Wildl. Dis.,* 15:351–358, 1979.

Paling, R.W. et al., The occurrence of infectious diseases in mixed farming of domesticated wild herbivores and livestock in Kenya. II. Bacterial diseases, *J. Wildl. Dis.,* 24:308–316, 1988.

Parris, R. and Child, G., The importance of pans to wildlife in the Kalahari and the effect of human settlement on these areas, *J. Southern Afr. Wildl. Manage. Assoc.,* 3:1–8, 1973.

Payne, A.L. and Jarman, P.J., Macropod studies at Wallaby Creek. X. Responses of eastern grey kangaroos to cattle, *Wildl. Res.,* 26:215–225, 1999.

Peter, T.F. et al., The distribution of heartwater in the highveld of Zimbabwe, 1980–1997, *Onderstepoort J. Vet. Res.,* 65:177–187, 1998.

Peterson, A.T. and Brisbin, I.L., Genetic endangerment of wild Red Junglefowl *Gallus gallus*? *Bird Conserv. Int.,* 9:387–394, 1999.

Plowright, W., Rinderpest virus, in *Virology Monographs 3,* Gard, S., Hallauer, C., and Meyer, K.F., Eds., Springer-Verlag, New York, 1968.

Plowright, W., The effects of rinderpest and rinderpest control on wildlife in Africa, *Symp. Zool. Soc. London,* 50:1–28, 1982.

Rasmussen, G.S.A., Livestock predation by the painted hunting dog *Lycaon pictus* in a cattle ranching region of Zimbabwe: a case study, *Biol. Conserv.,* 88:133–139, 1999.

Reid, R.S. et al., Land-use and land-cover dynamics in response to changes in climatic, biological and socio-political forces: the case of southwestern Ethiopia, *Landscape Ecol.,* 15:339–355, 2000.

Reynolds, J.C. and Tapper, S.C., Control of mammalian predators in game management and conservation, *Mammal Rev.,* 26:127–156, 1996.

Rhymer, J.M. and Simberloff, D., Extinction by hybridization and introgression, *Annu. Rev. Ecol. Syst.,* 27:83–109, 1996.

Roberts, D.H., Determination of predators responsible for killing small livestock, *South Afr. J. Wildl. Res.,* 16:150–152, 1986.

Rodgers, W.A. and Homewood, K.M., Cattle dynamics in a pastoralist community in Ngorongoro, Tanzania, during the 1982–1983 drought, *Agric. Syst.,* 22:33–52, 1986.

Rodriguez-Estrella, R. et al., Status, density and habitat relationships of the endemic terrestrial birds of Socorro Island, Revillagigedo Islands, Mexico, *Biol. Conserv.,* 76:195–202, 1996.

Rossiter, P., Jessett, D., and Karstad, L., Role of wildebeest fetal membranes and fluids in the transmission of malignant catarrhal fever virus, *Vet. Rec.,* 113:150–152, 1983.

Sauer, C.O., *Agricultural Origins and Dispersals,* George Grady Press, New York, 1952.

Schaller, G.B. and Wulin, L., Distribution, status and conservation of the wild yak, *Bos grunniens*, *Biol. Conserv.,* 76:1–8, 1996.

Scherf, B.D., Ed., *World Watch List for Domestic Animal Diversity,* FAO, Rome, 2000.

Schwartz, C.C. and Ellis, J.E., Feeding ecology and niche separation in some native and domestic ungulates on shortgrass prairie, *J. Appl. Ecol.,* 18:343–353, 1981.

Sere, C., Steinfeld, H., and Groenwold, J., *World Livestock Production Systems: Current Status, Issues and Trends,* FAO, Rome, 1996.

Simon, N., *Between the Sunlight and the Thunder. The Wildlife of Kenya,* Collins, London, 1962.

Singh, H.S. and Kamboj, R.D., Predation pattern of the Asiatic lion on domestic livestock, *Indian For.,* 122:869–876, 1996.

Skinner, J.D., Monro, R.H., and Zimmermann, I., Comparative food intake and growth of cattle and impala on mixed tree savanna, *South Afr. J. Wildl. Res.,* 14:1–9, 1984.

Smith, F.D.M. et al., Estimating extinction rates, *Nature,* 364:494–496, 1993.

Srivastava, K.K. et al., Food habits of mammalian predators in Periyar Tiger Reserve, South India, *Indian For.,* 122:877–883, 1996.

Steinfield, H., De Haan, C., and Blackburn, H., *Livestock and the Environment: Issues and Options,* World Bank/FAO, U.K., 1997.

Sutmoller, P. et al., The foot-and-mouth disease risk posed by African buffalo within wildlife conservancies to the cattle industry of Zimbabwe, *Prev. Vet. Med.,* 44:43–60, 2000.

Thomson, G.R., Alternatives for controlling animal diseases resulting from interaction between livestock and wildlife in southern Africa, *South Afr. J. Sci.,* 95:71–76, 1999.

Tilman, D. et al., Forecasting agriculturally driven global environmental change, *Science,* 292:281–284, 2001.

Ucko, P.J. and Dimbleby, G.W., Eds., *The Domestication and Exploitation of Plants and Animals,* Duckworth, London, 1969.

Van Soest, P.J., *Nutritional Ecology of the Ruminant,* 2nd ed., Cornell University Press, Ithaca and London, 1994.

Veeramani, A., Jayson, E.A., and Easa, P.S., Man-wildlife conflict: cattle lifting and human casualties in Kerala, *Indian For.,* 122:897–902, 1996.

Vitousek, P.M. et al., Human domination of Earth's ecosystems, *Science,* 277:494–499, 1997.

Voeten, M., Living with wildlife: coexistence of wildlife and livestock in an East African Savanna system, in *Tropical Resource Management Papers 29,* Wageningen University and Research Centre, Netherlands, 1999.

Voeten, M. and Prins, H.T., Resource partitioning between sympatric wild and domestic herbivores in the Tarangire region of Tanzania, *Oecologia,* 120:287–294, 1999.

Waghela, S. and Karstad, L., Antibodies to *Brucella* spp. among blue wildebeest (*Connochaetes taurinus*) and African buffalo (*Syncerus caffer*) in Kenya, *J. Wildl. Dis.,* 22:189–192, 1986.

Wayne, R.K., Conservation genetics in the Canidae, in *Conservation Genetics: Case Histories from Nature,* Avise, J.C. and Hamrick, J.L., Eds., Chapman & Hall, New York, 1996.

Weigund, S. et al., Conception for conserving animal genetic resources in poultry in Germany, *Arch. Geflügelkunde,* 59:327–334, 1995.

Wells, E.A. et al., Mammalian wildlife diseases as hazards to man and livestock in an area of the Llanos Orientales of Colombia, *J. Wildl. Dis.,* 17:153–162, 1981.

Wheeler, J.C., Evolution and present situation of the South American Camelidae, *Biol. J. Linnean Soc.,* 54:271–295, 1995.

WHO, Anthrax control and research, with special reference to national programme development in Africa: memorandum from a WHO meeting, *Bull. WHO,* 72:13–22, 1994.

Wilson, C.J. et al., Effects of land-use and tsetse fly control on bird species richness in southwestern Ethiopia, *Conserv. Biol.,* 11:435–447, 1997.

Wilson, R.T., Animal genetic resources and domestic animal diversity in Nepal, *Biodiversity Conserv.,* 6:233–251, 1997.

Yeo, J.J. et al., Influence of rest-rotation cattle grazing on mule deer and elk habitat use in east-central Idaho, *J. Range Manage.,* 46:245–250, 1993.

CHAPTER 9

Tropical Agriculture and Human Disease: Ecological Complexities Pose Research Challenges

Mark L. Wilson

CONTENTS

0-8493-1581-6/03/$0.00+$1.50

INTRODUCTION

This chapter addresses an apparent contradiction: adequate, nutritious food is essential for human health, yet contemporary agricultural production in many tropical settings is linked to increased risk of disease. This irony involves fundamental production and ecological processes. It is evident that agroecosystems derive from complex social and economic interactions that influence the manner in which humans alter and exploit their environments. Many of these transformations produce unintended modifications that affect important ecological links among people, wild and domestic animals, nondomesticated plants, natural and synthetic toxins, or diverse microorganisms. Often, such changes produce consequences for human well-being that counteract possible benefits from increased food production. Certain of these unintended ill-health effects are well understood and easily prevented, such as the direct toxic effects of agricultural pesticides on people. Others are less obvious and not easily anticipated without undertaking careful observation and thoughtful analysis. In particular, many infectious diseases that result from agricultural development and environmental change are poorly understood or easily overlooked.

The following discussion first considers the diverse pathways by which agricultural production may affect health. Then we address basic principles of environmental change and how they may alter human disease patterns. The manner in which agroecosystem transformations foster new infectious diseases in humans, or lead to their increased pathogenicity, distribution, or incidence is explored. These principles are illustrated with specific examples. Finally, we examine knowledge needs and research strategies that ultimately should reduce the likelihood of unexpected outcomes.

PATHWAYS OF AGRICULTURE EFFECTS ON HEALTH

Most of the world's agricultural production is driven by economic decisions designed for return on investments made by private capitalist enterprises. Underlying this economic activity is the biological need of humans to eat food and maintain health. Consumption of adequate nutrients in products that are free of toxins and pathogens is essential to good health. However, food production often has many, diverse disease outcomes (Kaplan, 1997; Lipton and deKadt, 1998; Ruttan, 1994).

The pathophysiology of these effects is generally well understood and represents the main focus of most academic discussion and study. This is particularly true for ill health produced by simple or direct causal pathways. Less often addressed are the more complex mechanisms underlying ecological and economic causation. Some examples of these pathways are briefly discussed below (summarized in Table 9.1).

Exposure to Toxins

One direct effect on health that may result from agriculture practices anywhere, but particularly in the impoverished tropics, involves exposure to toxins. These could be insecticides or fertilizers, and they may harm both agricultural workers and consumers. Intensification of production in most tropical environments is associated with increased use of potentially toxic chemicals. Technologies to prevent such well-recognized effects on health exist, yet disease still occurs (e.g., Weil et al., 1990). Expansion of tropical agriculture combined with features of this chemical-intensive technology has increased disease risk.

Agricultural Production and Potable Water

Another pathway of impact on human health passes through the availability of adequate, safe drinking water (e.g., Birley, 1995). Poorly managed runoff of pesticides or fertilizers can pollute surface waters that serve as the primary source of drinking water in the tropics. The rural poor, in particular, rarely have access to treated drinking water; in their case, agricultural toxins may be more frequently consumed. In addition, fertilizer-enhanced eutrophication of rivers or reservoirs diminishes water quality and access. As forests are cut to expand cropping area, less plant cover and resulting erosion lead to surface water that becomes less stable, with more rapid runoff reducing sustained fresh water availability and increasing risk of floods.

Table 9.1 Examples of Pathways by which Agricultural Production Changes Can Lead to Increased Human Disease

Disease or Condition	Mechanism or Pathway
Exposures to toxic chemicals	Inhalation or skin contact with insecticides or fertilizers by agricultural workers, consumers
Inadequate or polluted potable water	⇑ water diversion to irrigation, fertilizer/insecticide runoff
Insufficient food, inadequate nutrients	Conversion to cash crops and ⇓ in production for local consumption
Loss of natural medicinals	Deforestation for crop production and loss of plant biodiversity
Labor migration and increased disease	⇑ social disruption, poor housing, less immunity
International trade, food-borne pathogens	⇑ transfer of fresh produce harboring microbes
Infectious diseases and environmental change	Diverse pathways for increased risk of disease (see Table 9.3)

Food Availability and Nutrition

Diverse, less direct health effects can result from pathways that include other variables. One such indirect pathway involves health as mediated through nutrition. Where cash crops for export are replacing food crops for local consumption, diminished availability or increased cost of locally produced commodities may limit food access (Weil et al., 1990). Whether new employment will produce adequate income to purchase food is uncertain. Undernutrition and associated disease may result.

Natural Sources of Prevention and Treatment

Another unintended effect of tropical agricultural expansion may be loss of plants or animals used as traditional medicines. Deforestation that reduces or eliminates certain animal or plant populations can diminish availability of natural sources used to treat diseases (e.g., antibiotics, diuretics, analgesics, etc.). Similarly, reduced species diversity that intensifies with the growth of many tropical agroecosystems diminishes opportunities for identifying new, undiscovered natural products that will enhance health (Grifo and Rosenthal).

Migration and Disease

Large-scale movement of poor people toward new agricultural employment often is intended as an outcome of agricultural development schemes; wage labor that produces income may relieve some of the pressures toward urbanization. Such migration to rural areas, however, may increase risk of disease through various avenues. The social disruption that often accompanies such population shifts initially places people under less sanitary conditions, increasing the risk of particular infectious diseases. Often, these migrants have not had previous exposure to locally transmitted agents, hence they lack antibodies that might otherwise confer protection. Risk of disease in these new settings would be further increased by limited pharmaceutical access in small villages or rural areas.

International Trade and Food-Borne Pathogens

Globalization has increased the amount and diversity of tropical agricultural products that are shipped internationally, especially to developed countries of the North. Normally, these foods (especially fresh fruits and vegetables) would improve the year-round access to diverse and nutritious food among people in temperate regions. However, such products may also harbor infectious microbes, thus introducing food-borne pathogens to areas distant from their origin. Although this probably goes unrecognized when only a few people become ill, recent, larger outbreaks of food-borne disease linked to tropical agriculture products have been recognized. Outbreaks in the U.S., for example, due to *Salmonella* on Brazilian mangos and *Cyclospora* on Guatemalan raspberries represent two recent, well-documented examples.

These examples involve infectious diseases linked to tropical agricultural production methods and sales. Indeed, microbial pathogens represent the main cause of morbidity and mortality in most tropical countries. The links between tropical agroecology, environmental change, and pathogen transmission represent a particularly important influence on disease risk in the tropics. The remainder of this chapter focuses on health effects due to ecological changes that result from agricultural changes.

ENVIRONMENTAL LINKS TO INFECTIOUS DISEASE DYNAMICS

Infectious disease epidemiology considers patterns of ill health to result from interactions among the causative microbes, human hosts, and relevant characteristics of the environment (Webber, 1996). This triad simplifies diverse interactions among many variables that typically produce indirect and nonlinear effects over differing temporal scales, often involving time lags and various feedbacks (Wilson, 2001). For this reason, risk of disease is complex, and the presence of a pathogenic microbe in a region is necessary but not sufficient for disease occurrence. Many other factors influence which and how many people become infected, and whether or not they become ill (Anderson and May, 1991). Because only some infections produce disease, it is important to understand how agroecology influences these factors, and others such as previous exposure, nutrition, age, and immune status. Some of these variables have links to environmental change.

The environment affects different kinds of transmission differently. Human diseases can be divided into those for which the pathogen has a life cycle involving only humans as a vertebrate reservoir host (anthroponotic diseases) and others for which nonhuman animals are the main vertebrate reservoirs, with occasional transfer to people (zoonotic diseases). Transmission may be direct (involving close physical contact) or indirect (via physical objects such as air, food, water, etc. or animal vectors such as blood-feeding arthropods). Depending on the natural mode of transmission of each microbe, the importance of particular characteristics of the environment will vary. Generally, directly transmitted anthroponoses are less affected by the biophysical environment, while directly transmitted zoonoses and indirectly transmitted anthroponoses are more often influenced. The most important role of the environment involves transmission of vector-borne zoonoses, but the direction of effect is sometimes difficult to predict.

Exposure is largely a function of the abundance and time–space distribution of each infectious organism and its transfer characteristics. The environment, therefore, essentially defines transmission characteristics of *water*-borne, *air*-borne, and *food*-borne disease agents, whether anthroponotic or zoonotic. The environment determines the abundance of nonhuman vertebrate hosts harboring microbes that cause human zoonoses. Environmental variables influence the reproduction and survival of arthropod vectors (e.g., mosquitoes, sand flies, kissing bugs, ticks, etc.) that transmit many pathogenic agents, both anthroponotic and zoonotic. Indeed, through various pathways, many characteristics of the environment represent fundamental cornerstones of the transmission of most human infectious disease agents. Environmental change, whether or not linked to tropical agriculture change, is increasingly recognized as a primary determinant of the emergence and reemergence of infectious diseases in both developed and underdeveloped countries.

ENVIRONMENTAL CHANGE AND EMERGING DISEASES

Within the past few decades, infectious diseases have begun to reappear as a major threat to human health in most parts of the world (Lederberg, Shope, and Oaks, 1992). This has been particularly true in the tropics. Identified as emerging, reemerging, or resurgent, these diseases share one or more epidemiological characteristics that may include a newly recognized pathogen or syndrome, increased severity or reduced treatment capacity, greater local incidence, or expanding distribution into new areas (Wilson, 1994). More than one of these criteria may apply to any specific emerging disease. While there are many diverse forces underlying these processes (CDC, 1994; Morse, 1994; Wilson et al., 1994; Krause, 1998), environmental change or new human–environment interactions have been signaled as very important to most. As shifts in the environmental mosaic occur, changes in pathogen abundance or distribution should be expected. Similarly, as humans alter their behavior in relation to existing environments, new exposures will occur. This will alter exposure patterns. Examples included increased incursion into less exploited habitats, or rapid long-distance travel.

Of particular concern are emerging diseases that are linked to tropical agroecology where natural habitats have been transformed to permit or improve crop production (Table 9.2). Such environmental exploitation not only creates new exposures to previously unencountered pathogens, but often increases contact with pathogens that were already present. Previously unrecognized pathogens that have been linked to agriculture include, for example, those that cause hemorrhagic fevers (Junin, Machupo, and Guanarito viruses) (Morse, 1994). Another example, linked to agricultural fertilizer-based eutrophication, is the dinoflagellate *Pfiesteria piscicida* that is highly lethal to fish in Atlantic estuaries, while producing a toxin that afflicts people (Pinckney et al., 2000).

Reemergence may involve greater transmission of existing agents, increased abundance of animal reservoirs of certain pathogens, or greater exposure to transfer via water, air, or arthropod vectors (discussed below). Examples that may not involve recent shifts in agricultural production include hunting-associated Ebola hemorrhagic fever in eastern Africa (Leirs et al., 1999) or increased malaria in Brazil linked to gold mining (Vosti, 1990; Singer and de Castro, 2001). Interestingly, however, examples of raccoon rabies and Lyme disease in North America have emerged as human disease risks during the past few decades. These changes appear to be due to *reduced* agricultural activities (Wilson, 2001). The regional abandonment of farms during the past century has led to reforestation of much of the eastern U.S. and Canada, with accompanying population increases of specific vertebrate reservoirs and vectors.

Other emerging/reemerging diseases are linked to increased transmission resulting from different environmental changes. Urbanization and urban decay in the Third

Table 9.2 Some Examples of Emerging Diseases Newly Recognized since the 1950s that Are Related to Changing Agricultural Production

Disease	Year Recognized	Pathogen	Agricultural Link
Argentine hemorrhagic fever	1958	Junin virus	Agriculture favors rodent reservoirs
Bolivian hemorrhagic fever	1959	Machupo virus	Rodents near houses, agriculture
Ebola hemorrhagic fever	1966	Ebola virus	Animal reservoir contact in forest
Lassa fever	1969	Lassa virus	Rodent abundance near houses
Cryptosporidiosis	1976	*Cryptosporidium parvum*	Cattle contamination drinking water
Campylobacter enteritis	1980s	Quinolone-resistant *Campylobacter jejuni*	Antibiotic use in animal husbandry
Hemolytic uremic syndrome	1982	*Escherichia coli* O157-H7	Fecal contamination of food
Pfiesteriosis, toxin	1988	*Pfiesteria piscicida*	Agricultural runoff to estuaries
Venezuelan hemorrhagic fever	1989	Guanarito virus	Rodents in cultivated fields
Avian influenza	1997	Influenza virus H5N1	Integrated pig–duck farming (China)

World have greatly increased the number of people living under conditions of poor hygiene, thus enhancing transmission of specific diseases such as dengue fever and cholera. Civil strife and refugee movements have placed millions of people in environments unsuited for their basic hygiene and health needs. These and other politically or economically mediated changes in human–environment relations are producing many more emerging diseases, especially in poor countries.

Climate variability, perhaps global climate change, represents another type of environmental change that is not simply agriculturally driven, even though worldwide agroeconomic activities affect regional and global environments. These climate/weather effects, in turn, alter transmission of infectious disease agents. For example, it has been suggested that agricultural practices have increased temperatures in highland areas of Africa, thereby affecting transmission of malaria (Lindblade et al., 2000), but there are few studies that document this effect. Most climate-related studies have suggested that the risk of vector-borne diseases such as malaria and dengue will become more widespread or increasingly transmitted as global warming gradually occurs (Patz et al., 2001). Long-term forecasts, however, are controversial (e.g., Reiter, 2001) and generally assume that socioeconomic factors and prevention will remain constant.

Other food production-linked risks would appear to have fewer environmentally governed processes presenting threats to health in tropical and subtropical settings. The international commerce and transport of food commodities is internationally regulated, but accidental or illegal long-distance shipment of pathogens or vectors is frequent and well recognized. Examples of direct and indirect pathways of infectious disease transmission related to agroecological changes are addressed below.

CHANGING PATTERNS OF AGRICULTURE AND ECOLOGICAL EFFECTS ON HEALTH

Agricultural activities require environmental modification. Such changes inevitably affect the intensity, seasonality, spatial heterogeneity, and regional distribution of human pathogen transmission. The manner in which these fundamental epidemiological processes operate depends heavily on patterns of human contact with these microbes (Table 9.3). Exposure is the main component of risk for most pathogens, even though nutrition, existing immunity, and other factors are related variables. Most spatial and temporal patterns of infectious diseases are based in the evolutionary ecology of each pathogen's mode of transmission, since this largely determines exposure. Understanding how such characteristics govern transmission is critical to reducing associated infectious disease.

Changing Water Patterns

Consider, for example, how modern agricultural practices typically alter the distribution of water in existing cropping fields (temporal abundance, land surface covered, etc.). Such changes may or may not affect transmission of human disease agents. Although intensified irrigation schemes generally increase soil moisture and atmospheric humidity (encouraging aerosol- and soil-borne diseases), the human impact will depend on local dynamics involving animal population abundance, arthropod vectors, human density and behavior, and even human susceptibility and immune responses. Malaria, discussed in detail below, is an example of a disease that is water-associated because the ecologies of the *Anopheles* mosquito species

TABLE 9.3 Examples of Agricultural Changes and Associated Effects on Infectious Diseases

Agricultural Changes	Example Diseases	Pathway of Effect
Irrigation, dams, canals	Malaria	⇑ Breeding sites for mosquitoes
	Schistosomiasis	⇑ Snail host habitat, human contact
	Helminthiases	⇑ Moist soil increases larval contact
	River blindness	⇓ Blackfly breeding, ⇓ disease
	Rift Valley fever	⇑ Pools for mosquito breeding
Agricultural intensification	Malaria	Crop insecticides and ⇑ vector resistance
	Venezuelan hemorrhagic fever	⇑ Rodent abundance, ⇑ contact
Deforestation and new cropping	Malaria	⇑ Breeding sites and vectors, immigration of susceptible people
	Oropouche	⇑ Contact, breeding of vectors
	Visceral leishmaniasis	⇑ Contact with sand fly vectors
Urbanization, urban slums	Cholera	⇓ Sanitation, hygiene, ⇑ water contamination
	Dengue	Water-collecting trash, ⇑ *Aedes aegypti* mosquito breeding sites

Source: Modified from Wilson, 2001.

that transmit the causative protozoan are based on water for egg-laying. Other water-associated diseases might increase or decrease with agricultural change. Schistosomiasis, for example, which is associated with particular dam/canal schemes, may become a greater problem where introduction of irrigation alters stream volume, turbidity, or flow (discussed below).

Still other diseases transmitted by arthropod vectors whose reproduction depends on water volume, constituents, turbidity, or flow (e.g., mosquitoes, blackflies, etc.) also might increase or decrease with cropping-associated water development effects (Service, 1991; Sharma, 1996). Rift Valley fever (RVF), an *Aedes* mosquito-borne viral infection in sub-Saharan Africa, has been linked to extreme precipitation (Linthicum et al., 1999), but also to water impoundment that created epidemic RVF transmission when dams have been closed (Ksiazek et al., 1989). Onchocerciasis (river blindness) has forced people from agricultural land near rivers in much of sub-Saharan Africa where vector blackflies (*Simulium* spp.) require flowing water for larval development. Interestingly, where dams have reduced the flow of water, larval blackfly survival may be diminished (except near spillways), thus reducing risk of river blindness. In each case, details of the water-related transmission ecology become important in understanding risk (Hunter et al., 1993).

Expansion into Forests

Another type of agricultural intensification, expansion of production into previously forested areas, may alter disease transmission in yet other ways. While encroachment into forests and accompanying deforestation could reduce the risk of some diseases (e.g., anthroponoses that depend on direct contact and human crowding), such changes are likely to increase the risk of exposure to zoonotic pathogens that normally are transmitted among animals of the region. This process has been observed with greater contacts with the forest fringe involving elevated cases of Leptospirosis (*Leptospira* spp. infect various forest mammals and are transmitted from urine via water), normally to others of the same species. Humans are unnatural hosts that experience febrile, sometimes life-threatening disease. Similarly, New World Leishmaniasis has been linked to agriculture expansion into forests where sand flies (*Lutzomyia* spp.) transmit *Leishmania* spp. protozoa among forest mammals and can also transfer this parasite to people (e.g., Morrison et al., 1993). Yet another New World tropical example involves expansion of dwellings into forest edges where common midges (*Culicoides* spp.) are suspected vectors of Oropouche virus. People in the Amazon basin in particular are at risk of Oropouche fever after being infected by these biting flies (Baisley et al., 1998). In general, encroachment into forest ecologies that were previously not part of modern agroecological practice has increased exposure to various pathogens, hence elevating the human infectious disease burden among these people.

Ironically, agricultural expansion into forests can also have unfortunate effects on people other than those who are part of this change. People who already live in these sites are at enormous risk of diseases introduced by the pioneers, diseases to which they have no previous immunity (e.g., dengue, malaria, influenza, etc.). In this unusual way, agricultural expansion into forests produces victims of infectious disease who often do not benefit from these ecological and population changes.

Increased Human Concentration

Other, less direct effects of agricultural development function through yet different environmental pathways. In general, human-to-human transfer of microbes should generally decrease as crowding/contact decreases, but some agricultural development may create greater contact among people where housing may be limited. Even though one rationale of many tropical agriculture projects is to reduce migration to urban centers by people in search of employment, these programs may also create unusual concentrations of people near farming sites under conditions of poor hygiene. Such ecological effects operate via changes in human behavior or population density interacting with local sanitation and are primarily a potential problem with anthroponotic diseases. Thus, as human populations move into areas of intensified aquatic tropical agriculture (e.g., rice cultivation, fish farms), risk of helminth infections such as hookworm increases. Humans infected with *Ancylostoma duodenale* produce eggs in feces that when deposited in moist soil will hatch into larvae that can infect other people through skin contact; thus, both human concentration and poor hygiene are part of this health risk. Another example involves mosquito-borne dengue fever, which is considered primarily an urban/periurban disease. However, villages that emerge around new cropland can quickly provide suitable breeding habitat for *Aedes aegypti* mosquitoes, the vector of dengue and yellow fever viruses. This mosquito species evolved to lay eggs in water-filled tree holes, but reproduction is now largely based on availability of artificial water-filled containers (e.g., cans, bottles, jars, tires, and other trash) (Gubler, 1998). Throughout most of the tropics and subtropics, dengue risk is directly linked to the abundance of *Aedes aegypti* vectors that flourish in these human-created sites.

Resistance to Treatment

Most contemporary agricultural production relies heavily on synthetic chemicals that kill plant pests (insects, mites, helminths, etc.), animal vectors or pests (flies, ticks, mosquitoes, etc.), or animal infectious agents (bacteria, protozoa, etc.). While many of these chemicals successfully increased agricultural production, diverse, indirect, longer term effects on human health are appearing. Even though pests of plants rarely harm people directly, toxins used against these arthropods or helminths also kill nontarget species. Thus, nontarget arthropods that transmit human disease agents (vector mosquitoes, for example) are experiencing increased selective pressure that appears to have increased resistance to these or related insecticides. Indeed, certain agrochemicals may result in yet other indirect effects on human health by reducing the effectiveness of insecticides that also are being used to reduce human diseases.

Another indirect effect involves animal production that uses large amounts of antimicrobial compounds to kill bacteria, fungi, or protozoa in cattle, pigs, poultry, fish, etc. These antibiotics are used extensively for growth enhancement and infection prophylaxis in diverse animal-based food production efforts. Commercial animal feed for aquatic farm-raised animals (shrimp, fish) typically contains such antibiotics. A growing body of evidence suggests that such widespread practices are increasing the rate of antibiotic resistance of many pathogens that also infect people and that these compounds are becoming less effective in treating human infections (e.g., Engberg et al., 2001; Gorbach, 2001). There are other indirect pathways that could be considered, but the remainder of this review will explore in more detail how environmental change associated with tropical agroecology appears to alter human disease risk.

SOME CASE STUDIES

A few specific examples illustrate the complexity of ecological, demographic, and behavioral pathways by which infectious disease risk is altered by agricultural activities. What follows is illustrative of diverse processes that give rise to these and other related human diseases.

Malaria

In some ways, malaria represents the quintessential example of how tropical agricultural activities can affect disease risk. First, there are important agroenvironmental links (*Anopheles* mosquito reproduction and survival are strongly water, humidity, and temperature dependent). Second, human exposure to infectious mosquitoes is environmentally influenced (we are the only vertebrate reservoir of *Plasmodium* spp. that cause malaria in us), elevating risk where people live nearby each other in simple housing. Third, tropical agriculture sometimes involves expansion into forested or grassland areas where suitable habitats that did not exist before are being created. Fourth, domestic animals (cattle, sheep, goats, etc.) may serve as hosts for blood-meals that allow mosquito populations to be maintained even where people are protected with screens and repellants. In general, most tropical agricultural practices do little to reduce malaria transmission, indeed often elevating risk. This relationship between agroecology and malaria, however, also represents opportunities for intervention.

Various ecological approaches to food production intended to reduce malaria transmission have been proposed and studied (e.g., http://www.cgiar.org/iwmi/sima/examples.htm for a summary), most of which function by reducing vector abundance. The success of such interventions depends heavily on local characteristics of the principle *Anopheles* vectors in an area and how widely these practices are adopted (Singh and Mishra, 2000). Rice field flooding, for example, may increase availability of breeding sites for competent vectors; yet if intermittent irrigation were introduced, it might actually reduce egg-laying while increasing rice yields (Ijumba and Lindsay, 2001). As already mentioned, domestic animals are fed upon by some female mosquito species, thus increasing their reproduction, but these hosts neither transmit nor become infected with the human malaria parasite. The role of these animals is complicated, but could lead to reduced transmission to humans (Mutero et al., 1999).

Among domestic animals, cattle are particularly implicated in creating microhabitats for *Anopheles gambiae* breeding when their hoofprints fill with rain, but even this unintended risk could be minimized with changes in herding or use of focused breeding site treatment.

A different agriculturally induced malaria risk comes from insecticides, particularly as used in cash crop production, which increase development of insecticide resistance among malaria-transmitting mosquitoes. The challenge is to use other agricultural pest control methods that will reduce selective pressure toward resistance in these nontarget mosquitoes. Widespread, open-ditch irrigation has been practiced in many settings where malaria has become a major health problem, resulting in the unintended creation of breeding sites and greater malaria risk. Canals and feeder lines that are carefully constructed could reduce this problem and, if combined with simple bucket-kit drip irrigation systems with treadle pumps, might increase water efficiency, create fewer puddles and pools, cost less, and allow more malaria prevention and treatment. Another example involves synthetic fertilizers, used for rice production, since they also may enhance larval mosquito food resources (algae, etc.). More efficient use of these fertilizers could lead to fewer and less long-lived adults, hence less *Plasmodium* transmission. These are but a few of the pathways by which an enlightened tropical agricultural might simultaneously address associated malaria risk while enhancing sustainable production.

Schistosomiasis

Another example of how agricultural practices can increase risk of infectious diseases involves human schistosomiasis. This indirectly transmitted, water-associated anthroponotic infection is caused by trematode worms or blood flukes of the genus *Schistosoma*. Transmission is primarily in tropical or subtropical regions, involving certain snail species, primarily of the genera *Biomphalaria, Bulinus,* and *Onchomelonia,* that are required during part of each schistosome's life cycle. Disease results primarily from eggs that lodge in specific human organs. These eggs are produced by adult worms that live for years inside chronically infected people. Thus, people and snails play a role in *Schistosoma* biology, and water-associated influences affect both. Humans become infected as they work or bathe in water infested with *Schistosoma* larvae (cercaria) released by snails. These cercaria penetrate the skin of people and travel to internal organs where they mature to adult worms that mate and reproduce. Eggs are then released when people urinate or defecate and, if this is into snail-infested water, more snails become infected, hence perpetuating the transmission cycle.

Various kinds of tropical agriculture can enhance or diminish transmission, depending on how water patterns are altered. The role of environmental factors in the epidemiology of schistosomiasis is considerable, as determinants such as precipitation, water salinity and flow, vegetation, and water use all affect risk. Environmental changes such as the creation of dams, the diversion of water, canalization and irrigation, or diminished disposal of human wastewater all may affect risk of schistosomiasis. The exact manner in which these changes affect transmission of schistosomiasis will depend on details of the local ecology and human exposure.

Venezuelan Hemorrhagic Fever

In another example human health is affected through a very different pathway involving agricultural intensification in Venezuela. During the past few decades, forests in the Llanos region of the west-central part of the country have been cut to permit expansion of cropping (mostly grains) there. In 1992, a new, sometimes fatal disease named Venezuelan hemorrhagic fever (VHF) appeared, the result of a newly recognized virus in the Arenaviridae (Tesh et al., 1993). Arenaviruses are normally transmitted among rodents, with considerable species specificity between virus and host. Typically, they do not cause disease or reduced fitness in their natural reservoir hosts. Both rodent reservoirs (*Sigmodon alstoni* and *Zygodontomys brevicauda)* that were implicated as primary hosts to this causative agent (Guanarito virus) are granivorous inhabitants that prefer oldfield and cropped habitats (Fulhorst et al., 1999).

Apparently, as forests were being replaced by fields, other small mammal species declined while these two species became more abundant. More agricultural work in these fields, combined with simple rural house construction that allowed easy access by rodents, led to increased human contact with infectious rodent urine and feces. Many people had moved to the region with the hope of employment in this emerging sector, which further increased the likelihood human–rodent virus contact. As a result, cases of VHF began to appear and a local epidemic was identified. Although much is now known about the source of infection and methods for reducing exposure, people each year continue to be infected.

Rift Valley Fever

As a final and still different example, Rift Valley fever (RVF) risk results from two other kinds of agricultural links. RVF virus is transmitted by flood-water species of *Aedes* mosquitoes throughout most of sub-Saharan Africa. It is believed that native ungulates had evolved as the primary vertebrate reservoirs and that transmission occurred as adult mosquitoes feed on blood from these animals. As the ungulate fauna was gradually replaced with domestic ungulates (cattle, sheep, goats, etc.), RVF became an infection of predominantly these animals. Less coadapted to infection, RVF virus in these animals can cause abortion in pregnant females and death to very young animals. Humans may also be bitten by many of these competent *Aedes* mosquito species, and thereby may become infected. Severe, sometimes fatal, disease can result. Thus, RVF is a threat to domestic animals and humans alike, and where large herds of animals exist near human settlements, epizootic transmission occasionally leads to epidemic spillover transmission to people.

Agricultural activities can affect RVF risk in various ways. First, water impoundment for irrigation often creates suitable breeding habitat for vector mosquitoes, as was observed in Senegal/Mauritania during 1989 when an epidemic resulted from flooding and rice cultivation along the Senegal River (Ksiazek et al., 1989). Second, increased animal production has resulted in larger herds that enhance the risk of epizootics, particularly when animals are brought to graze in open areas where natural breeding sites abound. Third, the illegal transport of animals that have not been quarantined has been implicated in the introduction on at least three occasions of RVF to northern Africa (Arthur et al., 1993) and the Middle East (Ahmad, 2000).

This brief summary of a few examples illustrates the various pathways by which tropical agricultural changes can affect infectious disease risk. For some diseases, new insights are just now becoming available to inform future development schemes. For others, we have understood the underlying mechanisms that increase transmission and have had the means to design agricultural activities such that the risk of ill health is minimized. Most often, this has not been undertaken because of a lack of understanding of the potential costs in relation to the economic gains. Sometimes, there has been inadequate information to anticipate problems. Regardless, research generally has been inadequate to develop accurate predictions of how changes in tropical agriculture are likely to increase disease incidence.

FUTURE RESEARCH NEEDS

This review has outlined some of the links between tropical agriculture and human health, with particular emphasis on infectious diseases. The pathways are often complex, rarely linear, and frequently contradictory. Research aimed at understanding and prediction must recognize this reality. Many of the pathways involve basic biological mechanisms, as well as processes that are economic, social, behavioral, and cultural. Thus, one aspect of a new research strategy must involve cross-disciplinary scholarship. Evidence suggests that particular strategies of research are likely to be more productive that others.

Analysis of Agriculture Change and Disease Risk

Future changes in agricultural activities, whether in practice, location, or extent, must consider how corresponding environmental changes may alter human disease incidence. Many such effects are now occurring or are shortly planned. Research should be undertaken to understand what unintended health effects might result, and how they can be reduced or avoided. We should no longer accept that tropical agriculture projects and practices must inevitably produce unexpected outcomes or unanticipated effects. This will require field studies of past projects that resulted in unintended ill health, which analyze how the biological, social, or behavioral changes produced such effects. Understanding more than just the ecology or economics of changing agricultural patterns should be part of the research agenda of development projects. This addition to the normal research and planning agenda is in recognition of the importance of prospective investigations of how such changes affect disease risk. As projects are initiated, data on various indirect effects and changes in exposures need to be gathered. Such surveillance, particularly if it covers longer time periods, will provide an information base that has research value and can also lead to changes in plans before much harm is done.

Ecoepidemiological Data

Historically, long-term surveillance has been indispensable in demonstrating how environmental changes affect the distribution and incidence of disease. Such systematically gathered epidemiological records provide the basic information needed to track and analyze changes in diseases during and after agroecological changes. Greater emphasis on simple, rapidly published experiments has reduced such long-term prospective observations. This is particularly true in the underdeveloped tropics, where inadequate surveillance has diminished our ability to compare time–space patterns of disease with agroecological patterns. There must be renewed effort placed on naturalistic observations and ecoepidemiological experiments to understand mechanisms and processes of agricultural effects on infectious diseases. Prospective, multiyear monitoring designed to test hypotheses concerning changing ecological and health patterns should be encouraged.

Understanding Dynamics

Another research need is improved understanding of the dynamics of agricultural development projects, including how complex, indirect, and nonlinear feedback can lead to unintended outcomes. Thus, theoretical investigations on the dynamics of a large number of variables could be extremely valuable. Such research could be quantitative or qualitative and could explore interactions among diverse factors that include ecological, demographic, immunologic, societal, and other processes. The goal of such investigations could be to understand the general behavior of such systems or how specific kinds of tropical agroecology in particular settings interact with other processes to produce unintended outcomes. The long-term objective of such efforts would be to develop a predictive science of how agricultural development is likely to positively or negatively affect health and well-being.

Cross-Disciplinary Inquiry

Implied in the recognition that health risks and benefits are determined by many factors interacting in a complex manner is a need for analysis that crosses traditional academic disciplines. Contributions from investigators in the social and behavioral sciences need to be included in studies that historically have involved environmental or medical researchers. Most training of scientists today continues to be very disciplinary, but a new cadre of generalists could add enormously to understanding and solving problems of such breadth. In addition, research that integrates methods and perspectives from anthropology, microeconomics, demography, and development planning, for example, with agroecology would surely improve understanding how changing patterns affect people's well-being. New research strategies must be designed with the initial intent of integrating analyses from many disciplines, not as an add-on after less inclusive study has failed to produce accurate results.

Analysis of Inequalities

New research into who benefits and who pays in tropical agricultural projects is needed. Most studies of the economic effect of tropical agroecology consider the average or total result on regional production. Rarely are careful studies undertaken that explore how different classes of people are affected differently. Even then, most analysis of inequality has focused on the economic benefits or losses, with less attention to other measures of well-being. Analysis of whether peasants, migrant workers, herders, or food processors are at greater risk of disease, even if they have increased income, would help guide policies on the kinds of projects to be undertaken, and possible negative effects that should be monitored. This is especially important for the poor, who stand to be most harmed by modern and expanding agriculture in the tropics.

Understanding New Technologies

Health effects of new technologies, such as genetically engineered crops, growth hormones, and antibiotics in animal feeds, are poorly understood. Might engineered plants be less nutritious or in other ways produce ill health in people who consume them? How likely is it that hormones used to increase animal growth alter human hormonal responses or immune capacity? Mounting evidence suggests that antibiotics in animal feed, intended as growth promoters, is increasing resistance in microbes that also infect people. What would be the effect of halting the use of such antibiotics in animal feed, and what are the larger economic and societal costs that extend beyond the profit margins of the animal production industry?

Finally, human health should be placed first on the agenda of development planners rather than as an afterthought to consider only if it does not conflict with an economic development scheme. Research that initially anticipates indirect pathways, which would increase disease risk, should start with the premise that there will be unintended health effects. Then the research strategy involves how to plan a larger agroecological program that considers human well-being in a broader realm. Existing knowledge must be used in planning future change, because not doing so may lead to failure if improving the human condition is the larger goal. Can we apply the many past lessons of how tropical agroecology has produced ill health to future endeavors? Will the previous patterns of unexpected side effects be considered as new policies are being developed? Can we design alternative techniques that actually decrease risk of human disease while enhancing sustainable food production? Ultimately, can we develop a science of food production and economic growth that considers indirect effects and complex dynamics, such that all people in a region benefit from agricultural change? There are no valid excuses for not answering "yes" to each of these questions.

ACKNOWLEDGMENTS

Inspiration for this summary of how tropical agroecosystems affect health comes from Richard Levins. By sharing his understanding of the multiple causal pathways in complex systems, the interdependence of the biological and social, and the importance of analyzing contradiction as a fundamental process of life, Professor Levins has provided the conceptual framework for this review, and for the scholarship of countless others. This work was supported by the Global Health Program at the University of Michigan.

REFERENCES

Ahmad, K., More deaths from Rift Valley fever in Saudi Arabia and Yemen, *Lancet,* 356:1422, 2000.

Anderson, R.M. and May, R.M., *Infectious Disease of Humans: Dynamics and Control,* Oxford, University Press, New York, 1991.

Arthur, R.R. et al., Recurrence of Rift Valley fever in Egypt, *Lancet,* 342:1149–1150, 1993.

Baisley, K.J. et al., Epidemiology of endemic Oropouche virus transmission in upper Amazonian Peru, *Am. J. Trop. Med. Hyg.,* 59:710–716, 1998.

Birley, M.H., *The Health Impact Assessment of Development Projects,* HMSO Publications, London, 1995.

Centers for Disease Control and Prevention, *Addressing Emerging Infectious Disease Threats: A Prevention Strategy for the United States,* Centers for Disease Control and Prevention, Atlanta, GA, 1994.

Engberg, J. et al., Quinolone and macrolide resistance in *Campylobacter jejuni* and *E. coli*: resistance mechanisms and trends in human isolates, *Emerg. Infect. Dis.,* 7:24–34, 2001.

Fulhorst, C.F. et al., Natural rodent host associations of Guanarito and Pirital viruses (Family Arenaviridae) in central Venezuela, *Am. J. Trop. Med. Hyg.,* 61:325–330, 1999.

Gorbach, S.L., Antimicrobial use in animal feed — time to stop, *New Engl. J. Med.,* 345:1202–1203, 2001.

Grifo, F. and Rosenthal, J., *Biodiversity and Human Health,* Island Press, Washington, D.C., 1997.

Gubler, D.J., Dengue and dengue hemorrhagic fever, *Clin. Microbiol. Rev.,* 11:480–496, 1998.

Hunter, J.M. et al., *Parasitic Diseases in Water Resources Development,* World Health Organization, Geneva, 1993.

Ijumba, J.N. and Lindsay, S.W., Impact of irrigation on malaria in Africa: Paddies paradox, *Med. Vet. Entomol.,* 15:1–11, 2001.

Kaplan, C., *Infection and Environment,* Butterworth-Heinemann, Oxford, 1997.

Krause, R.H., Ed., *Emerging Infections,* Academic Press, San Diego, CA, 1998.

Ksiazek, T.G. et al., Rift Valley fever among domestic animals in the recent West African outbreak, *Res. Virol.,* 140:67–77, 1989.

Lederberg, J., Shope, R.E., and Oaks, S.C., Eds., *Emerging Infections, Microbial Threats to Health in the United States,* National Academy Press, Washington, D.C., 1992.

Leirs, H. et al., Search for the Ebola virus reservoir in Kikwit, Democratic Republic of the Congo: reflections on a vertebrate collection, *J. Infect. Dis.,* 179(Suppl. 1):S155–S163, 1999.

Lipton, M. and de Kadt, E., *Agriculture-Health Linkages,* WHO Offset Publication 104, World Health Organization, Geneva, 1988.

Lindblade, K.A. et al., Land-use change alters malaria transmission parameters by modifying temperature in a highland area of Uganda, *Trop. Med. Int. Health,* 5:263–274, 2000.

Linthicum, K.J. et al., Climate and satellite indicators to forecast Rift Valley fever epidemics in Kenya, *Science,* 285:397–400, 1999.

Morrison, A.C. et al., Dispersal of the sand fly *Lutzomyia longipalpis* (*Diptera: Psychodidae*) at an endemic focus of visceral leishmaniasis in Colombia, *J. Med. Entomol.,* 30:427–435, 1993.

Morse, S.S., Ed., *Emerging Viruses: Evolution of Viruses and Viral Diseases,* Princeton University Press, Princeton, NJ, 1994.

Mutero, C.M. et al., Livestock management and malaria prevention in irrigation schemes, *Parasitol. Today,* 15:394–395, 1999.

Patz, J.A. et al., The potential health impacts of climate variability and change for the United States. Executive summary of the report of the health sector of the U.S. National Assessment, *J. Environ. Health,* 64:20–28, 2001.

Pinckney, J.L., Responses of phytoplankton and *pfiesteria*-like dinoflagellate zoospores to nutrient enrichment in the Neuse River Estuary, North Carolina, *Marine Ecology-Progress Series*, 192:65–78, 2000.

Reiter, P., Climate change and mosquito-borne disease, *Environ. Health Perspect*., 109(Suppl. 1):141–161, 2001.

Ruttan, V.W., Ed., *Health and Sustainable Agricultural Development,* Westview Press, Boulder, CO, 1994.

Service, M.W., Agricultural development and arthropod-borne diseases: a review, *Rev. Saude Publica,* 25:165–178, 1991.

Sharma, V.P., Ecological changes and vector-borne disease, *Trop. Ecol*., 37:57–65, 1996.

Singer, B.H. and de Castro, M.C., Agricultural colonization and malaria on the Amazon frontier, *Ann. NY Acad. Sci.,* 954:184–222, 2001.

Singh, N. and Mishra, A.K., Anopheline ecology and malaria transmission at a new irrigation project area (Bargi Dam) in Jabalpur (Central India), *J. Am. Mosq. Control Assoc.,* 16:279–287, 2000.

Tesh, R.B. et al., Field studies on the epidemiology of Venezuelan hemorrhagic fever. 1. Implications of the cotton rat *Sigmodon alstoni* as the probable source of infection to humans, *Am. J. Trop. Med. Hyg.,* 49:227–235, 1993.

Vosti, S.A., Malaria among gold miners in southern Para, Brazil: estimates of determinants and individual costs, *Soc. Sci. Med.,* 30:1097–1105, 1990.

Webber, R., *Communicable Disease Epidemiology and Control,* CAB International, Oxon, U.K., 1996.

Weil, D.E.C. et al., *The Impact of Development Policies on Health,* World Health Organization, Geneva, 1990.

Wilson, M.E., Levins, R., and Spielman, A., Eds., *Disease in Evolution: Global Changes and the Emergence of Infectious Diseases,* New York Academy of Sciences, New York, NY, 1994.

Wilson, M.L., Developing paradigms to anticipate emerging diseases: transmission cycles and the search for pattern, *Ann. NY Acad. Sci.,* 740:418–422, 1994.

Wilson, M.L., Ecology and infectious disease, in Aron, J.L. and Patz, J., Eds., *Ecosystem Change and Public Health: A Global Perspective,* Johns Hopkins University Press, Baltimore, MD, 2001.

INDEX

D

E

F

G

H

I

J

K

L

M

N

O

P

Q

R

S

T

U

V

W